Natural Computing Series

Series Editors: G. Rozenberg
Th. Bäck A.E. Eiben J.N. Kok H.P. Spaink

Leiden Center for Natural Computing

Advisory Board: S. Amari G. Brassard K.A. De Jong
C.C.A.M. Gielen T. Head L. Kari L. Landweber T. Martinetz
Z. Michalewicz M.C. Mozer E. Oja G. Păun J. Reif H. Rubin
A. Salomaa M. Schoenauer H.-P. Schwefel C. Torras
D. Whitley E. Winfree J.M. Zurada

Joshua Knowles · David Corne ·
Kalyanmoy Deb (Eds.)

Multiobjective Problem Solving from Nature

From Concepts to Applications

With 178 Figures and 53 Tables

 Springer

Editors

Joshua Knowles
Manchester Interdisciplinary Biocentre
University of Manchester
131 Princess Street
Manchester M1 7DN, UK
j.knowles@manchester.ac.uk

David Corne
Room G39
Earl Mountbatten Building
Heriot-Watt University
Edinburgh EH14 4AS, UK
dwcorne@macs.hw.ac.uk

Kalyanmoy Deb
Dept. of Business Technology
Helsinki School of Economics
P.O. Box 1210
FIN-00101 Helsinki, Finland
kalyanmoy.deb@hse.fi
and
Deva Raj Chair Professor
Dept. of Mechanical Engineering
Indian Institute of Technology Kanpur
Kanpur, PIN 208016, India
deb@iitk.ac.in

Series Editors

G. Rozenberg (Managing Editor)
rozenber@liacs.nl

Th. Bäck, J.N. Kok, H.P. Spaink
Leiden Center for Natural Computing
Leiden University
Niels Bohrweg 1
2333 CA Leiden, The Netherlands

A.E. Eiben
Vrije Universiteit Amsterdam
The Netherlands

ACM Computing Classification (2008): F.2, G.1, I.2, J.6

ISSN 1619-7127
ISBN 978-3-662-50119-1   Springer Berlin Heidelberg New York
DOI 10.1007/978-3-540-72964-8

This work is subject to copyright. All rights are reserved, whether the whole or part of the material is concerned, specifically the rights of translation, reprinting, reuse of illustrations, recitation, broadcasting, reproduction on microfilm or in any other way, and storage in data banks. Duplication of this publication or parts thereof is permitted only under the provisions of the German Copyright Law of September 9, 1965, in its current version, and permission for use must always be obtained from Springer. Violations are liable for prosecution under the German Copyright Law.

Springer is a part of Springer Science+Business Media

springer.com

© Springer-Verlag Berlin Heidelberg 2008
Softcover re-print of the Hardcover 1st edition 2008

The use of general descriptive names, registered names, trademarks, etc. in this publication does not imply, even in the absence of a specific statement, that such names are exempt from the relevant protective laws and regulations and therefore free for general use.

Typesetting: by the Editors
Production: Integra Software Services Pvt. Ltd., Pondicherry, India
Cover Design: KünkelLopka, Heidelberg

Printed on acid-free paper   45/3100/Integra   5 4 3 2 1 0

To Julia and Luca
(JK)

To Mervyn Joseph Corne
(DC)

To Debjani
(KD)

Preface

To those unfamiliar with the field of evolutionary computation (EC), its problem-solving achievements must seem as magical, nearly, as the products of natural evolution itself. Air traffic control in four dimensions and robot teams that perform co-operative navigation; billion-transistor microchips and expert-level poker playing: these are not the future, but just some of the past trophies of the computer scientist's version of descent with modification.

Of course, behind these achievements lurks some human ingenuity, and liberal amounts of human perspiration. Practitioners of EC know that it does not do its magic at the mere twitch of a wand — and there is much work still ahead to understand how the next step-changes in capability will be reached.

But it remains true that EC demands relatively little from the practitioner in order to function with at least moderate success. Three ingredients, only, are needed: a way to express a solution as a data-structure, a way to modify instances of that data-structure, and a way to calculate the relative quality of two solutions. These are often simple things to design and implement, and consequently EC enjoys the labels 'generic' and 'flexible', able to tackle a huge diversity of problems.

In at least one important respect, however, the flexibility of EC was not fully realized until the emergence, in the 1980s, of *evolutionary multiobjective optimization* (EMO), now a burgeoning sub-discipline. Handling problems with multiple (conflicting) objectives *the way EMO does* can be profoundly useful. Consequently, EMO has spread rapidly, with some three- or four-thousand scientific papers on the subject being published since its inception, sprinkled among the literature of many disciplines.

Straightforward explanations of EMO's growth and appeal typically refer to the extra information it provides when it yields a diverse set of solutions. However, it turns out that EMO has *many* more feathers in its cap. We propose, in this book, a characterization of EMO that accounts more for recent innovations, and which shows where we think much of the future growth in EMO and its applications will be.

The view we adopt, and that the contributed chapters here make concrete, stems from the observation that, alongside 'vanilla' EMO research there has been a parallel development in terms of the *ways that multiple objectives can be used to help solve problems* in general. With notable and often remarkable effectiveness, we find, for example, that EMO techniques can accelerate the search process (for single objective problems), provide novel methods for machine learning, and reliably address dynamic optimization tasks. Similarly we see that EMO techniques can uncover novel design principles, help us to better understand natural complex systems, lead to better solutions even for problems that are unashamedly single-objective, and more.

Some of the ideas presented in this book have become apparent to one or other of the editors, in gradual degrees over the past half-dozen years or so. For JK, the idea of objective function decomposition, explored with Richard A. Watson and DC, in 2000–2001, is one of his earlier memories of thinking more flexibly about how EMO would be used in the future. JK has also been inspired by the recent work he did with Julia Handl on multiobjective clustering, particularly her innovative ideas about objectives as proxies for fundamentally unmeasurable criteria. For DC, an ever-present interest in the link between landscape topology and search dynamics (partially seeding the aforementioned 'multiobjectivization' — objective function decomposition — work) underpins his view that *every* realistic problem is a many-objective one, and he has come to see EMO as a way to help in understanding the 'true' structure of a problem while, or before, solving it. For KD, the concept of using EMO principles for other kinds of problem-solving tasks came to him in 1999, while working on another book. His earlier experiences with single-objective optimization algorithms had taught him that the dogged pursuit of a single specified goal often leads to a rapid loss of solution diversity, with many potentially powerful solutions being discarded; the possibility of using helper objectives to prevent this effect was thus intriguing. KD is also excited with the possibility of using EMO-found trade-off solutions for knowledge discovery in real-world problem-solving tasks.

Our combined interest in this area was piqued again, most recently, by the contributions to the MPSN workshop we co-chaired at PPSN in Reykjavik in 2006, where many of the ideas in this book finally came together and 'brushed shoulders' for the first time.

It has been a lot of work; if only science could disseminate itself. Since it can't, we are most grateful to Ronan Nugent, the Springer editor, for his general support as well as his careful checkign of some of the txet. But, all in all, we have had great pleasure in compiling this book, and we do hope readers will find in it some exciting challenges for their future work. We hope so, or the various sacrifices and injustices imposed by us on our families during the book's production will be wasted. So, enjoy it or else!

Manchester, Edinburgh, Helsinki
August 2007

Joshua Knowles
David Corne
Kalyanmoy Deb

Contents

List of Contributors

Hussein A. Abbass
The Artificial Life and Adaptive
Robotics Laboratory
School of ITEE, ADFA, University
of New South Wales
Canberra, Australia
h.abbass@adfa.edu.au

Johnson A. R. Abraham
Level E Ltd
ETTC, The King's Buildings,
Mayfield Road
Edinburgh EH9 3JL, UK
johnson@levelelimited.com

Johannes Bader
Computer Engineering and Networks
Laboratory (TIK)
ETH Zurich, Switzerland
bader@tik.ee.ethz.ch

Stefan Bleuler
Computer Engineering and Networks
Laboratory (TIK)
ETH Zurich, Switzerland
bleuler@tik.ee.ethz.ch

Jürgen Branke
Institute AIFB
University of Karlsruhe, 76128
Karlsruhe, Germany
branke@aifb.uni-karlsruhe.de

Dimo Brockhoff
Computer Engineering and Networks
Laboratory (TIK)
ETH Zurich, Switzerland
brockhoff@tik.ee.ethz.ch

Anthony Bucci
DEMO Laboratory,
Michtom School of Computer
Science,
Brandeis University
415 South St.
Waltham, MA 02454
abucci@cs.brandeis.edu

Lam T. Bui
The Artificial Life and Adaptive
Robotics Laboratory
School of ITEE, ADFA, University
of New South Wales
Canberra, Australia
l.bui@adfa.edu.au

Carlos A. Coello Coello
CINVESTAV-IPN (Evolutionary
Computation Group)
Departamento de Computación
Av. IPN No. 2508, Col. San Pedro
Zacatenco, México D.F. 07360
ccoello@cs.cinvestav.mx

David Corne
School of Mathematical and Computer Sciences (MACS)
Heriot-Watt University
Edinburgh, UK
dwcorne@macs.hw.ac.uk

Vincenzo Cutello
Department of Mathematics and
Computer Science
University of Catania
Viale A. Doria 6, 95125 Catania
Italy
vctl@dmi.unict.it

Edwin D. de Jong
Institute of Information and
Computing Sciences,
Utrecht University
PO Box 80.089
3508 TB Utrecht, The Netherlands
dejong@cs.uu.nl

Kalyanmoy Deb
Department of Mechanical
Engineering
Indian Institute of Technology
Kanpur
Kanpur, PIN 208016, India
deb@iitk.ac.in

Richard M. Everson
School of Engineering, Computer
Science and Mathematics
University of Exeter, EX4 4QF, UK
R.M.Everson@exeter.ac.uk

Sevan Gregory Ficici
Harvard University
School of Engineering and Applied
Sciences
Maxwell-Dworkin Laboratory
33 Oxford Street, RM 242
Cambridge, MA 02138 USA
sevan@eecs.harvard.edu

Jonathan E. Fieldsend
School of Engineering, Computer
Science and Mathematics
University of Exeter, EX4 4QF, UK
J.E.Fieldsend@exeter.ac.uk

Peter J. Fleming
Automatic Control and Systems
Engineering
University of Sheffield, Mappin
Street, Sheffield S1 3JD, UK
p.fleming@sheffield.ac.uk

Julia Handl
Faculty of Life Sciences
University of Manchester
Manchester, UK
j.handl@manchester.ac.uk

Evan J. Hughes
Dept. Aerospace, Power and Sensors
Cranfield University
DCMT, Shrivenham
Swindon, UK
ejhughes@theiet.org

Hisao Ishibuchi
Department of Computer Science
and Intelligent Systems
Graduate School of Engineering,
Osaka Prefecture University
1-1 Gakuen-cho, Naka-ku, Sakai,
Osaka 599-8531, Japan
hisaoi@cs.osakafu-u.ac.jp

Yaochu Jin
Honda Research Institute Europe
Carl-Legien-Str. 30
63073 Offenbach, Germany
yaochu.jin@honda-ri.de

Joshua Knowles
School of Computer Science
University of Manchester
Manchester, UK
j.knowles@manchester.ac.uk

Isao Kuwajima
Department of Computer Science
and Intelligent Systems
Graduate School of Engineering,
Osaka Prefecture University
1-1 Gakuen-cho, Naka-ku, Sakai,
Osaka 599-8531, Japan
kuwajima@ci.cs.osakafu-u.ac.jp

Azahar Machwe
ACDDM Lab, University of the West
of England
Bristol, UK
azahar.machwe@uwe.ac.uk

Efrén Mezura-Montes
Laboratorio Nacional de Informática
Avanzada (LANIA A.C.)
Rébsamen 80, Centro, Xalapa,
Veracruz, 91000 México
emezura@lania.mx

Amiram Moshaiov
School of Mechanical Engineering
The Iby and Aladar Fleischman
Faculty of Engineering
Tel Aviv University, Ramat Aviv,
Tel Aviv 69978, Israel
moshaiov@eng.tau.ac.il

Giuseppe Narzisi
Courant Institute of Mathematical
Sciences
New York University
715 Broadway #1010, New York
NY 10003, USA
narzisi@nyu.edu

Frank Neumann
Algorithms and Complexity,
Max-Planck-Institut für Informatik
Department 1: Algorithms and
Complexity
Building 46.1, Room 317
Stuhlsatzenhausweg 85
66123 Saarbrücken, Germany
fne@mpi-inf.mpg.de

Minh-Ha Nguyen
The Artificial Life and Adaptive
Robotics Laboratory
School of ITEE, ADFA, University
of New South Wales
Canberra, Australia
m.nguyen@adfa.edu.au

Giuseppe Nicosia
Department of Mathematics and
Computer Science
University of Catania
Viale A. Doria 6,
95125 Catania, Italy
nicosia@dmi.unict.it

Yusuke Nojima
Department of Computer Science
and Intelligent Systems
Graduate School of Engineering,
Osaka Prefecture University
1-1 Gakuen-cho, Naka-ku, Sakai,
Osaka 599-8531, Japan
nojima@cs.osakafu-u.ac.jp

Ian C. Parmee
ACDDM Lab, Faculty of Comput-
ing, Engineering and Mathematical
Sciences
University of the West of England,
Frenchay Campus, Coldharbour
Lane, Bristol, BS16 1QY, UK
ian.parmee@uwe.ac.uk

Katya Rodríguez-Vázquez
Instituto de Investigaciones en
Matemáticas Aplicadas y en Sis-
temas
Universidad Nacional Autónoma de
México, México, D.F. 04510
katya@uxdea4.iimas.unam.mx

Dhish Kumar Saxena
Indian Institute of Technology
Kanpur
Kanpur, PIN 208016, India
dksaxena@iitk.ac.in

Bernhard Sendhoff
Honda Research Institute Europe
Carl-Legien-Str. 30
63073 Offenbach, Germany
bernhard.sendhoff@honda-ri.de

Aravind Srinivasan
Indian Institute of Technology
Kanpur
Kanpur, PIN 208016, India
aravinds@iitk.ac.in

Edward Tsang
Department of Computer Science,
University of Essex
Wivenhoe Park
Colchester CO4 3QS, UK
edward@essex.ac.uk

Ingo Wegener
FB Informatik, LS 2
Univ. Dortmund, 44221 Dortmund,
Germany
ingo.wegener@uni-dortmund.de

Qingfu Zhang
Department of Computer Science,
University of Essex
Wivenhoe Park
Colchester CO4 3QS, UK
qzhang@essex.ac.uk

Aimin Zhou
Department of Computer Science,
University of Essex
Wivenhoe Park
Colchester CO4 3QS, UK
azhou@essex.ac.uk

Eckart Zitzler
Computer Engineering and Networks
Laboratory (TIK)
ETH Zurich, Switzerland
zitzler@tik.ee.ethz.ch

Introduction: Problem Solving, EC and EMO

Joshua Knowles[1], David Corne[2], and Kalyanmoy Deb[3]

[1] School of Computer Science, University of Manchester, UK
   j.knowles@manchester.ac.uk
[2] School of Mathematical and Computer Sciences (MACS), Heriot-Watt
   University, Edinburgh, UK   dwcorne@macs.hw.ac.uk
[3] Kanpur Genetic Algorithms Laboratory (KanGAL), Indian Institute of
   Technology, Kanpur, India   deb@iitk.ac.in

Summary. This book explores some emerging techniques for problem solving of a
general nature, based on the tools of EMO. In this introduction, we provide background material to support the reader's journey through the succeeding chapters.
Given here are a basic introduction to optimization problems, and an introductory
treatment of evolutionary computation, with thoughts on why this method is so
successful; we then discuss multiobjective problems, providing definitions that some
future chapters rely on, covering some of the key concepts behind multiobjective
optimization. These show how optimization can be carried out separately from subjective factors, even when there are multiple and conflicting ends to the optimization
process. This leads to a set of trade-off solutions none of which is inherently better than any other. Both the process of multiobjective optimization, and the set
of trade-offs resulting from it, are ripe areas for innovation — for new techniques
for problem solving. We briefly preview how the chapters of this book exploit these
concepts, and indicate the connections between them.

1 Overview

> Intellectual activity consists mainly of various kinds of search.
> *Alan M. Turing, 1948*

When we say that computers can solve problems, it is a sort of half-truth, to
be taken with a medium-sized pinch of salt. It is manifestly true that computers
solve problems when they almost autonomously carry out everyday tasks involving
communication, auditing, logistics and so forth, and computers even act somewhat
more 'intelligently' when they do such tasks as controlling an automatic transmis-
sion, or making a medical diagnosis, in which they may even exhibit a computerized
form of learning. But it is also true that computers are not very autonomous in
solving *new* problems. Much human input and human innovation still goes into the
process of solving difficult problems (designing a cable-stayed bridge, finding novel
drug interventions, brokering international peace initiatives), a state of affairs that
is likely — and desirably so — to continue for some time to come.

In this book, we consider an area in computer science which is at the forefront of techniques for solving difficult problems. Evolutionary computation — that is, methods that resemble a process of Darwinian evolution — can be used as a way to make computers 'evolve' solutions to problems, which we, as humans, do not ourselves know how to solve. By giving computers the capability of searching for their own answers to problems — through huge spaces of possibilities, in a very flexible way that goes beyond numerical methods — innovative and intelligent-seeming solutions and actions can be produced.

Much of evolutionary computation is concerned with optimization. For a human to use evolutionary computation to solve a new optimization problem, very little is required. This is where the flexibility of EC comes from: one must only provide the computer with (i) some way of representing or even 'growing' candidate solutions to the problem, and (ii) some function (or method) for evaluating any candidate solution, estimating its goodness on a numerical scale. Enormous varieties of problems can be stated succinctly in this way, from electronic circuits to furniture designs, and from strategies for backgammon to spacecraft trajectories. And the boon of evolutionary computation (though not one hundred percent realized) is that it turns computers into almost universal problem solvers which can be used by anyone with even minimal computational/mathematical competence. Of course, many problems are fundamentally intractable, in the sense that we cannot hope to find truly optimal solutions, but this does not really limit the uses of EC, but makes it more useful, since its strength is in finding the best solution possible given the time allowed.

But, while we just said that optimization is widely applicable, it is but a subset of a larger and even more flexible method of problem solving: multiobjective optimization (MOO). The drawback of standard optimization (let's call it single-objective optimization or SOO) is the requirement for a function that can score each and every candidate solution in terms of a single objective number. Many problems exist that are not so easy to state in terms of a single function like this. Humans are generally much more comfortable with, and used to, thinking in terms of aims and objectives *plural*, when stating a problem. And, while it is possible, in principle, for people to combine their aims in some over-arching function by weighting or ordering them by importance, in practice, different aims and objectives are often not measured in the same units, on the same scales, and it is often nigh-impossible to state the importance of different objectives when one has seen no solutions yet! More fundamentally, many of the methods for combining functions together into a single one, necessarily miss potentially interesting solutions. And many methods are very difficult to use because a small change in weights, gives a totally different solution. Therefore, it would be great if one could exploit the power of evolutionary algorithms, but use them to search for solutions even when the aims and objectives cannot be boiled down to a single function.

We are being purposely obtuse here, of course, because such methods already do exist in the field of *evolutionary multiobjective optimization (EMO)*, and they have been growing more and more effective over the past twenty years or so. An EMO algorithm is, loosely-speaking, one in which objectives are treated independently, and a *set* of optimal trade-offs (called Pareto optima) is sought, rather than a single optimum. But we introduced the field in this roundabout way to emphasise the point, made above, that computers solve problems — difficult problems at least — only in concert with humans. Humans are still the ones who generally own the problems and understand something about what they desire and hope for in a solution. So, one of the key hurdles to problem solving with computers is to be able to formulate a problem in such a way that a computer can actually solve it, and a human is happy with the solution. EMO has this ability in spades, so rather than being a mere branch of EC, it actually represents a major step forward generally in computer problem solving.

Where this book advances further in terms of problem solving with computers, and problem solving specifically using the tools of evolutionary multiobjective optimization, is in examining the critical area of how to exploit the greater flexibility of search afforded by a multiobjective optimization perspective. While other books and articles on EMO [6, 7, 5] have given a thorough grounding in the development of EMO techniques, and have been formidable advocates of its benefits, it is only relatively recently that a groundswell in terms of researchers confidently exploiting EMO tools to new and innovative ends has really been apparent. It is on this we concentrate.

The new uses of EMO do not represent a step-change, but a gradual realization that there are few hard-and-fast rules in solving problems with the technique. Thus, researchers have begun to ask themselves such things as what would happen if I took away a constraint and treated it as an objective, what would happen if I had a problem where I had one objective but it seemed possible to decompose it into several, what would happen if I had some different functions which were inaccurate proxies for a true ideal objective function? In this book, we see how current research is dealing with these questions and further we see valuable products of this exploration. We see here that, ironically, EMO is very useful in *coevolution*, an area characterized by problems that have *no* formal objective function at all (evaluation occurs only by competitions). We see it helps in traditional SOO problems, where it speeds up search. We see it put to numerous uses in ill-posed problems, especially those in machine learning. Along the way, the chapters also consider the important issue of how to analyse and exploit the *sets* of solutions that are obtained from EMO, both in terms of decision making (i.e., usually choosing one final solution) or of learning from the set of trade-offs. And the development of EMO methods with respect to their scalability to larger problems with more objectives is considered, and supports the ideas proposed throughout the book.

In this chapter, we seek to do two jobs. First, to preview the book, which we have partly done, but which we continue in Sec. 4. Secondly, to cover some bases for any readers who might be unfamiliar with the fundamental concepts whose knowledge is assumed in some of the chapters, we give some appropriate introductory material. To these ends, Sec. 2 recalls the formal definition of a problem, including, in particular, an optimization problem. Hard problems, evolutionary algorithms, and the use of the latter on the former, are then discussed. Sec. 3 deals with multiobjective optimization, giving definitions for Pareto optimality and related issues. Then, Sec. 4 is a rundown of the four parts the book is divided into, and provides summaries of each of the chapters that make it up. Finally, Sec. 5 briefly concludes this introduction chapter.

2 Problems and Solution Methods

2.1 Optimization Problems

Problems, in computer science, are both abstract and precise. They are abstract in the sense of describing a whole class of *instances*; they are nevertheless precise in the sense that both the inputs and the solution of a problem instance are members of well-defined (mathematical) sets.

Specifically, a problem consists of: (i) a set of instances, where this set can be defined by either listing it exhaustively (enumerating it), or, much more usually, by specifying all the givens that define the form of an instance; and (ii) a set of solutions, being a definition of the entities that comprise a valid solution and the criteria for accepting it.

For example, we could define a sorting problem. A valid instance could be defined as any finite set of positive integers; a valid solution as an ordering of the input set that is strictly increasing.

An *optimization problem* is just a problem where the solution part of the problem is defined in terms of a function (the objective function), which is to be maximized or minimized.

Definition 1 *An optimization problem is specified by a set of problem instances and is either a minimization problem or a maximization problem.*

From here onwards we will consider only minimization problems. In mathematical parlance, this is done 'without loss of generality', i.e., everything we say is true of both minimization and maximization problems, as long as we replace 'smallest' with 'largest' and such like.

Definition 2 *An instance of an optimization problem is a pair (X, f), where the solution set X is the set of feasible solutions and the objective function f is a mapping $f : X \rightarrow \Re$. The problem is to find a globally optimal solution, i.e., an $i^* \in X$ such that $f(i^*) \leq f(i)$ for all $i \in X$. Furthermore, $f^* = f(i^*)$ denotes the optimal cost, and $X^* = \{i \in X \mid f(i) = f^*\}$ denotes the set of optimal solutions.*

In this definition, 'solution' is being used in a broad sense to mean any well-formed answer to the problem that maps to a cost through the function f. The objective is not to find *a* solution, but to find a *minimum cost* one. In this case, therefore, it is meaningful to talk about an approximate solution to the problem, i.e., one that is *close* to optimal.

Once the problem is defined, we commonly express the task of optimization – the fact that we wish to minimize our cost function over a particular set of potential solutions, as follows:

$$\text{minimize } f(\mathbf{x}), \text{ subject to } \mathbf{x} \in X. \tag{1}$$

Here, f is the objective function, and it maps any solution $\mathbf{x}$ that is a member of the set of feasible solutions X to the set of real numbers, $\Re$. We are asked to find an x, such that f is minimized.

2.2 Hard Optimization Problems

Some optimization problems are easy to solve, and fundamentally so. An easy problem is one where there exists a method that always works — that solves every valid instance — finding an optimal solution in a reasonable amount of time, or number of steps. An example of an easy problem is that of finding a shortest path between two nodes in a network, where the nodes are separated by links with certain lengths. This problem can be solved using Dijkstra's famous algorithm (dynamic programming), an ingenious method that greedily constructs the optimal solution, by considering partial paths through the network and keeping track of the competing alternatives.

Unfortunately, a great many problems that we encounter — and almost all of those in science and industry — are hard. That is to say, there is no known method for solving instances of them exactly and reliably. More than that, these problems are fundamentally hard in that they can be shown to belong to a set of problems, all of which are essentially equivalent in their difficulty. The equivalence means that finding quick and reliable methods for solving one of the problems would result in quick and reliable methods for all of them.

However, no such method has yet been found, and many think that such a method does not exist. Computer scientists call optimization problems that are fundamentally hard in this way, NP-hard [17].

Consider the following list of problems:

- Find a competent, or good, strategy for playing the game, Othello
- Allocate resources for flood protection of the UK
- Define a taxonomy of prokaryotic genes, by their functional activities
- Design a suspension bridge
- Schedule the jobs in a factory, as orders and raw materials arrive periodically.

When defined more precisely, each of these is an example of a fundamentally hard problem. There is no way to go directly to an optimal solution, or to organize a search in a super-efficient fashion that gives reliable results for all instances.

Instead, for problems like these, we can only hope to find good (or approximately optimal) solutions, by a process of searching through alternatives. Further, being fundamentally *hard*, this process is likely to take significant time, even using modern hardware, and even just to find an acceptable, rather than good, solution.

Fortunately, however, we rarely need to resort to a blind, random search through the set X. Each of these problems, and realistic problems in general, have some inherent *structure* that can be exploited as we try to devise a strategy for seeking good solutions. This structure is usually apparent in the way that similar solutions are related in terms of the cost function(s). For example, if two suspension bridge designs are very similar, but differing slightly in the distance between the points of suspension, then the performance characteristics of these two bridges are likely to be very similar too. Such structure is partly captured by the *search landscape*; once we have settled on the details of X (having decided how we will encode potential solutions to the problem), the details of f, and the specifics of how we will generate new solutions from others previously sampled, this *landscape* comes to life as a high-dimensional mathematical structure. A fundamental aspect of all modern search strategies is to sample points in this landscape (i.e., points in X), and attempt to discover (or simply assume) certain properties of the landscape in an attempt to navigate a path through it towards good solutions. One class of such strategies, called evolutionary computation, is recognized as being particularly successful (in comparison to other methods, at least) at handling the landscapes found in most real-world problems.

2.3 Evolutionary Computation

> Algorithms are conceived in analytic purity in the high citadels of academic research, heuristics are midwifed by expediency in the dark corners of the practitioner's lair ... and are accorded lower status. *Fred Glover*

Origins

Problem solving by simulated evolution has been invented independently several times. Its pre-history can be traced back at least to Butler, who pronounced that that machines might 'become as complicated as us' by evolutionary processes, long before general purpose computers had even been conceived (see Dyson [9], chap. 2). Actual computerized simulations began in the 1950s, one of the early notable examples of problem solving being work by Baricelli [1] to 'evolve' fragments of code that cooperated to play a game called Tac-Tix. Fraser [16], Bremmerman [3], Rechenberg [30], Schwefel [33] and Fogel [14] all experimented with models of evolution in independent work conducted in the 50s and 60s, and were joined by many others, notably Holland [19] in the 70s. These researchers had very different ends in mind, and emphasised different elements of the neo-Darwinian principles of evolution in the models that they investigated (for detailed accounts of this early, pioneering work see [13] and [9]). For some time, the differences were cemented, and the distinct methods known as evolution strategies, genetic algorithms and evolutionary programming developed in isolation. Today, and since the 1990s, evolutionary computation is inclusive of all these areas, and innovations cross the boundaries, using common concepts and abstractions from nature.[4]

The Basic Evolutionary Algorithm

Despite the high-falutin ideals of evolutionary computation to model and abstract from the complexity and richness of nature's wandering adaptive walks, the basic evolutionary algorithm for optimization has much more in common with a process of selective breeding (as used by Mendel) than it does with adaptation. The optimization problem provides a static goal to which the algorithm is directed, and a *population* of solutions are improved towards this by rounds of evaluation, biased selection, reproduction, variation and replacement. One round or cycle is called a *generation*, and the pseudocode in Fig. 1 captures the central process in practically *every* evolutionary algorithm:

Typically, in evolutionary algorithms, there is made a distinction between the genotype and the phenotype of a candidate solution, with the genotype being the medium of reproduction and variation (step 3.2 in Fig. 1), but with selection being based on the evaluation of the phenotype (steps 2 and 4). There are several alternative schemes for selection, but the basis for them is usually a relative ranking

---

[4] We might say there is panmictic (all-mixing) evolution.

1. Generate a population of candidate solutions to the problem
2. Evaluate the fitness of each candidate in the population
3. Produce some new solutions from this population:

while not done **do**

    3.1 Select (preferring the fitter ones) some to be 'parents'

    3.2 Produce 'children' (new candidate solutions) from the parents

end while

4. Evaluate the fitness of each of the children.
5. Update the population, by incorporating some of the new children, and removing some of the incumbents to make way for them
6. Until there is a reason to stop, return to step 3.

Fig. 1. Pseudocode for an evolutionary algorithm

of the solutions; the *fitness* of a solution then refers to the expected reproductive opportunity afforded it by selection.

Reproduction occurs by a combination of replication and mutation events, or recombination and mutation — both lead to variation in the children produced. In recombination, genetic material from two or more solutions are crossed over to yield one or more recombinant offspring. Mutation refers to a small change that is made to the genotype of an offspring, following a recombination or replication event. Evolutionary algorithms are always stochastic, and in most cases, selection, recombination, and mutation are all based on dice-rolls. Moreover, the starting or *initial population* (step 1 of Fig. 1) is most often created by a stochastic process. The offspring population, once created, replaces the population of the previous generation, becoming the new *current* (parent) population.

Typically, the population in an evolutionary algorithm is of a fixed size from one generation to the next. Replacement of the old by the new can again be based on competition (i.e., selection), but it can also be entirely random. One particularly popular replacement scheme is for the best few individuals (the elites) of the parent population to be protected from replacement, so that they survive across generations. This is called elitism, and it is implemented in most evolutionary algorithms because it ensures non-retardation of the best solution(s).

Generally, there is immense variety in the way that individual evolutionary algorithms will carry out each of these steps. To a large degree, the reasons for and nature of this variety relate to the precise problem being addressed — this will dictate, for example, the *encoding* used (how candidate solutions are represented as data structures), which in turn affects the way that recombination and mutation may be done. The more problem-independent aspects of EAs are selection for breeding (step 3.1) and so-called 'environmental selection' (step 5). Many techniques have been tried and tested, but the general lessons that are clear from practice are that a 'low pressure' strategy tends to do well on the more interesting and difficult problems. That is, though we may be tempted to strongly prefer to breed exclusively from the most fit solutions in step 3.1, and be sure to be rid of the poorest solutions in step 5, it turns out that we get more capable and reliable algorithms when we ensure that these steps are *gently* influenced by fitness.

Why EC works, and the benefits of EC

Evolutionary computation is very successful, but why? There is continued debate about what underpins its general degree of success, but we first need to clarify what we mean by 'success' in order to address this question properly. What appears to be the case, at least for some important problems, is as follows:

Performance For some problems, EC is capable of much better solutions (and achieved in better, or reasonable, time) than all or most other known methods.

Sufficiency For some problems, EC is capable of solutions competitive with solutions achieved by all or most other known methods.

Applicability EC is applicable to almost *any* optimization problem.

Accessibility For some problems, EC has been applied, and works fine, but no comparisons with other methods have been done.

Opportunity For some problems, EC is the *only* approach that can be used with any chance at all of success — in other words, with EC we can solve problems that we couldn't solve before.

The *Performance* statement is what most people would assume is meant by "EC is successful". Indeed it is true, but it must be stressed that this situation exists in relatively few cases, usually those in which an EC practitioner has worked hard in configuring the key components of the method – the encoding, the operators, and perhaps other features (such as using a heuristic to provide seed solutions in the initial population). The reason for EC's success in such cases is tied up in the fact that EC provides a framework within which a new approach can be engineered. At heart, it seems plausible that the use of the central evolutionary concepts (Darwinian selection from a population, coupled with a means of variation) is a key element in the success here — i.e., it is a fundamentally powerful all-purpose landscape search strategy. However, it is worth noting that much work (usually) needed to be done to craft the landscape, turning it into a problem more amenable to this strategy.

The *Sufficiency* point is true for a great many problems, and it doesn't seem to characterize EC in a particularly exciting light, yet it speaks to EC's 'dependability'. The key point here is that the 'other' methods tend to have a much larger variance in their success than EC. Given, for example, a set of 100 different real-world problems to solve, an EC approach, crafted with no undue effort in each case, will probably do at least 'OK' on each of them. In contrast, an alternative method such as Simplex, that does well in a few cases, may be inapplicable for all other cases; a graph-based search technique that does well on a certain problem may perform terribly in other cases, and so on. Sometimes, use of EC may be over the top, like using an electron microscope to read the small-print in an insurance policy, and a rival technique will do just as well in far less time. However, that rival may have abysmal performance elsewhere in this set of problems. And so it goes on. As we noted with regard to *Performance*, it seems safe to attribute the success of EC to the notion that the Darwinian principles of evolution comprise a good all-purpose strategy for navigating the kinds of landscapes that spring up once we start to solve a real-world problem. The style of success inherent in *Sufficiency*, adds some weight to this. We can understand the high variance in the performance of other methods in this context, by suggesting that such an alternative method will be ideally tuned to aspects of the landscape structure of some (maybe very few) problems, while being a manifestly hopeless strategy for other problems. The well-known gradient-descent approach is an obvious example. Essentially, with gradient-descent search, your problem will be solved quickly and optimally if the landscape's structure is that of a single, smooth multidimensional bowl (the optimum being the lowest point in the centre of the bowl); but on almost any other landscape this strategy will miss the optimum, by perhaps a great distance.

Bearing much relation to the last point, the *Applicability* of EC is well-known, and this, in its own right, is a type of success that EC enjoys in abundance. Almost by definition, if we have an optimization problem to solve, then we already have to hand some notion of candidate structures for X and f. We need very little more than that before we are then able to at least make a first attempt at using EC to solve that problem. By contrast, other optimization techniques may require additional elements that are either unforthcoming, or painful to arrange — such as necessary features of the differentiability of f, or a sensible way to assess the quality of *partial* solutions, or a requirement that candidate solutions be real-valued vectors of a fixed length. An additional feature of EC's general applicability, important to some, is the great ease with which EC can exploit parallel computing resources.

Riding on its ready applicability, combined with its essential simplicity (requiring no particular mathematical or programming prowess, for example), EC is successful partly through its *Accessibility*. This in itself has led to many applications in which EC has been used, pronounced 'good', 'fine' (or whatever), but not actually evaluated in comparison with any alternative approaches. That is, some practitioners, given some problem that they needed to solve, have chosen EC (for one or more of the reasons already discussed), used it, and left it at that. Such cases are valid examples of 'EC successes', and some are in commercial use, but that is not to say that some alternative method wouldn't be (perhaps much) faster, and/or produce higher-quality solutions.

So, many of the applications of EC that we see in the literature, or even in the popular press, may only provide evidence that EC is a highly accessible algorithm, rather than contribute to the evidence that it is the best choice for the problem at han. Nevertheless there is ample evidence that for many important problems it is indeed an appropriate choice; while, in some cases it is arguably the *only* choice. The fact that EC imposes no constraint at all on the nature of the structures in X, the set of candidate solutions, leads to some notable achievements for EC when researchers exploit the *Opportunity* this provides. Essentially, there is no candidate in the list of (non-EC) potential methods for optimization that is able to be applied to the problem of finding ideal strategies for a fighter pilot to use during a dogfight. However, EC has been used for this, with notable success [35]. Similarly, though one can think of antenna design as a problem in which standard parameter vectors are manipulated to achieve variants on standard designs, EC provides the opportunity to think of optimizing antennae in a much wider sense; thus, Lohn [23] used EC to optimize a set of *algorithms* for constructing antennae (such an approach, Genetic Programming [21], is a large subfield of EC), enabling a search through a space of possible antenna designs in which existing styles was just a tiny, imperceptible corner. In this and many other cases EC becomes a way (perhaps the only successful, and automated, way) to discover innovative solutions, rather than simply optimize around standard, prior designs. Many other such examples, as well as examples of more conventional successes, may be found on a visit to the 'HUMIES' awards website at http://www.genetic-programming.org/hc2007/cfe2007.html

We haven't yet quite answered the question of 'why' EC works. But there is no great mystery there. The well-known 'No Free Lunch' theorem [39] tells us that, given an entirely random collection of landscapes (so, think in terms of *all conceivable landscapes*) no single approach is capable of the type of success that we have claimed here for EC. The flip-side of this is that, given some *non-random* collection of problems — a collection in which there is a bias towards certain elements of general structure in the problem landscapes, say — a method may well exist which is generally better than others. It is highly plausible to suggest that the collection of *real-world* problems is highly biased in such a sense. In particular, once we have gone through the process of formalising a problem sensibly, and defined X, f and the operators we will use to move within X, landscapes that we construct are invariably correlated, in the sense that nearby elements of X tend to have similar cost. Thus real-world problems are highly biased towards correlated landscapes. Meanwhile, the essential Darwinian strategy used by EC, which is to follow 'clues' in a landscape under the assumption of such correlation (apples not falling far from the tree), yet not to overcommit too soon to any particular path or region in a landscape (everyone has some chance to reproduce, rather than only the very fittest), seems well suited to most of the landscapes in this class, while rarely being a particularly *poor* approach.

The basic theory behind the EC search strategy is well known, and exemplified in Price's theorem [27], which basically expresses part of the above in formal terms. In the EC field, specialisations of Price's theorem have been derived [19], [29], [26], which express nuances of the central idea, tied to specific kinds of solution structure X. Given these highly general statements, we can be satisfied that the prowess of EC is not a magical ability, but explainable. Meanwhile, similarly general statements for several classes of EC algorithm enable us to be satisfied that an EA will generally make progress in reasonable time [12, 32, 31, 37]. Beyond this, which we need not (and choose not) go into here, the EC literature is replete with incremental steps in our understanding of the many aspects involved in how to best configure an EA for a particular problem class. There remains very much to discover about that very point, but one key theme, which we certainly *will* develop further here, is one which also further evidences EC's traits of *Applicability* and *Opportunity*. Sometimes, with other optimization methods, a problem with two or more objectives can only be addressed if it is first simplified to a single-objective (and hence, a different) problem. But, as we will see, this hurdle is not present with EC.

3 Multiobjective Optimization: Why Many Are Better Than One

We have seen in the last section that amongst the benefits of EC is its general applicability in optimization. Yet, an optimization problem defined by a single objective function is itself a restricted class of problems. More generally, an optimization problem may have multiple objectives.

A multiobjective optimization problem (MOP) is typically formalized like this:

$$\begin{array}{ll} \text{minimize} & \{f_1(x), f_2(x), f_3(x), \ldots, f_k(x)\} \\ \text{subject to} & x \in X \end{array} \tag{2}$$

expressing the fact that we want, ideally, a single solution x that minimizes *each* of k distinct cost functions (also called objective functions). These functions may well be conflicting to various degrees, in the sense that a solution a for which $f_1(a)$ is particularly good, may be such that $f_2(a)$ is particularly bad.

At this point, some terminology is necessary. A solution structure $x \in X$ is often called a decision vector (and X is called the decision space), since the values in x tend to encapsulate the design decisions that we need to make. Meanwhile, the vector $(f_1(x), f_2(x), ..., f_k(x))$ is referred to as the vector of objectives, which inhabits the so-called objective space, typically but not necessarily $\Re^k$.

The quality of a candidate solution x is now no longer measured as a scalar, but as a vector. This makes necessary a new way to assess whether or not some solution x is better than some other solution y. Previously we might say either "x is better than (worse than) y" or "x and y are equally fit". Now, we can still say that they are equally fit, to describe cases in which the objective vectors for x and y are identical, but there are two distinct ways in which x and y's performance on the task may be different. First, we might have "x is better than y", as before, to cover cases in which x's objective vector is better than y's in at least one objective, and no worse in all the others. This is called *dominance* and we say that x dominates y. Secondly, we may have a case in which x is better than y on some objectives, but y is better than x on other objectives. In this situation we say that x and y are *incomparable*, or we say that they are *nondominated*.

Given a set of multiobjective solutions (such as the current population of solutions during an EMO algorithm run), some of this set will be dominated by others in this set. Those that are not dominated by any others in that set (which may be a single solution, or the whole set) form what we call the *Pareto set*. In objective space, the set of objective vectors corresponding to the Pareto set is called the *Pareto front*.

Commonly, the true, optimal solution to a real multiobjective problem is such a set, containing more than one, and perhaps hundreds or thousands, of nondominated points. Put another way, no single solution in X dominates (or is equal to) all other solutions in X; instead, the minimization task is satisfied by a set of distinct, nondominated solutions. Strictly speaking, this set is the Pareto set (and the corresponding objective vectors are the Pareto front), while all other sets of nondominated solutions that we may form from elements of x (such as the nondominated points of the current population during an EMO run) are, at best, 'approximations' to the Pareto set. Commonly, however, papers in the field refer to "Pareto front of the current population", and the precise meaning is usually clear from the context.

Any single point on the Pareto front is called *Pareto optimal*; it is (usually) not optimal in the single-objective sense, since it (usually) does not minimize each of the objectives; however, it represents a compromise, such that if any solution exists that improves upon it on one objective, then that solution will be worse on at least one other objective. Clearly, the Pareto set for any problem contains, for each objective, a point that 'truly' minimizes that objective. That is, if we are trying to find a bridge design that has minimal mass, minimal cost, and whose construction would involve a minimal carbon footprint, then we can expect three of the solutions in the Pareto set to be, respectively, the best possible solutions in these respects among those members of X that are feasible bridge designs. It is almost always too much to expect, of course, that any single solution will do particularly well on any pair of objectives, or all three at once, especially where there is such obvious conflict. For example, cheaper designs will invariably use less optimal materials in terms of strength/mass ratio, and will typically exploit mass-produced, environmentally questionable sources. Nevertheless, the vector of scores for these three points on each objective, representing the best attainable for each objective, is itself a useful reference point in multiobjective optimization, known as the *ideal point*. Some multiobjective optimization methods use an estimation of this point in order to set target directions for the search. Similarly, the so-called *nadir point* represents the vector of worst values for each objective, for points in the Pareto front (note that, for a problem whose Pareto front shrinks to a single point, the ideal and nadir points are the same).

These concepts are illustrated very simply in Fig. 2, where we see contrived examples of Pareto fronts for two problems. The white circles are supposed to represent the Pareto optimal solutions, plotted in objective space, for a two-objective minimization problem. The circles correspond to actual points (designs, decision vectors, etc.) in the Pareto set. The white square locates the ideal point — a solution that we cannot actually achieve in this problem, but showing the best attainable result for each objective individually. Notice that this particular Pareto front 'bulges' *towards* its ideal point — this is called a *convex* front. More generally, convexity is present in a Pareto front if we can generally draw straight lines between two different solutions, and find that there are solutions on the Pareto front that dominate the points on the line. Alternatively, fronts in some problems may display much concavity — this is the case with the Pareto front represented by black circles in Fig. 2 (these are also used in the figure to illustrate the concept of a nadir point). A particularly interesting aspect of problems with concavities in the Pareto front is that the solutions in the concavity are *not* the optima of any simple weighted sum of the objectives. That is, these may be points that the decision maker (see later in this section) will choose, since they may form an ideal trade-off given various considerations. However, they will invariably be missed in a search based on a single objective weighted sum, since, on such a unidimensional view of fitness, these so-called *unsupported* solutions are bested by other points on the front.

Now, from the viewpoint of the 'owner' of the optimization problem we are trying to solve, we seem to have a difficulty. In the more common approach to optimization, we will typically combine our different objectives into one (for example, adding up a bridge design's scores for cost, mass and carbon footprint) and concentrate on minimizing their sum. This eventually yields a single result – which is the bridge design that achieved the best combined score. Alternatively we may find several solutions that achieve the same best score, but, when using single objective methods, these will invariably turn out to be quite similar, and effectively the same design. Hence, the problem has been defined, the optimization has been done, and we can provide the solution, and move on to the next job. However, if we treat this as a multiobjective problem, and perform a multiobjective search, our tactics for the end game are not immediately clear. The outcome of our search is now a *set* of solutions, and these will typically contain quite a variety of different designs. What do we deliver as the single best design?

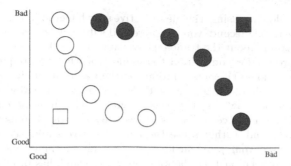

Fig. 2. Examples of possible Pareto fronts, convex (light) and concave (dark), show-ing the ideal point for the convex front (white square) and the nadir point for the concave front (dark square)

Before we answer that, certain notes will be instructive. First, whatever was the optimal point in the single-objective (added up) formulation of the problem is sure to be on the Pareto front of the multiobjective version of the problem. That is, an adequately designed multiobjective search will deliver the 'best' single-objective solution as one of the contents of the returned set of solutions. This is trivial to see, by noting that the solution that truly optimizes a single objective 'added up' formulation must be a nondominated solution. Second, notice that the usual practice, when combining many objectives into one, is to attempt to weight them suitably. So, if cost is more important to us than anything else, we will give this a much heftier multiplier than mass or carbon footprint, in the hope that this will guide our (single-objective) search towards cheaper solutions, but which still show some consideration for the other objectives. However, for the same reasons as before, the optimal solution to this 'new' single objective problem will also be on the Pareto front of our (unaltered) multiobjective search. Indeed, the Pareto front contains the optima for *every* possible weighted-sum based single objective search for this problem.

Thus, one way to view a multiobjective search is as a way to free the problem-solver from the need to specify weights for each objective. It is notoriously difficulty to decide on the correct relative weights in the first place, but you can be sure that the returned Pareto set will contain (at least a good approximation to) the solution that optimizes the 'correct' weighted-sum single-objective formulation (as well as all others), without ever needing to specify the weights.

Suppose, then, that you are an engineer who normally casts your problem as a single-objective weighted sum, but has been convinced, by one of the authors of this chapter, to do a multiobjective search instead. Suppose, too, that beforehand you specify a set of weights, W, for each objective, just as you would normally do. Now, proceed to do a multiobjective search, but without actually making use of W. At the end, you have a set of solutions — a set of bridge designs, or factory production schedules, or whatever. Faced with this choice of possible solutions, and looking for an easy, swift, automatic way to make the choice, you can simply take the one that minimizes the single objective sum specified by W. So, why do it multiobjective at all? Well, the difference is potentially twofold. First, via the multiobjective search you may have found a better result, in terms of the single-objective score found by

W, than you would have using a single-objective search method. This is a commonly observed phenomenon. Second, you are presented with a diverse set of solutions that provides information about the trade-offs available to you. Even though the weight set W may represent for you a robust statement of what you require in a design (though usually it doesn't), some solutions on the Pareto front that don't optimize this particular weighted sum may nevertheless grab your attention. You may well discover, for example, that an unexpectedly good saving in mass may be possible for just a slight increase in cost. True, what we are suggesting here is that a decision needs to be made, and in that sense the multiobjective search seems not to have automatically solved your problem for you. But, on anything more than cursory inspection, it becomes clear that the multiobjective search has provided everything that the single-objective search would have provided for you, *plus more*, so this is not an extra decision to be begrudged, it is an extra opportunity, to grasp or ignore as you see fit.

Multiobjective search is therefore viewed as a way of providing the opportunity for a *decision maker* to make informed decisions about the solution based on information about the solutions that inhabit the Pareto front. In contrast, a single-objective formulation and search, when applied to an inherently multiobjective problem, provides a solution that may look appealing in the absence of alternatives, but is otherwise potentially far from what the decision maker may choose given a better supply of possibilities.

When we therefore decide to face a multiobjective problem on its own terms, and apply a search method that supplies a variety of different but equally 'optimal' solutions for a decision maker to consider, there are various ways we can respond to this opportunity. As noted above, if we have a preferred weight vector at hand, we can use that to pick the 'best' one. If instead we are skeptical about this, or any, weight vector, other approaches are available to us, from the long-established field of *multicriterion decision making*.

3.1 A Note on Multicriterion Decision Making

> The man who, though exceedingly hungry and thirsty, [is] both equally, being equidistant from food and drink, is therefore bound to stay where he is.
>
> Aristotle, On the Heavens (Book II)

Given that we have used an approach that generates an approximation to the Pareto front, the decision maker is provided with this as both a collection of different solutions to the problem, and a source of information about the conflicts between the objectives, and other aspects of the space of possible solutions. If the decision maker is an expert in the problem domain (which should normally be the case!), she may go into a dark room, and emerge some time later having made her choice, based on perhaps deep consideration of the information at hand as well as other, unformalized (maybe unformalizable) aspects of the probable performance characteristics of the various potential solutions.

But, such decision makers are expensive, and it is therefore desirable to have more formal, automated ways to help decision makers minimize their effort. These are generally ways to use additional information about the problem or problem domain, which may have been difficult to include in the original search that led to the Pareto set. There are many standard such methods, and the reader may refer to any textbook on multicriterion decision making for further information on the many existing techniques for selecting a final preferred solution (e.g see [10, 25, 34, 38]).

To provide a flavour of the type of method in use, however, we mention first the idea of 'preference articulation'. When an expert is at hand who is able to provide authoritative views on how to balance conflicting measures and goals, this can be exploited by using preference articulation techniques [8, 2, 20], whereby a series of concrete questions about preferences are asked to the decision maker. The answers then determine if it is possible to build one or other type of consistent model of the decision maker's internal utility function; if so, then an automated procedure can potentially be developed for solution evaluation/selection. Note that far more complicated types of model exist for this than a simple weighted sum over the objectives.

3.2 Visualization Methods

When tackling an optimization problem, visualization may be used to present various features revealed about the problem, or to present information about the search method being used. Amongst other things, the purposes of visualization include estimating the optimal solution value, monitoring the progress or convergence of an optimization run, assessing the relative performance of different optimizers (including stochastic optimizers whose results form a distribution), and surveying features of the search landscape.

In multiobjective optimization, the above purposes of visualization remain important, but the set-valued nature of the results and the conflicts that exist between objectives mean that additional or dual aspects come into play. These will often include gaining an appreciation of the location and range of the Pareto set/front, assessing conflicts and trade-offs between objectives, and selecting preferred solutions.

In the following, we briefly present some of the visualization techniques used by contributors in this book. For more information on visualization techniques that go beyond those used in this book, the reader is referred to [25] (pp. 239–249), [28] and [24].

A basic task in MOO visualization is to illustrate the Pareto front, or the approximation of it found by an optimizer. A raw plot of the Pareto front approximation for bi-objective (or sometimes three-objective) problems (see Fig. 3) is thus very common. In this book, several of the chapters use this visualization technique to present results or concepts. When used carefully, it is an intuitive and straightforward method which can yield much information in a small amount of space. It is worth remembering, however, that the eye's tendency to interpolate between points is usually not to be trusted when considering points shown in such a plot. The boundary of the region dominated by a set of points is represented by its attainment surface, as shown in Fig. 3. This is the representation used in the chapter by Handl and Knowles in their visualization of the Pareto fronts obtained by multiobjective clustering.

Many problems of interest go beyond two or three objectives, of course. To gain an understanding of the range of the Pareto front and the conflicts between

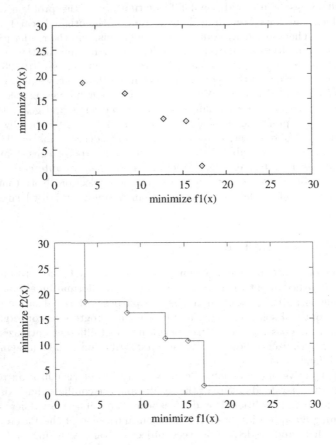

Fig. 3. (Top) A standard two-objective plot of a Pareto front approximation. (Bottom) The corresponding attainment surface represents the family of tightest goals that are known to be attainable as a result of the points found

objectives (and to help select preferred solutions), a parallel axis plot (also known as a value path) [18] is one method that has much to recommend it (see Fig. 4): it can handle relatively many objectives; all objectives are represented simultaneously; the value of each objective is shown by position along a standard numeric axis; and the conflicts between pairs of adjacent objectives is represented by the angles of lines. By adding interaction, allowing a user to change the order of presentation of the objectives and to set acceptable values for objectives, it is possible to learn much about the Pareto set. Parallel axes plots, or variations of them, can be seen in use in the chapters of Rodriguez and Fleming, Brockhoff et al, and Parmee et al in this book.

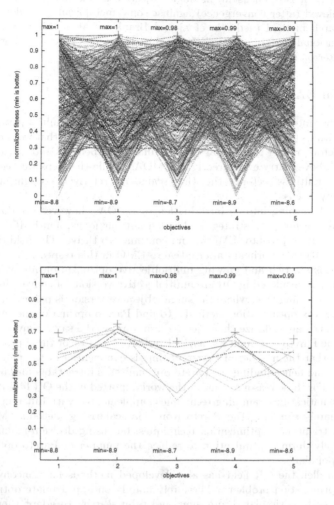

Fig. 4. (Top) A parallel axes plot showing a whole approximation set. (Bottom) A subset of the approximation set, arrived at by interactively setting acceptable levels for each objective (shown by the cross-hairs)

To assess the progress of an optimizer, a plot of best function value against iteration (or time) is the standard method used in single-objective optimization. In multiobjective optimization, this simple and very useful visualization device cannot be used. However, some authors in the book use an appropriate one-dimensional measure of progress in place of best function value, and plot this instead. Cutello et al do this in their chapter, showing how progress towards a known gold standard is made over time.

Similarly, it is often useful in single-objective evaluation to plot best function value achieved (after convergence) against some metaparameter, like a parameter of the search method used. Jin et al and Branke et al in this volume both plot one-dimensional performance metrics against meta-parameters to show the effects of the latter on search performance.

3.3 From EAs to EMOAs

Our primary interest in this book is the use of evolutionary multiobjective optimization algorithms (EMOAs) — that is, evolutionary algorithms that work directly with fitness vectors, rather than scalar fitness values. The benefits and merits of EAs, as discussed above, carry over directly to EMOAs, while the steps we need to take to operate with fitness vectors rather than scalars are relatively straightforward, as we shall briefly discuss.

Before launching into an introduction to EMOAs, however, we do not dismiss alternative optimization strategies. Though not our focus, non-EMOA multiobjective optimization predates EMOA, and continues to thrive. The field of operations research (OR) is the primary alternative to EMO in this respect.

OR boasts a rich and growing multiobjective optimization literature. Many of the problems considered in OR are multiobjective versions of convex problems, e.g., minimum spanning tree, where the single-objective version is polynomial-time solvable by convex optimization methods; to find Pareto optima of the multiobjective problem, one can scalarize the objectives, and apply the same convex methods (so finding one Pareto optimum is 'easy'). However, for most of these problems, it can be shown that the number of Pareto optima is exponential in the number of input variables, and hence finding the Pareto optimal set is intractable for large problems (see [11]). For these reasons, some of the work reported in the OR literature is based on mathematical programming techniques supplemented with different schemes for approximately sampling the Pareto front via scalarizing methods. Methods also based on traditional optimization techniques, but using decision-making before or interactively during optimization to reduce the number of Pareto optima to find, are also popular.

In parallel, the OR field has also developed methods for nonconvex multiobjective optimization problems. These rely mainly on approximate optimization algorithms such as simulated annealing and tabu search. To adapt them for use in multiobjective optimization, two main approaches are possible. One is to adjust the core functions of the algorithm, such as the acceptance function, so as to base decisions on dominance relations between solutions (or other factors, such as proximity of solutions in objective space). The other is to leave the basic algorithm in its original form, and rely on scalarizing techniques (as used in the mathematical programming approaches discussed above) to build up an approximation of the Pareto optimal front.

Meanwhile, in recent years, some OR researchers have begun to experiment with evolutionary algorithm approaches, encouraged by the rapid development of the field of evolutionary multiobjective optimization. This brings us to our main topic.

There are comprehensive and authoritative recent texts available providing reviews of and descriptions of the field of EMO (e.g., [5, 15, 7]). The central difference to the single objective case is the assignment of fitness — obviously, each individual in the population now has its performance characterized by a fitness *vector*, rather than a single value. Perusal of Fig. 1 suggests that the only way this need affect the operation of an evolutionary algorithm is in the selection steps — i.e., steps 5 and 9, in which we are making decisions that require us to compare candidate solutions in terms of relative fitness. This is indeed the case, and EMO algorithms tend to be characterized by how these steps are performed.

The primary style of approach, often referred to (see several of the coming chapters) as Pareto ranking, is to give each point in the population a single score based on the degree to which it dominates, or is dominated by, other points in the population. In one of the most celebrated such approaches, *nondominated sorting* [36], the score assigned to a candidate solution reflects the 'depth' to which it dominates other candidates in the current population. Specifically, all of the nondominated solutions in the population are given the best rank (say, rank 1). These are then marked as 'ranked', and we proceed to find the nondominated front among the *unranked* remainder. These are ranked 2, and the process continues until all candidates have been ranked. An individual's rank is then converted into a selective probability in any one of numerous ways. Often, for example, practitioners will borrow the simple *Tournament Selection* method common in single objective EAs, in which some (usually) small number from the population are randomly picked, and the best of these (highest ranked, breaking ties randomly) becomes selected as a parent. As for the other selection step (step 5 in Fig. 1), the typical approach in EMOAs is to ensure that the Pareto front of the merged parent and child population is preserved, and any remaining space in the population is filled by relatively highly ranked members. It is common, however, for the Pareto front size to be larger than the population size — when this happens (and more generally too, e.g., in considering which of the non-front candidates to keep, when this is an issue) — selection decisions are often based on density in objective space. So, we might prefer to maintain points that are in relatively sparsely populated parts of the front, and don't mind losing solutions that are in crowded sections. These are certainly not the only approaches to selection, and meanwhile we have skipped over many issues that are the topic of hot research subareas in the field. However, the reader is again referred to any of several accessible papers and texts that present the main techniques, and others that describe ongoing research areas.

Meanwhile, the remainder of the book attempts to characterize and focus on a certain category of subareas, in which EMO is considered more widely, as a framework in which a variety of novel approaches to problem solving can be devised and implemented.

4 New Directions in Problem Solving with MOO Techniques

Rather than regarding problems as immutably divided into single-objective and multiobjective, and basing such distinctions purely on the properties required of a solution to the problem, EMO scientists, in practice, are finding reasons to blur these distinctions, co-opting the EMO mechanisms for handling individual objectives for related but distinct purposes that enhance the optimization process. This book is largely about the most prominent and successful of these new problem-solving approaches. It is not a book of EC techniques or of isolated applications, but rather concentrates on the concepts guiding the use of EMO in *broad* problem classes. In particular, it shows how EMO

- can be used to understand and resolve ill-defined problems;
- helps in dynamic optimization environments, in problems with constraints, and in various learning problems where quality is not always directly measurable or free from biases;
- can eliminate a measurement bias or other confounding factor in the optimization of an objective;
- can combine information from multiple sources;
- can change the landscape of a problem, making it easier to search;
- supports proper progress and convergence in search problems where the objective is not explicit, but instead based on tests or competitions (e.g., in co-evolution);
- and, may be employed in the reverse-engineering of artificial and natural systems, where it can contribute to the quest for new principles of design.

Among these and other lessons, we also learn more about the decision-making step for choosing a 'final' solution that is associated with more traditional uses of multiobjective optimization, and explore how this need not be an entirely subjective matter. In some uses of EMO presented in this book — for example, when applying it to traditional 'single-objective' problems, like constrained optimization — the 'best' solution is not a function of decision-maker preferences, and its identification can be automated.

4.1 Chapter Summaries

Part I — Exploiting Multiple Objectives: From Problems to Solutions

Part I of the book is a collection of chapters about problem formulation. It shows how *broad classes* of problem, usually formulated with a single objective to optimize, can be re-cast as multiobjective problems, with various beneficial and sometimes even dramatic effects. In some cases, the effect achieved is an improvement in the efficiency of searching the problem space. In other cases, there is a more profound effect, so that different and better solutions can be accessed asymptotically, i.e., the ultimate potential outcome of the search is improved. In yet other cases, the new problem formulation yields a greater understanding of a problem, with its competing goals and objectives, and this can help to re-evaluate the problem, possibly leading to a more conscious refinement of it.

Using several objectives to help solve what are traditionally considered 'single-objective' problems may raise the spectre of 'relaxing' the problem in some people's minds, resulting in a set of trade-offs, when only one solution is really wanted. However, as is shown throughout the chapters in Part I, this does not turn out to be a difficulty: the single-objective formulation (if it exists in well-defined form) can always be invoked *post hoc* to select the best solution; and where this is found inappropriate, it is because the multiobjective problem formulation has revealed a weakness, or a hidden assumption, in the original problem definition — one that should be externalized and dealt with appropriately. This issue of selecting the final solution is discussed at some level in most of these chapters, and what is found is an interesting contrast to the usual view that multiobjective optimization always implies a phase of (human) decision making.

Ficici (Chap. 2) considers co-evolutionary algorithms (CEAs), a thriving area of research that has the potential to make valuable contributions to the breadth of the whole problem-solving domain. The defining characteristic of CEAs is that the fitness of an individual (a candidate solution) is defined, implicitly, by interactions with other individuals. From this, it follows that CEAs can be used to solve problems for which no known (explicit) objective function exists: problems that would be impossible to tackle using traditional optimization methods, such as finding optimal strategies in two- or multi-player games. Ficici's chapter shows how Pareto optimality can be used as an organizing principle in CEAs, with each individual being viewed as a potential objective for optimization. From this idea, several long-standing difficulties associated with CEAs can be better understood and largely circumvented. Moreover, the multiobjective framework allows the general co-evolutionary learning problem to be handled in such a way that monotonic improvement of solutions is ensured. This idea relates to elitism in MOEAs, and to the use of archives of nondominated solutions, a theme which is touched on in several other chapters in the book, particularly de Jong and Bucci's. The issue of decision making is raised in the chapter, and its relationship to the concept of *refinement*, and the *equilibrium selection problem* in game theory is described, with some tantalizing prospects for future developments.

One of the earliest uses of multiobjective methods for solving single-objective problems was their application in constrained optimization, an area thoroughly reviewed in Mezura-Montes and Coello's work (Chap. 3). Constrained optimization problems represent a large and important chunk of real-world problems, especially in engineering, but they still pose a challenge to traditional evolutionary algorithm methods. In this chapter, the common EA approach of using penalty functions is compared, conceptually, with multiobjective formulations of the problem. Two benefits of the latter are posited: that weights do not have to be selected in order to balance the different importance or ranges of the constraints; and, that the number of improving paths to the optimum is much greater, which increases the possibility of approaching good solutions. Work that both supports and criticizes these assertions is considered, and an empirical study is used to compare some of the current state-of-the-art methods. In constrained optimization, it is shown that decision making never enters as a matter of DM preferences, even in the multiobjective formulation. In other words, there is an automatic way of selecting the objectively best solution from the Pareto front in all cases.

Another considerable area of research in problem solving concerns optimization in dynamically changing contexts, as explained in Chap. 4., Bui et al. investigate using a secondary helper objective for this class of problems, aimed at maintaining diversity in the evolving population, and thus a readiness for sudden or periodic changes in the optima. The use of a secondary objective, and application of standard MOEAs, is a simple approach to handling dynamism, and what is more it does not introduce any issues related to decision making. Comparisons made with existing EA mechanisms for dynamic problems, namely hypermutation and random immigration, show the MOEA approach already gives the most consistently good performance, while there remains much room for further development of the technique.

Cutello et al (Chap. 5) apply multiobjective EAs to the traditionally single-objective problem of predicting a protein's native structure, a pressing and massively significant problem in the biological sciences. The rather specialized nature of this problem belies the fact that it may serve as an archetype for others in which the objective function is not really objective or final, but a proxy used to help find solutions. In structure prediction, it is an energy function that is minimized, and this is essentially a guess made up of several components of energy; the ultimate arbiter of quality, however, is not the objective function, but the distance to the observed, real structure (which is not available at the time of optimization in real instances of the problem, however). Taking a multiobjective approach to the problem (here by decomposing the energy function into its components) is a process of learning how to *align* the objective functions with the ultimate measure of solution quality. Here, the flexible nature of multiobjective search is being used as a way to improve the models on which the optimization is based.

Neumann and Wegener's interest (Chap. 6) is in the possibility that for well-defined problems a multiobjective formulation could be straightforwardly faster to solve for an evolutionary algorithm than a single-objective one. Taking two classic problems from combinatorial optimization, the single source shortest path problem and the minimum spanning tree problem, they demonstrate that this possibility is not a fiction. For the MST, the asymptotic expected optimization time is derived for simple multiobjective and single-objective EAs, indicating the superiority of the former. Experimental results on different instance types also show a performance advantage of multiobjective algorithms for some classes of minimum spanning tree. In all cases, the problem formulations used here directly yield unique solutions to the original problem and no extra step of decision making is involved.

Handl and Knowles (Chap. 7) express and develop a view of problem-solving-via-MOO that concords with some of the ideas expressed in the preceding five chapters. They believe that in practical problem-solving applications, MOO is used in a variety of subtly different ways, called *modes*. The modes capture the specific reason why the problem has been formulated with multiple objectives, and what job each of the objectives is doing. Handl and Knowles identify five different modes and provide examples of each from their own research. They also show how some modes require no decision making for solution selection, while others reveal useful trade-off information that would normally be hidden, but which must be accounted for to select a final operational solution. For the latter case, however, some automatic and semi-automatic methods of decision making have been successfully devised, notably those based on the shape of the Pareto front and the consideration of control distributions.

Part II — Multiple Objectives in Machine Learning

Part II of the book concerns the application of MOO to different problems in machine learning. The chapters collected here, like those in part I, emphasise the reasoning behind the multiobjective formulations presented, and demonstrate that core difficulties in machine learning can be understood and alleviated by the multiobjective approach. Common themes in machine learning, and in these chapters, are the trade-off between accuracy and model complexity; conflicts between training, validation and testing errors; and the combining of rules or classifiers. More unusual issues that are also highlighted include competing errors in multi-class problems, program bloat in genetic programming, and the value of promoting models that humans can understand in system identification.

Fieldsend et al (Chap. 8) consider the supervised learning paradigm, in which the output class or value of a datum must be predicted from its inputs, following a period of training on a random i.i.d. sample of example data. It is well known that the supervised learning problem is about generalization performance, which is difficult to assess during training, and hence different terms are often added to the basic training error objective to achieve regularization or model selection. Fieldsend et al consider multiobjective approaches to this central issue, and show some graphical methods for identifying solutions that best balance accuracy vs model complexity. The chapter also identifies a number of supervised learning problems where competing error terms are inherent, and a balance must be struck between them. One such is the different costs of misclassifications in multi-class data, most notably in disease diagnosis. Some groundbreaking methods in this problem area are presented.

The first of two consecutive chapters on genetic programming is Bleuler et al's (Chap. 9). Genetic programming is a form of computer program induction, based on evolutionary algorithm principles (see [22]). Bleuler et al focus on the problem of 'bloat' in GP, whereby evolved programs have the tendency to grow larger and larger, containing more and more useless code. This problem with GP has been a bugbear for several years, and several methods for counteracting it have been proposed and studied. In recent years, several multiobjective approaches have been tried, with considerable success. In this chapter, the reason behind the success of the Pareto-based approach to reducing bloat is investigated, following a thorough review of this area.

Rodriguez-Vázquez and Fleming (Chap. 10) concentrate on the use of genetic programming in system identification, specifically for non-linear dynamical systems. They show how a multi-stage process, which involves going back and forth between steps of structure selection, parameter estimation and validation can be compressed into a one-step process through the use of a multiobjective formulation. Moreover, human understanding of generated models is identified as an important issue which can be further enhanced by including objectives that control the type and complexity of model components used.

Rule mining is a method of classification, often for large databases, based on two processes, (i) extracting useful rules and (ii) combining them. Ishibuchi et al (Chap. 11) investigate both processes, exploring what is meant by a Pareto optimal rule and a Pareto optimal rule set, and how these can be approximated. They uncover interesting relations between accuracy and complexity, which echo the 'switchback effect' shown in Fieldsend et al's earlier chapter. They also show that Pareto optimal rule sets are not necessarily comprised entirely of Pareto optimal rules, and this is more the case as the ruleset size is allowed to grow.

Part III — Multiple Objectives in Design and Engineering

In design and engineering, it is quite widely understood and accepted that problems invariably have competing objectives, and that problem solving is about finding good balances, or spotting niches where a different type of solution might be attractive for the first time. Part III of the book is about this area, particularly open-ended design, where it is almost a given that problems are multiobjective, when viewed at some level. Instead of explaining why and how multiple objectives arise here, the chapters rather focus on how to support understanding, learning and invention in a multiobjective space, and also how the same principles that are used for design might also help when analysing and seeking to understand existing natural systems, which have inevitably evolved under several and various selection pressures.

Deb and Srinivasan (Chap. 11) suggest a systematic procedure of using two or more conflicting objectives (usually minimization of size and maximization of performance) to unveil salient knowledge about properties which when present in a solution would make it an optimal solution corresponding to the underlying objectives. The argument works as follows. Since Pareto-optimal solutions are no ordinary solutions in the search space, but rather correspond to optimal solutions of certain trade-offs among objectives, a series of such solutions is expected to possess some common properties that can provide a practitioner with important knowledge about 'what makes a solution optimal?'. This process of 'innovization' — the creation of innovative knowledge through multiobjective optimization — is illustrated through a number of engineering design problems.

Parmee's focus (Chap. 12) is on methods to support the human designer as she goes about her business, particularly in the area of conceptual design. Advanced methods of visualization, interactive evolution and machine learning are described, all aimed at taking away the drudgery of evaluation, and freeing the designer to make more insightful and high-level choices and inferences, based on an understanding of the multiobjective nature of the problem space.

Moshaiov is interested in analogies that can be drawn between artificial and natural systems (Chap. 13). He explores how and why such analogies have been useful because of what they can tell us about the design process, and about natural (evolved) phenomena. From this historic background, he moves on to consider why, in cybernetics, artificial life and evolutionary biology, the concept of trade-off is known, but a multiobjective view is rarely taken. Moshaiov explains how such a view could be made acceptable to both biologists and engineers, and considers what the consequences of this broader outlook might be.

Part IV — Scaling Up Multiobjective Optimization

Much evidence for the potential of multiobjective optimization to deliver new and powerful solutions to problems, from classic combinatorial optimization problems to open-ended design problems, is provided in the first three parts of the book. To turn this into a reality, there is, of course, a continuing need for the development of effective multiobjective optimization methods. Of great concern to the field in recent times has been the scalability of the algorithms and concepts we use — scalability to increasing numbers of objectives and to larger design spaces. Part IV of the book presents some of the latest developments in the design of scalable multiobjective evolutionary algorithms.

Jin et al (Chap. 15) consider how the relatively low-dimensional manifold in which Pareto optimal solutions reside can be modeled and projected back into the much higher dimensional parameter space. Such an approach promises to achieve great scalability in parameter space dimension, provided certain base assumptions are valid. Jin et al show excellent performance of their techniques for problems with up to 100 real-valued parameters.

The first of three chapters concerned with methods capable of handling problems with many objectives is provided by Hughes (Chap. 16). He provides a background to the issue and reviews the capabilities of several existing Pareto and non-Pareto multiobjective EAs at handling problems with four or more objectives. He considers both the issues of convergence to the Pareto front, and of controlling the distribution of points along it.

De Jong and Bucci (Chap. 17) are concerned with a particular class of problems that results in optimization problems with many objectives, easily tens or hundreds of them. This class is where the objectives are defined in terms of 'tests', an approach that may be taken when other methods of evaluation are not possible. Although the space of tests is usually very large, de Jong and Bucci show that tests need not all be arranged on orthogonal axes, but they can be grouped together, depending on how they affect the ordering of candidate solutions. This enables a significant reduction in the number of effective objective dimensions considerably.

Brockhoff et al (Chap. 18) also seek to reduce the number of objectives presented, to a lower number, which is easier to handle effectively. The first method for doing this has similarities with de Jong and Bucci's method for tests: it depends on inspection of the orderings induced by the objectives on sampled sets of solutions, and whether or not these orderings can be preserved when some objectives are removed from consideration. The second method does not consider orderings per se, it is based on principal components analysis on the objectives to establish the redundant ones. Both techniques are demonstrated on test instances, and unifying concepts are discussed.

5 Concluding Remarks

There are many interesting topics vying for inclusion in a book on multiobjective problem solving. Our choice has been guided by the desire to present the emerging concepts and methods that are being used to tackle important and long-standing classes of problems; and as a result some things have necessarily been left out. We have not touched on the use of other natural analogies in multiobjective problem solving, such as methods based on social insect behaviour, or flocking — methods that can trump evolutionary algorithms in some domains. We have not considered multiobjective optimization for problems where the number of solution evaluations possible is severely restricted, an area which is gaining rapid prominence because of the take-up of EMO in engineering and science, where evaluations are often costly. In machine learning, we covered several topics, but left out the contribution MOO is beginning to make in searching for ensembles of neural networks [4], though ensembles of rules are considered in the chapter by Ishibuchi et al.

And you, the reader, may well consider that we have overlooked something else — and you are surely right. But what remains in the book is, we hope, the beginnings of a synthesis that shows the contributions MOO is making to the core activity of problem solving with computers. Rather than an esoteric technique at the fringes of evolutionary algorithm research, or a specialist's area in operations research, MOO is now being used to help people solve all sorts of problems by offering genuinely new approaches to them. It goes beyond a method that allows engineers to balance out different criteria — though that is an important aspect of it. It can transform a problem, reaching solutions that were not possible before, or allowing monotonic progression where cycling previously occurred. It can take us to solutions, even to classic problems, more quickly than before. And we have seen that the algorithms for achieving this are being developed anew, with fresh concepts to take us on to a new scale of problem-solving ability.

Acknowledgment

JK gratefully acknowledges the support of the Biotechnology and Biological Sciences Research Council (BBSRC), UK.

References

[1] N. A. Barricelli. Numerical testing of evolution theories Part II: Preliminary tests of performance, symbiogenesis and terrestrial life. *Acta Biotheoretica*, 16(3–4):99–126, 1963.

[2] V. Belton and T. J. Stewart. *Multiple Criteria Decision Analysis: an Integrated Approach*. Springer-Verlag, Berlin, Germany, 2002.

[3] H. J. Bremermann. Optimization through evolution and recombination. In *Self-Organizing Systems*, pages 93–106. Spartan Books, Washington, DC, 1962.

[4] A. Chandra and X. Yao. Ensemble Learning Using Multi-Objective Evolutionary Algorithms. *Journal of Mathematical Modelling and Algorithms*, 5(4):417–445, 2006.

[5] C. A. Coello Coello, D. A. Van Veldhuizen, and G. B. Lamont. *Evolutionary Algorithms for Solving Multi-Objective Problems*. Kluwer Academic Publishers, New York, May 2002. ISBN 0-3064-6762-3.

[6] D. Corne, K. Deb, P. Fleming, and J. Knowles. The Good of the Many Outweighs the Good of the One: Evolutionary Multi-Objective Optimization. *IEEE Connections Newsletter*, 1(1):9–13, 2003.

[7] K. Deb. Evolutionary Algorithms for Multi-Criterion Optimization in Engineering Design. In K. Miettinen, M. M. Mäkelä, P. Neittaanmäki, and J. Periaux, editors, *Evolutionary Algorithms in Engineering and Computer Science*, chapter 8, pages 135–161. John Wiley & Sons, Ltd, Chichester, UK, 1999.

[8] J. S. Dyer, P. C. Fishburn, R. E. Steuer, J. Wallenius, and S. Zionts. Multiple criteria decision making, multiattribute utility theory: the next ten years. *Management Science*, 38(5):645–654, 1992.

[9] G. Dyson. *Darwin Among the Machines*. Penguin Books, 1999.

[10] M. Ehrgott. *Multicriteria Optimization*. Number 491 in Lecture Notes in Economics and Mathematical Systems. Springer, Berlin, 2000.

[11] M. Ehrgott and X. Gandibleux. An Annotated Bibliography of Multi-objective Combinatorial Optimization. Technical Report 62/2000, Fachbereich Mathematik, Universität Kaiserslautern, Kaiserslautern, Germany, 2000.

[12] D. Fogel. Asymptotic convergence properties of genetic algorithms and evolutionary programming: Analysis and experiments. *Cybernetics and Systems*, 25(3):389–407, 1994.

[13] D. B. Fogel. An introduction to simulated evolutionary optimization. *IEEE Transactions on Neural Networks*, 5(1):3–14, 1994.

[14] L. J. Fogel. Autonomous automata. *Industrial Research*, 4(1):14–19, 1962.

[15] C. M. Fonseca and P. J. Fleming. An Overview of Evolutionary Algorithms in Multiobjective Optimization. *Evolutionary Computation*, 3(1):1–16, Spring 1995.

[16] A. S. Fraser. Simulation of genetic systems by automatic digital computers. *Australian Journal of Biological Sciences*, 10:484–491, 1957.

[17] M. R. Garey and D. S. Johnson. *Computers and Intractability: A Guide to the Theory of NP Completeness*. W. H. Freeman and Company, San Francisco, 1979.

[18] A. M. Geoffrion, J. S. Dyer, and A. Feinberg. An Interactive Approach for Multi-Criterion Optimization, with an Application to the Operation of an Academic Department. *Management Science*, 19(4):357–368, 1972.

[19] J. H. Holland. *Adaption in Natural and Artificial Systems*. University of Michigan Press, 1975.

[20] R. L. Keeney and H. Raiffa. *Decisions with Multiple Objectives: Preferences and Value Trade-Offs*. Cambridge University Press, Cambridge, UK, 1993.

[21] J. R. Koza. *Genetic Programming: On the Programming of Computers by Means of Natural Selection*. MIT Press, 1992.

[22] W. B. Langdon and R. Poli. *Foundations of Genetic Programming*. Springer, 2002.

[23] J. Lohn, G. Hornby, and D. Linden. An Evolved Antenna for Deployment on NASA's Space Technology 5 Mission. In *Genetic Programming Theory and Practice II*, pages 13–15. Springer, 2004.

[24] A. V. Lotov, V. A. Bushenkov, and G. K. Kamenev. *Interactive Decision Maps: Approximation and Visualization of Pareto Frontier*. Kluwer Academic, 2004.

[25] K. Miettinen. *Nonlinear Multiobjective Optimization*. Kluwer Academic Publishers, 1999.

[26] R. Poli and W. Langdon. Schema Theory for Genetic Programming with One-Point Crossover and Point Mutation. *Evolutionary Computation*, 6(3):231–252, 1998.

[27] G. R. Price. Selection and covariance. *Nature*, 227:520–521, 1970.

[28] A. Pryke, S. Mostaghim, and A. Nazemi. Heatmap visualization of population based multi objective algorithms. In *Evolutionary Multi-Criterion Optimization (EMO 2006)*, volume 4403 of *LNCS*, pages 361–375. Springer, 2006.

[29] N. Radcliffe. The algebra of genetic algorithms. *Annals of Mathematics and Artificial Intelligence*, 10(4):339–384, 1994.

[30] I. Rechenberg. Cybernetic solution path of an experimental problem, 1965. Library Translation 1122, Royal Aircraft Establishment, Farnborough, UK.

[31] G. Rudolph. Convergence analysis of canonical genetic algorithms. *IEEE Transactions on Neural Networks*, 5(1):96–101, 1994.

[32] G. Rudolph. Convergence of evolutionary algorithms in general search spaces. In *Proceedings of the IEEE International Conference on Evolutionary Computation*, pages 50–54, 1996.

[33] H.-P. Schwefel. Kybernetische Evolution als Strategie der experimentellen Forschung in der Stromungstechnik. *Master's thesis, Hermann Föttinger Institute for Hydrodynamics, Technical University of Berlin*, 1965.

[34] Y. Siskos and A. Spyridakos. Intelligent multicriteria decision support: Overview and perspectives. *European Journal of Operational Research*, 113(2):236–246, 1999.

[35] R. E. Smith, B. A. Dike, B. Ravichandran, A. El-Fallah, and R. Mehra. Discovering Novel Fighter Combat Maneuvers in Simulation: Simulating Test Pilot Creativity. In *Creative Evolutionary Systems*, pages 467–486. Morgan Kaufmann, 2001.

[36] N. Srinivas and K. Deb. Multiobjective optimization using nondominated sorting in genetic algorithms. Technical report, Department of Mechanical Engineering, Indian Institute of Technology, Kanpur, India, 1993.

[37] D. Thierens and D. Goldberg. Convergence models of genetic algorithm selection schemes. In *Parallel Problem Solving from Nature — PPSN III*, volume 866 of *LNCS*, pages 119–129, 1994.

[38] P. Vincke. *Multicriteria Decision-aid*. Wiley, New York, 1992.

[39] D. H. Wolpert and W. G. Macready. No free lunch theorems for optimization. *IEEE Transactions on Evolutionary Computation*, 1:67–82, 1997.

Part I

Exploiting Multiple Objectives: From
Problems to Solutions

Multiobjective Optimization and Coevolution

Sevan Gregory Ficici

Harvard University
School of Engineering and Applied Sciences
Maxwell-Dworkin Laboratory
33 Oxford Street, RM 242
Cambridge, Massachusetts 02138 USA
sevan@eecs.harvard.edu
www.eecs.harvard.edu/~sevan

Summary. This chapter reviews a line of coevolutionary algorithm research that reframes coevolution as a form of multiobjective optimization. This shift in perspective towards coevolution as a form of multiobjective optimization has led researchers to new algorithmic and analytical formulations that have advanced the state of the art in coevolutionary algorithm research. We review relevant literature and discuss the basic concepts and issues involved in the application of multiobjective optimization to coevolutionary algorithms.

1 Introduction

Recent work in coevolutionary algorithm (CEA) research considers coevolution as a form of multiobjective optimization (MOO). This creates an unusual type of MOO in which the objectives, and optimizers of those objectives, are comprised of the coevolving individuals. Like any evolutionary algorithm (EA), a CEA is a stochastic population-based search method. A fundamental difference between a CEA and an "ordinary" (i.e., non-coevolutionary) EA concerns the process of evaluation. In an EA, each individual is evaluated with respect to some predefined objective function; the objective function provides a fixed standard that enables the direct comparison of individuals. In the CEA, by contrast, an individual is evaluated by having it interact with other coevolving individuals.[1] This interactive aspect of evaluation in coevolution is fundamentally game-theoretic in nature; hence, much recent research in coevolution takes a game-theoretic perspective. Thus, the outcome of evaluating an individual in a CEA depends upon the context of whom the individual interacts with. In one context, the evaluated individual may look superior, while in another context the same individual may appear inferior. This context sensitivity is

---

[1] In biology, the term *coevolution* refers to co-adaptation between two distinct populations; in evolutionary computation, we abuse the term somewhat by using it also to refer to the co-adaptation of individuals within a single population.

characteristic of coevolutionary systems and responsible for the complex dynamics for which coevolution is (in)famous.

The concept of MOO is easily imported into CEA research: each individual with which interaction can occur can be viewed as an objective for optimization. For example, let us say that individual X interacts with individuals A, B, and C. In this case, X's performance is measured with respect to three distinct optimization objectives. One optimization objective indicates how well an individual (e.g., X) performs when it interacts with A; another objective concerns performance with respect to B, and the third with respect to C. As with most MOO problems, the search space may be such that the optimal behaviour with respect to one objective (e.g., A) is suboptimal with respect to another objective (e.g., B); thus, a *trade-off surface* may exist. This trade-off surface underlies the context-sensitive nature of evaluation that has long been recognized in CEAs.

By viewing evaluation in coevolution as a form of MOO, we are able to invoke the concept of *Pareto optimality* as an organizing principle. This chapter will show how Pareto optimality can be used to articulate our goals in using a CEA, and thus serve as a *solution concept* for coevolution. In particular, we will consider how the MOO view of coevolution opens a new vista on the context sensitivity of evaluation in CEAs, pointing to improved understanding and more effective search algorithms.

This chapter is organized as follows. Section 2 provides an introduction to CEA research; it reviews some early and notable work in coevolution, defines the learner-teacher paradigm which CEAs often use, discusses various problems that can arise when using coevolution, and reviews the literature on the application of MOO to coevolution. Section 3 introduces the idea of solution concepts, and discusses the importance of a solution concept to evolutionary search and CEAs in particular. Section 4 introduces the fundamental concepts involved in using MOO in a coevolutionary algorithm, and discusses how an MOO perspective can help address the problems that can arise when using a CEA. Section 5 introduces the concept of monotonic progress in coevolution, and outlines how the Pareto optimality solution concept can have the property of monotonicity. Section 6 reviews issues that can arise in practice when you treat coevolution as a multiobjective optimization process. Finally, Section 7 offers some concluding remarks.

2 Coevolutionary Algorithms

Before we explore the MOO view of coevolution, we need to consider prior coevolution research to understand what the MOO perspective contributes. In this section, we present a selective literature review to provide a helpful background; more thorough reviews of coevolution research can be found in Ficici [18] and Wiegand [59]. This section concludes with a review of how MOO has been used in coevolution.

2.1 Early Work

Coevolution is a form of machine learning through *self-play*, where the computer plays a game against itself and incrementally adjusts its game strategy in response to how the play unfolds. In the late 1950's, Samuel's ground-breaking work [51] used self-play to learn the game of checkers; though this work did not use an evolutionary strategy-adjustment method, the first coevolutionary algorithms would soon follow. Published in 1963, Barricelli's research [3] used coevolutionary dynamics to learn strategies for the *TacTix* game, which is similar to Nim; follow-up work appears in 1967 [48]. Some 20 years later, Axelrod [2] used coevolution to evolve strategies for the Iterated Prisoner's Dilemma.[2]

Two iconic applications of coevolution are those of Hillis [29] and Sims [54]. Hillis [29] used coevolution as an optimization procedure to evolve minimal sorting networks. Because the number of possible input sequences to a sorting network grows rapidly with the length of the sequence, exhaustive testing of the network quickly becomes impractical. Some form of sampling is required. Hillis showed that coevolutionary dynamics can provide an effective approach to perform this sampling. Working with input sequences of length 16, he obtained a sorting network that used 61 comparators—just one more than the smallest network known for this input size. In contrast, when Hillis used random sampling of input sequences, instead of coevolving them, the smallest network found that could correctly sort all inputs had 65 comparators. Hillis' study presents evidence that coevolving the input sequences can provide a more effective gradient for evolving minimal sorting networks.

In a more open-ended setting, Sims [54] used coevolution to evolve virtual creatures that played a competitive two-player game in an environment with simulated physics. The objective for each creature was to gain control over a cube placed in the centre of an arena. By allowing both creature morphology and control to evolve, Sims was able to obtain a variety of interesting behaviours for playing this game; for example, some agents tried to scoop the cube to obtain control, whereas others used long extremities to push opponents away. Over evolutionary time, strategies and counter-strategies were seen to arise. Such evolutionary dynamics are often likened to an "arms race," where competitors must discover increasingly sophisticated strategies to survive. The potential of coevolution to engender an arms race in complexity is the primary appeal of coevolutionary algorithms.

2.2 The Learner-Teacher Paradigm

Hillis' [29] work on sorting networks used two populations; one population contained candidate sorting networks while the other contained sets of input sequences that required sorting. These populations were placed in a competitive relationship that manifested an asymmetric zero-sum game. The sorting networks were rewarded for properly sorting input sequences and input sequences were rewarded for thwarting sorting networks. This arrangement has become a standard paradigm known as "learner-teacher," "host-parasite," or "test-based"[3] coevolution.

One feature of the learner-teacher paradigm is that it often has certain asymmetries. First, the game being played is usually asymmetric, meaning that the strategy spaces for the learners and teachers are different. Second, we are often interested primarily in evolving competent learners; we care about evolving challenging teachers only insofar as they help the learning process of the learners. A third common feature of this paradigm is that the difficulties of the two search spaces are asymmetric;

---

[2] These early papers on coevolution are reproduced in [25].

[3] "Test-based" is the term used by de Jong and Bucci in their chapter in this volume.

specifically, it is usually much more difficult to find competent learners (e.g., effective sorting networks) than difficult teachers (e.g., input sequences that are hard to sort). This fact has heavily influenced much research in coevolutionary algorithms, as we will discuss below. In situations where both populations are of equal interest, or where we have just one population, individuals can simultaneously be learners as well as teachers for others.

The work of Pagie and Hogeweg [46] uses the learner-teacher paradigm in a symbolic regression task. One population contains genetic program (GP) functions while the other population contains data points that the GP functions must fit. A space of 676 possible data points is obtained by discretizing the domain of a two-dimensional target function. A GP function is rewarded for minimizing the difference between itself and the target function over the data points on which the GP function is tested; data points are rewarded for exposing large differences between a GP function and the target function.

The space of possible data points is small enough that results obtained by coevolving data points can be compared to results obtained from GP evolution using complete evaluation (i.e., where all 676 data points are used to evaluate each genetic program). Pagie and Hogeweg [46] find that coevolved data points produce genetic programs that better approximate the target function than genetic programs evolved with full evaluation. This result is surprising because evaluation on the full set of test points should provide a more accurate assessment of fitness. In contrast, evaluation using coevolved data points is highly focused on a relatively small portion of the target function domain; as a result, the fitness measurement obtained for a genetic program depends strongly upon the subset of data points used to take the measurement. Both Hillis [29] and Pagie and Hogeweg [46] argue that coevolving tests provides better performance because this context sensitivity of evaluation in coevolution allows the learner population to escape local optima—local optima that would otherwise stall evolutionary learning in an ordinary EA using full evaluation.

Much of the interest in using MOO for coevolution revolves around the goals of better understanding how to conduct evaluation and evolve teachers, and thereby make learner evolution more effective. For example, a conventional CEA performs evaluation by aggregating the outcomes of a learner's interactions with teachers into a scalar, typically by averaging. Though their algorithms have some less conventional aspects, both Hillis [29] and Pagie and Hogeweg [46] average over a learner's outcomes. Such aggregation loses information. In contrast, the MOO approach keeps the outcomes with each teacher separate, producing a vector rather than a scalar. Thus, multiobjective evaluation uses information differently, and promises to use it more advantageously.

2.3 Problems with Coevolution

Despite their many successful applications, coevolutionary algorithms are well known to exhibit a variety of undesirable behaviours that hinder problem solving [18, 59]. Among the various problems that can arise are *disengagement* and *cyclic dynamics*; both of these problems can further lead to *evolutionary forgetting*, where previously learned traits are lost and need to be re-learned. Yet another problem that can arise in coevolution is *overspecialization*.

Disengagement

Disengagement refers to a loss of fitness gradient in multi-population coevolution. Returning to our learner-teacher CEA paradigm, we note above that evolving a competent learner is typically much more difficult than evolving a challenging teacher. A naive coevolutionary algorithm will place learners and teachers in a strictly competitive framework, where learners obtain fitness for "solving" teachers and teachers obtain fitness for being difficult to solve. In such a framework, the more easily evolved population (typically the teachers) will quickly outpace the capabilities of the other population. This can eventually produce a state where none of the learners can solve any of the teachers. In such a situation, all of the learners will appear equally poor, and all of the teachers will appear equally challenging. Nevertheless, in reality, the learners will likely not be of equal competence—they will merely appear that way because teachers that can expose distinctions between the learners do not exist. Consequently, in each population, all individuals obtain the same fitness and are left subject to genetic drift; selection ceases to exert its influence. As Watson's numbers games show [58], we do not need asymmetric games for disengagement to occur; his experiments show that a loss of fitness gradient can also occur in symmetric games when members of one population can only interact with members of another population.

The problem of disengagement has been considered by many researchers (e.g., [50, 36, 47, 44, 10]). Though several methods have been used to ameliorate disengagement, they generally all provide a way to modulate the reproductive success of individuals that are deemed "too good" relative to others.

Cyclic Dynamics

Under the competitive dynamics of the conventional coevolutionary algorithm, *cyclic dynamics* can occur in both asymmetric (e.g., [36]) and symmetric games; in symmetric games, we can obtain cyclic dynamics in single-population (e.g., with Rock-Paper-Scissors (RPS) [30]) and multi-population (e.g., with the Penny-Matching game [26]) systems. In conventional coevolutionary algorithms, we have frequency-dependent selection; that is, the fitness of an individual depends not only upon what other strategies it interacts with, but also upon the frequency with which it interacts with these strategies. For example, in the RPS game, the fitness of an individual playing Rock depends upon how many Scissors and Paper players it interacts with; fitness improves as the proportion of Scissors increases, and fitness worsens as the proportion of Paper increases. Thus, if Paper is the majority of the population, then the proportion of Rock will increase, to the detriment of Scissors (which Rock beats) and the benefit of Paper (which beats Rock). Eventually, the fitness of Paper will surpass that of Rock, and so on in a cycle. The problem cyclic dynamics poses for coevolutionary search is that it can make the discovery of new areas of the search space more difficult; also, given the fact that real coevolutionary algorithms have finite populations, cyclic dynamics can drive a strategy to extinction during the portion of the cycle where the strategy is maladapted to the population; this leads to a loss of genetic diversity. Most importantly, whereas a search algorithm should converge onto a solution (assuming one exists), cycling is a divergent dynamic. Thus, cycling is generally a symptom of an improperly implemented solution concept; we discuss solution concepts further in Section 3.

Like the problem of disengagement, cycling has been examined by a number of researchers. Methods to avoid cycling generally work by diffusing selective pressure, for example by fitness sharing [49], using multiple genetically isolated populations [8, 31], or augmenting the environment of interaction to include non-agent elements [43].

Overspecialization

Overspecialization occurs when the evolutionary trajectory of learners concentrates on a sub-region of the search space where learner phenotypes exhibit only a narrow range of skills. For instance, if chess were our domain, overspecialization could result in strategies that, for example, only had good opening games but were otherwise poor. Often, a domain demands that learners acquire a breadth of competencies. The proper selection of teachers is required to shepherd the learner population towards good, well-rounded solutions.

2.4 Pareto Coevolution

The ideas of MOO and coevolution appear together at least as early as the work of Juillé and Pollack [34, 33, 32]. This work coevolves GP classifiers for the intertwined spirals problem, which originates from the neural networks community (e.g., [38]). Two intertwined spirals are composed of 97 data points each, and the classification task is to identify to which spiral each datum belongs. Juillé and Pollack [34, 33, 32] obtain effective GP classifiers that have an interesting modular structure that corresponds with a spatial division of the input space. They evaluate classifiers in a series of pairwise comparisons. Given a pair of classifiers L_a and L_b, a point is earned by L_a for each datum T_j where L_a correctly classifies T_j and L_b incorrectly classifies T_j, and *vice versa* for classifier L_b. This scoring method creates a form of niching that enhances genetic diversity during search. The 194 total data points of the spirals are described by Juillé and Pollack [32] as being independent optimization problems; hence a form of multiobjective optimization is being performed. Nevertheless, this work is distinguished from later work on Pareto coevolution by the fact that the objectives themselves are not subject to evolution.

Watson and Pollack [57] and Ficici and Pollack [21] introduce the idea that the coevolving entities themselves can be considered objectives for optimization. In Watson and Pollack [57], the criterion of Pareto dominance is used to implement selection in the SEAM algorithm; this allows the *shuffled HIFF* problem to be solved [56]. Pareto dominance is next used in algorithms by Noble and Watson [42] and Ficici and Pollack [23]. In Noble and Watson [42], Pareto coevolution is used to coevolve poker strategies; their method delivers encouraging results compared to those obtained with an ordinary EA. In Ficici and Pollack [23], Pareto coevolution is used to discover cellular automaton (CA) rules for the density classification task; a high-quality rule (84% correct classification) is obtained.

The density classification problem falls neatly into the learner-teacher paradigm. Two populations are involved: CA rules (learners) and automaton initial conditions (teachers). As demonstrated by Juillé and Pollack [35, 36], initial conditions that are difficult to classify are much easier to evolve than rules that are effective classifiers. This means that special effort is required to guide the evolution of the teacher population in order to provide a useful learning gradient for the learner population; without such guidance, problematic dynamics such as those described in Section 2.3 may arise and impede learning.

The need to create a useful gradient for learners affords Ficici and Pollack [23] the opportunity to explore not only the application of Pareto optimality to coevolution, but also the idea of *distinctions*. The concept of distinctions is applied to the teacher population; a teacher that causes a pair of learners to obtain different outcomes is said to *distinguish* the learners. For example, by causing two CA rules to perform differently, an automaton initial condition reveals a distinction in their performance. We may understand distinctions to be the dual of the Pareto concept: dominance and nondominance is determined by establishing differences in behaviour. The teachers' abilities to distinguish learners thus affects the learning gradient for the learner population.

As we will discuss below, an observation made early in Pareto coevolution research was that the number of possible objectives can quickly grow as search progresses. The work of Bucci and Pollack [6, 7] and de Jong and Pollack [16] establishes a theory of the minimal number of objectives required for Pareto coevolution to succeed. More recently, work by Bucci et al. [5] and de Jong and Bucci [15] develops a formalism to project this potentially high-dimensional space onto an alternative coordinate system that decomposes the search space into a minimal number of "behavioural" dimensions. One can then order individual phenotypes according to their competencies in these dimensions. This new ordering may allow for more effective evaluation and search.The chapter by de Jong and Bucci in this volume presents new work on using MOO to discover behavioural dimensions.

Finally, the concept of distinctions appears in the recent coevolutionary algorithm by Bongard and Lipson [4]; they present an online learning method for system identification that attempts to minimize the number of (sometimes costly) probes of the system being modeled. This work coevolves candidate system models (learners) with model tests (teachers). Tests are sought that cause the population of evolving models to differ in their predictions of what the real system would do given the input specified by the test; such tests distinguish the models. Once such distinctions are discovered, the system being modeled is probed on these tests to resolve model disagreements. This learning approach thus concentrates on probes of the real system that will reveal the learning gradient. Good results for a number of system identification problems are reported.

3 Solution Concepts

As we have seen above, many optimization problems can be recast as games. For example, in the case of sorting networks [29], an optimization process was transformed into a constant-sum two-player game between a sorting network and an input sequence. As with any EA used for optimization, when we use a CEA to "solve" a game, we are performing search. The process of search presupposes some notion of what it is we are searching for. The purpose of a *solution concept* is to articulate our goal for search.

3.1 Definition

A solution concept is a formalism that specifies which elements of a search space are *solutions* to a search problem. We borrow the term 'solution concept' from game theory [27]. Solving a game is a form of search; here, we are interested in applying the idea of a solution concept to search problems, in general.

We begin with a search domain $\mathcal{D}$. When we search a domain, we require a solution concept $\mathcal{O}$ to perform a binary partition of $\mathcal{D}$; one partition $\mathcal{D}_{\in \mathcal{O}}$ contains only solutions, while the other partition $\mathcal{D}_{\notin \mathcal{O}}$ contains only non-solutions. At most, one of these partitions may be empty. Thus, depending upon the contents of $\mathcal{D}$ and the definition of $\mathcal{O}$, a search problem may have one, many, or no solutions.

Typically, the elements in $\mathcal{D}_{\in \mathcal{O}}$ are distinguished in some systematic way from the elements in $\mathcal{D}_{\notin \mathcal{O}}$. Solution concepts that produce such systematic partitions are *intensional* in nature. Intensional solution concepts specify what (measurable) properties an element in the domain must have in order to constitute a solution to our search problem. The solution concept does not indicate what the actual solution(s) will be, if any—it only specifies what properties solutions will have if they exist. In contrast, an *extensional* solution concept lacks any underlying semantics, and can be used to obtain an arbitrary partition. Extensional solution concepts are rare in practice.

By applying a solution concept to a search domain, we obtain a specific *search problem*. We can imagine an infinity of search problems. Of course, if $\mathcal{D}$ is finite (which it need not be), the number of ways to partition it is finite, and exponential in its size (i.e., the possible solution sets are members of the power set $2^{\mathcal{D}}$).

The solution concept can also operate on any subspace $\mathcal{B} \subset \mathcal{D}$ such that $\mathcal{B}$ is partitioned. Note that solutions to $\mathcal{B}$ are not necessarily solutions to $\mathcal{D}$, and *vice versa*. One way to think of a subspace $\mathcal{B}$ is that it represents the portion of $\mathcal{D}$ that we have discovered so far. We discuss this idea more below.

3.2 Example Solution Concepts

Simple Examples

To illustrate how solution concepts may work, let us begin with a simple search space; let $\mathcal{D} = \{3, 4, 9, 12, 15, 21, 25, 27\}$. One solution concept may specify that solutions are prime numbers; in this case, we obtain the partition $\mathcal{D}_{\in \mathcal{O}} = \{3\}$ and $\mathcal{D}_{\notin \mathcal{O}} = \{4, 9, 12, 15, 21, 25, 27\}$. Another solution concept may specify that solutions are even numbers; in this case, we have $\mathcal{D}_{\in \mathcal{O}} = \{4, 12\}$ and $\mathcal{D}_{\notin \mathcal{O}} = \{3, 9, 15, 21, 25, 27\}$. If the solution concept specifies that the most positive number is a solution, then $\mathcal{D}_{\in \mathcal{O}} = \{27\}$; if we were aware only of the subspace $\mathcal{B} \subset \mathcal{D}$, where $\mathcal{B} = \{3, 9, 12\}$, then $\mathcal{B}_{\in \mathcal{O}} = \{12\}$. Thus, a solution concept can be applied to local states of knowledge; we go into more detail on this topic in Section 5.

Single-Objective Search

Moving to more realistic scenarios, say we have a search space that spans some continuous interval $\mathcal{D} = [-100, 100]$. If we are interested in maximizing some continuous function $f(x)$, then we have a single-objective search problem with a straightforward solution concept: any value of $x \in \mathcal{D}$ that maximizes $f(x)$ is a solution. If there is more than one solution, then we are indifferent as to which one we choose; if we have a preference of one solution over another, then we need to refine our solution concept, as we discuss in Section 3.3.

Multiobjective Search

Next, we may add a second continuous function $h(x)$, and construct a two-objective search problem where we seek to maximize $f(x)$ and simultaneously minimize $h(x)$. It could easily be that the value of x that maximizes $f(x)$ is not the value that minimizes $h(x)$, and vice versa; in this case, we have a trade-off curve that we need to explore. A sensible solution concept for this problem would be Pareto optimality; the solution would be the set of nondominated values of x. A value x is nondominated for our problem if there does not exist an alternative value x' such that $f(x') > f(x)$ and $h(x') \leq h(x)$, or $f(x') \geq f(x)$ and $h(x') < h(x)$.

When Pareto optimality is used as a solution concept, it is very often the case that we seek to locate more than just one, or a few, of the nondominated elements of the search space; indeed, we typically seek to identify as much of the trade-off surface as possible. That is, the desired output from the search process is the set of all nondominated elements of the search space, which is known as the *Pareto front*. Once this trade-off surface is obtained, we usually embark upon a separate decision-making process to select one element from the trade-off surface (we discuss this process more in Section 6.2). Thus, depending upon our needs, when we use Pareto optimality as a solution concept, we may define a solution to be either an element of the trade-off surface (less common) or the entire trade-off surface itself (more common).

Games

Arguably, the most sophisticated solution concepts are those applied to games. The mathematics of strategic reasoning is known as game theory [27]. A game G has N players; each player i has a set $\mathcal{S}_i$ of game strategies available to her. Each of these players chooses a strategy; for player i, this strategy choice may be either a *pure strategy* $s_i \in \mathcal{S}_i$, or a *mixed strategy*, which is a stochastic strategy defined by a probability distribution over the player's pure strategies $\mathcal{S}_i$. Note that a pure strategy is a degenerate mixed strategy where all the probability mass is on one pure strategy. Let $C = \langle m_0, m_1, \ldots, m_{N-1} \rangle$ be the *strategy profile* that represents the strategic choices of the N players, where m_i is the mixed strategy choice of player i. Given a strategy profile C, the game G also defines the *payoff* each player of the game receives. A solution concept, then, is a formal specification about the properties a profile C should have to constitute a solution to the game G.

The most famous solution concept in game theory is the Nash equilibrium [41]. Finding a Nash equilibrium is essentially a combinatorial search problem [11]. A profile $C^* = \langle m_0^*, m_1^*, \ldots, m_{N-1}^* \rangle$ is in Nash equilibrium when no player has incentive to deviate unilaterally from her mixed-strategy choice. That is, for each player i, no other mixed strategy $m_i \neq m_i^*$ that player i could use will give player i a higher payoff than m_i^*, given the strategic choices of the $N - 1$ other players.

A game may have more than one Nash equilibrium; indeed, the number of Nash equilibria may be exponential in the number of strategies [40]. For this reason, efforts to discover all of the Nash equilibria of a game are generally impractical; this is in contrast to the size of the Pareto front, which cannot be larger than the search space itself, and is often much smaller. Thus, the solution concept affects how the number of solutions relates to the size of the search space; this, in turn, affects our decision as to whether to locate one, some, or all solutions.

3.3 Refinements

A solution concept may yield more than one solution when applied to a domain, and these different solutions may have different properties; this is often the case, for example, in games that have multiple Nash equilibria. Given multiple solutions, we may prefer the properties of one solution over another. If we do have such a preference, then how do we formalize our preference to rationalize our choice? In game theory, this question is the *equilibrium selection problem* [52], and often entails *refining* our solution concept. A solution concept $\mathcal{O}_\beta$ is a (strict) refinement of another solution concept $\mathcal{O}_\alpha$ if and only if

$$\forall \mathcal{D} : \mathcal{D}_{\in \mathcal{O}_\beta} \subseteq \mathcal{D}_{\in \mathcal{O}_\alpha} \text{ and } \exists \mathcal{D} : \mathcal{D}_{\in \mathcal{O}_\beta} \subset \mathcal{D}_{\in \mathcal{O}_\alpha}.$$

That is, for $\mathcal{O}_\beta$ to strictly refine $\mathcal{O}_\alpha$, two conditions must be met. First, for every possible domain, all solutions to $\mathcal{O}_\beta$ must also be solutions to $\mathcal{O}_\alpha$. Second, there must exist at least one domain for which $\mathcal{O}_\beta$ specifies fewer solutions than $\mathcal{O}_\alpha$.

3.4 Implementation

Our definition of a search problem stipulates that the solution concept be intrinsic to the problem statement. Nevertheless, the solution concept itself must be implemented by our search algorithm. Of course, we have many search algorithms from which to choose. These algorithms may differ in terms of their computational demands and efficiency, but they must all implement the solution concept of our search problem if they are to be correct. When an algorithm implements a different solution concept, then it solves a different search problem. Solution concepts are therefore important because they link search problems and search algorithms. This basic observation is the foundation for the coevolutionary algorithm research described in [18].

Evolutionary game theory (EGT) [39] provides a convenient framework for illustrating how different algorithmic mechanisms may or may not implement a solution concept (e.g., Nash equilibrium) when used in a CEA. The EGT framework is a selection-only system (i.e., there is no variation) and has an infinitely large population. The canonical selection dynamics in EGT correspond to "fitness proportional" selection in evolutionary algorithms. In the EGT framework, all dynamical point attractors of proportional selection are Nash equilibria [30]. Nevertheless, evolutionary computation uses a wide variety of selection methods aside from proportional selection. Many of these alternative selection methods, however, have very different dynamical properties. In a certain class of games where proportional selection would have converged to Nash equilibrium, these alternative selection methods instead converge to non-Nash fixed points, or yield cyclic or even chaotic dynamics [20]. These are examples of how an algorithm can fail to implement a certain solution concept. Though all selection methods realize Darwinian selection in a broad sense, they are not interchangeable with respect to the solution concept; the implementation details of selection matter, as they affect the solution concept [20].

The EGT framework can also be used to study whether or not a certain solution concept is implemented by diversity maintenance methods [22, 18]. For example, both competitive fitness sharing [50] and similarity-based fitness sharing [28] fail to implement the Nash equilibrium solution concept in a class of variable-sum games. This result highlights the fact that the evolving population is essentially charged with two, sometimes conflicting duties. One duty of the population is to conduct search; the reason we use diversity maintenance methods is to improve the efficacy of the search. The other duty of the population is to contain the solution of the search process.

The difficulty arises when the solution to our search problem can only be represented by the state of the population as a whole, rather than by just an individual within the population. For example, the Hawk-Dove game [39] is a simple variable-sum game of two pure strategies (Hawk and Dove). Given a single population of pure-strategists, the Nash equilibrium for this game is represented by a *polymorphic population*, where the population contains both Hawks and Doves in a specific proportion; this proportion corresponds to the probabilities with which a Nash equilibrium mixed-strategy would play Hawk and Dove. Thus, while the solution demands one population state, a diversity maintenance method will likely demand some other population state (to maximize genetic diversity), bringing the two tasks charged to the population into conflict. This realization has led to the emergence of new algorithms where archives take over the duty of representing the solution obtained from search, leaving the population free to conduct the search itself (e.g., [24, 12]).

4 Pareto Optimality in Coevolution

4.1 Learners and Teachers

For purposes of exposition, let us assume that our search problem fits the learner-teacher paradigm that has so often been used in coevolutionary algorithms. We have already discussed two examples of this paradigm in Section 2.2. Another example is that of Juillé and Pollack [36], who coevolve a population of cellular automaton (CA) rules (learners) with a population of lattice initial conditions (teachers).

The solution concepts that interest us here are Pareto optimality and its "dual" concept, which concerns distinctions. The Pareto-distinctions duality fits neatly into the learner-teacher paradigm. As we note above, in this paradigm, the task of evolving a competent learner is typically much more difficult than evolving a challenging teacher. Thus, the problem is to evolve teachers that are of appropriate difficulty. Put another way, the goal of the teacher population is to create a gradient that the learner population can successfully follow such that learners improve.

4.2 Pareto Optimality

Domination, Mutual Nondomination, and Equivalence

Let $\mathcal{L}$ be a set of learners and $\mathcal{T}$ a set of teachers. Let G be a matrix of payoffs where $G_{i,j}$ is the payoff learner $L_i \in \mathcal{L}$ obtains from interacting with teacher $T_j \in \mathcal{T}$. Following the usual definition of Pareto domination, learner L_a *dominates* learner L_b with respect to $\mathcal{T}$, which we denote $L_a \overset{\mathcal{T}}{\succ} L_b$, if and only if

$$L_a \overset{\mathcal{T}}{\succ} L_b \equiv \forall T_j : G_{a,j} \geq G_{b,j} \text{ and } \exists T_j : G_{a,j} > G_{b,j}. \tag{1}$$

That is, learner L_a dominates learner L_b with respect to $\mathcal{T}$ when interaction with each teacher in $\mathcal{T}$ causes L_a to perform at least as well as L_b and interacting with some teacher in $\mathcal{T}$ causes L_a to perform strictly better than L_b.

Two learners L_a and L_b are *mutually nondominating* with respect to $\mathcal{T}$, denoted $L_a \overset{\mathcal{T}}{\sim} L_b$, if and only if

$$L_a \overset{\mathcal{T}}{\sim} L_b \equiv \exists T_j, T_k : G_{a,j} > G_{b,j} \text{ and } G_{a,k} < G_{b,k}. \tag{2}$$

That is, there exists some teacher with respect to which learner L_a outperforms L_b, and there exists some other teacher with respect to which learner L_b outperforms L_a.

Two learners L_a and L_b are *equivalent* with respect to $\mathcal{T}$, denoted $L_a \overset{\mathcal{T}}{=} L_b$, if and only if

$$L_a \overset{\mathcal{T}}{=} L_b \equiv \forall T_j \in \mathcal{T} : G_{a,j} = G_{b,j}. \tag{3}$$

That is, there exists no teacher in the set $\mathcal{T}$ that distinguishes the performance of L_a and L_b.

To reiterate, the relation that holds between a pair of learners is conditioned on the set of teachers; for example, it may be that $L_a \overset{\mathcal{T}}{\succ} L_b$ but $L_a \overset{\mathcal{T}'}{\sim} L_b$, for $\mathcal{T} \neq \mathcal{T}'$.

Dominated Set and Pareto Front

The set of dominated learners in $\mathcal{L}$ with respect to $\mathcal{T}$, denoted $D^0(\mathcal{L}, \mathcal{T})$, is

$$D^0(\mathcal{L}, \mathcal{T}) \equiv \{L \in \mathcal{L} : \exists L_x \in \mathcal{L}, L_x \succ L\}. \tag{4}$$

The *Pareto optimal* learners in $\mathcal{L}$ with respect to $\mathcal{T}$, denoted $F^0(\mathcal{L}, \mathcal{T})$, is therefore

$$F^0(\mathcal{L}, \mathcal{T}) \equiv \mathcal{L} - D^0(\mathcal{L}, \mathcal{T}). \tag{5}$$

That is, the Pareto optimal learners are those learners in $\mathcal{L}$ that are not dominated. The set $F^0(\mathcal{L}, \mathcal{T})$ is also known as the *Pareto front*. For any two learners in $F^0(\mathcal{L}, \mathcal{T})$, it must be the case that they mutually nondominate each other, or that they obtain identical payoffs across the set of teachers:

$$\forall L_x, L_y \in F^0(\mathcal{L}, \mathcal{T}) : L_x \overset{\mathcal{T}}{\sim} L_y \vee L_x \overset{\mathcal{T}}{=} L_y. \tag{6}$$

Each learner of a pair of mutually nondominating learners can do something better than the other. For example, if $L_a \overset{\mathcal{T}}{\sim} L_b$, then L_a possesses some competency that L_b lacks and vice versa. With respect to the set of teachers $\mathcal{T}$, a dominated learner is less capable, or general, than the learner (or learners) that dominates it. When $L_a \overset{\mathcal{T}}{\succ} L_b$, for example, it is the case that the competencies of L_a are at least as good as those of L_b in every tested objective and strictly better in at least one; thus, we can think of L_a as having more broadly applicable abilities than L_b with respect to $\mathcal{T}$.

Pareto Layers

If we have a finite set of learners, once we identify $F^0(\mathcal{L}, \mathcal{T})$, we can recursively examine $D^0(\mathcal{L}, \mathcal{T})$ to locate learners that would be nondominated if $\mathcal{L}$ did not contain the learners in $F^0(\mathcal{L}, \mathcal{T})$. Thus, we define

$$F^N(\mathcal{L}, \mathcal{T}) = F^0(D^{N-1}(\mathcal{L}, \mathcal{T}), \mathcal{T}), \tag{7}$$

$$D^N(\mathcal{L}, \mathcal{T}) = D^0(D^{N-1}(\mathcal{L}, \mathcal{T}), \mathcal{T}). \tag{8}$$

We can continue the recursion until every learner is identified as belonging to a particular *Pareto layer*. As with the Pareto front, learners within each layer N are either mutually nondominating or equivalent with respect to $\mathcal{T}$. Within a single layer, each learner of a pair of mutually nondominating learners can do something better than the other. Every learner in layer F^N must be dominated by some learner in layer F^{N-1}, and so is less general than its dominating learner(s) with respect to $\mathcal{T}$.

4.3 Distinctions

A set of teachers $\mathcal{T}$ specifies objectives for learner optimization. For each pair of learners L_x and L_y where $L_x \overset{\mathcal{T}}{\succ} L_y$ or $L_x \overset{\mathcal{T}}{\sim} L_y$, there exist specific teachers that *distinguish* the performance or behaviour of L_x from L_y. To identify the distinctions made by the teachers, we construct a new matrix H, where $H_{m,n} = 1$ if teacher T_m distinguishes the learners in the ordered pair $n = \langle L_i, L_j \rangle$ in favour of L_i; otherwise, $H_{m,n} = 0$. That is, $H_{m,n} = 1 \Leftrightarrow G_{i,m} > G_{j,m}$. This also means that if $H_{m,n} = 1$ for $n = \langle L_i, L_j \rangle$, then $H_{m,n'} = 0$ for $n' = \langle L_j, L_i \rangle$.

Equation (9) shows a simple example matrix G. In this example, we have three learners and three teachers. We see that learner $L_b \overset{\mathcal{T}}{\succ} L_c$; specifically, all three teachers distinguish L_b from L_c in favour of L_b. In contrast, $L_b \overset{\mathcal{T}}{\sim} L_a$ and $L_c \overset{\mathcal{T}}{\sim} L_a$. In the latter case, teacher T_α does not distinguish L_c from L_a, but the other two teachers do, one in favour of each learner.

$$G = \begin{array}{c|ccc} & T_\alpha & T_\beta & T_\gamma \\ \hline L_a & 1 & 1 & 3 \\ L_b & 4 & 3 & 2 \\ L_c & 1 & 2 & 1 \end{array} \tag{9}$$

From matrix G we construct matrix H, as shown in (10). We can now easily see the gaps in learner performance that each teacher exposes. In particular, we see that T_β makes all the distinctions that T_α makes, plus one other. T_γ makes distinctions that neither T_α nor T_β make, and *vice versa*. If we view the ordered learner pairs as optimization objectives for teachers, then we see that $T_\beta \succ T_\alpha$, $T_\gamma \sim T_\alpha$, and $T_\gamma \sim T_\beta$. Thus, we can use the idea of Pareto optimality to select teachers in a way that complements the use of Pareto optimality to select learners.

$$H = \begin{array}{c|cccccc} & \langle L_a, L_b \rangle & \langle L_a, L_c \rangle & \langle L_b, L_a \rangle & \langle L_b, L_c \rangle & \langle L_c, L_a \rangle & \langle L_c, L_b \rangle \\ \hline T_\alpha & 0 & 0 & 1 & 1 & 0 & 0 \\ T_\beta & 0 & 0 & 1 & 1 & 1 & 0 \\ T_\gamma & 1 & 1 & 0 & 1 & 0 & 0 \end{array} \tag{10}$$

4.4 Symmetric Games

In the learner-teacher CEA paradigm, learners and teachers are often different *types* of individuals. This is the case in domains like the sorting networks task [29] or the CA rules task [36]; these are two-player games where the type of one player, say the sorting network, is intrinsically different from that of the other player, say a sequence of unsorted numbers. The games being played in such domains are termed *asymmetric*, because each game role requires a different set of strategies. In contrast, a *symmetric* game is one where each role in the game shares the same strategy set and payoff function. Examples of such games are Rock-Paper-Scissors, the Iterated Prisoner's Dilemma [1, 2], and numbers games [58].

Though we introduce Pareto optimality and distinctions in the context of asymmetric games, nothing prevents us from applying these concepts in symmetric games. In a symmetric game, the game players share the same set of strategies, and we no longer have the situation where the learning task of one game player is intrinsically more difficult than that of another's. Thus, we no longer need to differentiate players as learners or teachers. Instead, players may simultaneously be learners and teachers; fitness may now be conferred onto an individual for being good at "doing" or for being good at "teaching", or for both.

4.5 Addressing Problems in Coevolution

In Section 2.3, we discuss various problems that can arise during the use of a coevolutionary algorithm. Here we discuss how the concepts of Pareto optimality and distinctions address these problems.

With respect to the issue of disengagement, Pareto optimality and distinctions help by providing an alternative to the naive competitive coevolutionary framework. By rewarding teachers for exposing gaps in learner competence (i.e., showing one learner to be able to "solve a problem" that another cannot), the teacher population is given incentive to remain engaged; that is, teachers obtain fitness by revealing a gradient structure for the learners to follow. This incentive structure prevents teachers from becoming uniformly too difficult for the learner population.

Next, cyclic dynamics can involve frequency-dependent fitness and intransitive structures (e.g., such as those found in RPS). Pareto optimality and distinctions are not frequency dependent; further, these concepts can detect intransitivities. For example, all three strategies in RPS are involved in an intransitive loop. Each strategy can beat some strategy that no other can; only Rock beats Scissors, and so on. As a result, each strategy is on the Pareto front (regardless of the frequency with which each strategy appears in the population). Each RPS strategy can also make a unique distinction; for example, only Rock can distinguish Paper and Scissors in favour of Paper, and so on. Thus, all three RPS strategies are nondominated teachers as well.

Finally, overspecialization occurs when an appropriate diversity of teachers is not used to guide learners. The problem of overspecialization is illustrated in de Jong's *compare-on-one* game [16]; in this game, conventional coevolutionary approaches easily lose teacher diversity, resulting in learners that are very good in one dimension of optimization, but very poor in another. One approach to address overspecialization is proposed by de Jong [13], who constructs an archive mechanism that uses both Pareto optimality and the concept of distinctions to maintain teacher diversity, and obtains more generalized learner progress.

5 Monotonic Behaviour

A desirable property for search algorithms, in general, is that the longer a search algorithm runs, the better its output should be. A similar statement would be to say that, the more we explore a search space, the more our estimation of a solution should improve. Here, we discuss whether a coevolutionary algorithm using Pareto optimality as a solution concept might have such properties. We will assume that we seek the entire Pareto front as our solution.

To explore these questions, we make some simplifying assumptions. First, we assume a single-population system; thus, each member of the population is evaluated by having it interact with members of the same population. Second, we make the strong assumption that the population may grow indefinitely as we discover more of the search space; that is, each newly discovered element of the search space is added to our population. This second assumption ensures that our knowledge of the search space will grow monotonically. Our questions then become the following: Given monotonically increasing knowledge of the search space, will we obtain monotonically improving estimations of our solution (i.e., the Pareto front)?

On the one hand, as knowledge of the search space increases, evaluation becomes more thorough. Thus, it may appear intuitive that monotonically increasing knowledge of the search space should imply monotonically improving estimates of the Pareto front. On the other hand, the outcome of evaluation is sensitive to whom an individual interacts with. Thus, it might be that when we add a newly discovered element of the search space to our evaluation process, the evaluations of previously discovered elements will change in dramatic and important ways, potentially upsetting a monotonic path towards the solution. We will see below that whether our estimations of the solution actually improve monotonically or not depends upon how we treat learners that appear equivalent. (For further discussion of monotonic progress in coevolution, see [18, 19, 14].)

5.1 Formalism

Let $\mathcal{W}^t \subseteq \mathcal{D}$ be our *state of knowledge* of the search space $\mathcal{D}$ at time t. With each new element of $\mathcal{D}$ that we discover, $\mathcal{W}$ increases. Thus, $\mathcal{W}^{t+1} \supseteq \mathcal{W}^t$. At each point in time t, we can apply our solution concept $\mathcal{O}$ to our state of knowledge $\mathcal{W}^t$. This gives us a sequence of states of knowledge and a corresponding sequence of solutions.

We can think of this latter sequence as a series of exact solutions to our states of knowledge or a series of estimated solutions to the full search space $\mathcal{D}$. Using Pareto optimality as our solution concept, we extract for each time point t the set of learners that are nondominated with respect to $\mathcal{W}^t$. We are interested in knowing how the nondominated set of $\mathcal{W}^t$ may change with time, and how it might relate to the true solution—the Pareto front of $\mathcal{D}$.

5.2 Example

Let us work through a simple example of how the Pareto front might change as we discover more of the search space. We assume a search space $\mathcal{D} = \{a, b, c, d, e, f\}$; the full payoff matrix for this space is

$$G = \begin{array}{c|cccccc} & a & b & c & d & e & f \\ \hline a & 5 & 3 & 2 & 4 & 2 & 2 \\ b & 0 & 3 & 1 & 3 & 2 & 1 \\ c & 2 & 1 & 1 & 2 & 0 & 1 \\ d & 5 & 3 & 2 & 4 & 1 & 3 \\ e & 0 & 3 & 2 & 1 & 1 & 2 \\ f & 1 & 0 & 0 & 2 & 0 & 1 \end{array}. \tag{11}$$

Matrix element $G_{i,j}$ is the payoff strategy i obtains when it interacts with strategy j. Let our initial state of knowledge be $\mathcal{W}^0 = \{a, b, c\}$. Knowing only the first three rows and columns of the payoff matrix (11), the Pareto front is comprised only of element a. Strategies b and c are each dominated by a; that is, $a \overset{\mathcal{W}^0}{\succ} b$ and $a \overset{\mathcal{W}^0}{\succ} c$. Next, say we discover a new strategy d, and our subsequent state of knowledge becomes $\mathcal{W}^1 = \{a, b, c, d\}$. Now we know the first four rows and columns of the payoff matrix. With this knowledge, we find that b and c are still dominated, but d appears to have the same payoff profile as a; that is, $a \overset{\mathcal{W}^1}{=} d$. Since a and d appear equivalent, we place them both on the Pareto front. Say we next discover strategy e, giving $\mathcal{W}^2 = \{a, b, c, d, e\}$. We find that e it distinguishes a and d in favour of a. As a result, we have $a \overset{\mathcal{W}^2}{\succ} d$, and only a is on the front. Finally, say we discover strategy f, which, although it is dominated, nevertheless distinguishes a and d in favour of d. This makes $a \overset{\mathcal{W}^3}{\sim} d$, and so the Pareto front is now a and d.

The above example gives us the following sequence of solution estimations as knowledge increases: $\{a\}, \{a, d\}, \{a\}, \{a, d\}$. Note that the first estimation in the sequence is discarded only to reappear as the third estimation; similarly the second estimation is discarded to reappear as the fourth. We call such behaviour *non-monotonic*. In contrast, what we call monotonic behaviour requires that, once we discard an estimation of the true solution due to an increase in knowledge, then we will never repeat that estimation for any future state of knowledge.

The source of the non-monotonic behaviour that we observe in our example occurs at time-step $t = 1$, where we discover strategy d. Our mistake was to add d to the Pareto front because it appeared equivalent to a. If we instead exclude from our estimation new strategies that appear equivalent to previously discovered strategies, then we will obtain monotonic behaviour from the Pareto solution concept; under this modified approach, our sequence of estimations will be $\{a\}, \{a\}, \{a\}, \{a, d\}$. As another possibility, we can include strategy d as an alternative estimation, along with a; that is, at time-step $t = 1$, we will estimate the true Pareto front to be either $\{a\}$ or $\{d\}$, instead of $\{a, d\}$. At time-step $t = 2$, the discovery of strategy e, which distinguishes a and d in favour of a, allows us to see that the true Pareto front cannot possibly be $\{d\}$. Note that this is different from saying that d cannot appear on the true Pareto front; we are merely saying that d alone is not the true Pareto front.

In general, the relation between any two strategies can only move from $=$ to $\succ$ or from $\succ$ to $\sim$ as knowledge increases. Once we transition from $x = y$ to $x \succ y$, we can never go back because we know there is at least one dimension of performance that distinguishes x and y; similarly, once we transition from $x \succ y$ to $x \sim y$, we can never go back because we know that x and y each has at least one dimension of performance that distinguishes it, in its favour, from the other.

Monotonic behaviour is desirable because it means that we can discard estimations with confidence; we will never change our minds and return to a previously discarded estimation in light of new knowledge of the search space. Further, Ficici [19] shows that the space of possible estimations can be arranged into a partial order; a monotonic sequence of estimations moves through the space of possible estimations such that the partial order is never violated. Thus, there exists a formal sense in which coevolutionary search can achieve directional progress (even when the search space is open-ended).

6 Issues in Practice

6.1 Managing the Nondominated Set

Most multiobjective optimization involves a modest number of objectives. In Pareto coevolution, however, the number of objectives can be on the order of the population size. Researchers investigating Pareto coevolution quickly found the potentially large number of objectives to be problematic, as it allowed the Pareto front to include most of the learner population [42, 23]. As a result, early work resorted to various heuristic methods to create a fitness gradient amongst the members of the Pareto front. The approach used by Noble and Watson [42] was to eliminate individuals that were nearly dominated; in Ficici and Pollack [23], fitness sharing schemes [50, 33] were used to obtain a gradient within a Pareto layer.

As we mention in Section 3.4, recent work on coevolutionary algorithms has begun to emphasize the use of structured archives as an antidote to the problem of expecting a population to simultaneously conduct search and represent the solution to the search effort. For example, Ficici and Pollack [24] introduce an archive that implements the Nash equilibrium solution concept. Work by de Jong [12, 13] introduces an archive based upon Pareto optimality. In each of these examples, the purpose of the archive is to organize (according to the solution concept in use) the elements of the search space that are discovered by a separate search mechanism (e.g., an evolving population). In this way, the task of performing search is made distinct from the task of representing the solution of the search process; this avoids

the potential conflicts discussed in Section 3.4 that can arise with diversity maintenance methods. The trend towards archive-based algorithms in coevolution mirrors
that in evolutionary multiobjective optimization (e.g., [37]).

6.2 Operationalizing the Result of Multiobjective Search

An important issue in multiobjective optimization concerns how the result of the
optimization process is to be used. Given a set of candidate solutions that approximate the true Pareto front, how do we choose a particular candidate as our final
"answer" to our search problem? Of course, the specifics of the search domain play
an important role in this decision making. For example, if we are evolving designs for
a bridge, we might care about the trade-off between material costs and durability.
Whatever trade-off we decide upon, it is clear that we will select exactly one design.
The situation for coevolution is typically more complicated because of the close
kinship between coevolutionary domains and game theory. The result of Pareto coevolution essentially yields the set of (pure) game strategies on the trade-off surface
of competence; each strategy will have its own strengths and weaknesses. Pareto
coevolution eliminates dominated strategies, which is very helpful, but otherwise
provides little assistance in operationalizing the result of search into an effective
strategy that we can field in an adversarial strategic setting. Indeed, Pareto coevolution essentially leaves us with the equilibrium selection problem of conventional
game theory [52]. Recall that many Nash equilibria involve mixed strategies, that is,
probability distributions over pure strategies; thus, Pareto coevolution is unusual,
as far as MOO goes, in that the ultimate solution can be a stochastic combination
of points on the Pareto front (this generally cannot be the case when building a
bridge, but the idea of stochastic combinations of Pareto-optimal points is intriguing to consider for other MOO domains). The elimination of dominated strategies
is a standard procedure in game theory to help find Nash equilibria [27]; thus, if we
are interested in constructing a Nash equilibrium mixed strategy, we know that the
Pareto front contains the appropriate set of pure strategies. Nevertheless, the issue
of operationalizing the result of Pareto coevolution largely remains an open issue.

7 Conclusion

A coevolutionary algorithm can only evaluate an individual by having it interact
with other individuals. This property ties coevolution to both game theory (the
mathematics of strategic interaction) and multiobjective optimization. This latter
connection is made clear once we view each individual that we can interact with as
an objective for optimization. The foundational concepts involved in the application
of MOO to coevolution are Pareto optimality and distinctions. These concepts help
address some of the difficulties that can be encountered when using a CEA, such
as loss of fitness gradient, cycling, and overspecialization. Viewing coevolution as a
form of MOO allows us to formalize coevolutionary search in new ways, and these
new formalisms suggest new insights and algorithmic approaches for solving games
and game-like problems.

References

[1] R. Axelrod. *The Evolution of Cooperation.* Basic Books, 1984.

[2] R. Axelrod. The evolution of strategies in the Iterated Prisoner's Dilemma. In L. Davis, editor, *Genetic Algorithms and Simulated Annealing,* pages 32–41. Morgan Kaufmann, 1987.

[3] N. A. Barricelli. Numerical testing of evolution theories Part II: Preliminary tests of performance, symbiogenesis and terrestrial life. *Acta Biotheoretica,* 16 (3–4):99–126, 1963.

[4] J. C. Bongard and H. Lipson. Nonlinear system identification using coevolution of models and tests. *IEEE Transactions on Evolutionary Computation,* 9(4): 361–383, August 2005.

[5] A. Bucci, J. Pollack, and E. de Jong. Automated extraction of problem structure. In Deb et al. [17], pages 501–512.

[6] A. Bucci and J. B. Pollack. Order-theoretic analysis of coevolution problems: Coevolutionary statics. In A. M. Barry, editor, *2002 Genetic and Evolutionary Computation Conference Workshop Program,* pages 229–235, 2002.

[7] A. Bucci and J. B. Pollack. A mathematical framework for the study of coevolution. In K. A. De Jong, R. Poli, and J. E. Rowe, editors, *Proceedings of the Foundations of Genetic Algorithms 2003 Workshop (FOGA 7),* pages 221–235. Morgan Kaufmann Publishers, 2003.

[8] S. Bullock. Co-evolutionary design: Implications for evolutionary robotics. Technical Report CSRP 384, School of Cognitive and Computing Sciences, University of Sussex, 1995. Presented as a poster at the Third European Conference on Artificial Life (ECAL 1995).

[9] Cantú-Paz et al., editors. *Proceedings of the 2003 Genetic and Evolutionary Computation Conference,* Lecture Notes in Computer Science 2723–2724, 2003. Springer.

[10] J. Cartlidge and S. Bullock. Learning lessons from the common cold: How reducing parasite virulence improves coevolutionary optimization. In D. B. Fogel, M. A. El-Sharkawi, and X. Yao, editors, *Proceedings of the 2002 Congress on Evolutionary Computation,* pages 1420–1425. IEEE Press, 2002.

[11] C. Daskalakis, P. Goldberg, and C. Papadimitriou. The complexity of computing a Nash equilibrium. In J. Kleinberg, editor, *Proceedings of the 38th Symposium on Theory of Computing (STOC),* pages 71–78. ACM Press, 2006.

[12] E. D. de Jong. The incremental Pareto-coevolution archive. In Deb et al. [17], pages 525–536.

[13] E. D. de Jong. Towards a bounded Pareto-coevolution archive. In Deb et al. [17], pages 2341–2348.

[14] E. D. de Jong. The maxsolve algorithm for coevolution. In O'Reilly et al. [45], pages 483–489.

[15] E. D. de Jong and A. Bucci. DECA: Dimension extracting coevolutionary algorithm. In M. Cattolico et al., editors, *Proceedings of the 2006 Genetic and Evolutionary Computation Conference,* pages 313–320. ACM Press, 2006.

[16] E. D. de Jong and J. B. Pollack. Learning the ideal evaluation function. In Cantú-Paz et al. [9], pages 277–288.

[17] K. Deb et al., editors. *Proceedings of the 2004 Genetic and Evolutionary Computation Conference,* Lecture Notes in Computer Science 3102, 2004. Springer.

[18] S. G. Ficici. *Solution Concepts in Coevolutionary Algorithms*. PhD thesis, Brandeis University, May 2004.

[19] S. G. Ficici. Monotonic solution concepts in coevolution. In O'Reilly et al. [45], pages 499–506.

[20] S. G. Ficici, O. Melnik, and J. B. Pollack. A game-theoretic and dynamical-systems analysis of selection methods in coevolution. *IEEE Transactions on Evolutionary Computation*, 9(6):580–602, 2005.

[21] S. G. Ficici and J. B. Pollack. A game-theoretic approach to the simple coevolutionary algorithm. In Schoenauer et al. [53], pages 467–476.

[22] S. G. Ficici and J. B. Pollack. Game theory and the simple coevolutionary algorithm: Some results on fitness sharing. In R. Heckendorn, editor, *2001 Genetic and Evolutionary Computation Conference Workshop Program*, pages 2–7, 2001.

[23] S. G. Ficici and J. B. Pollack. Pareto optimality in coevolutionary learning. In J. Kelemen and P. Sosík, editors, *Sixth European Conference on Artificial Life (ECAL 2001)*, pages 316–325. Springer, 2001.

[24] S. G. Ficici and J. B. Pollack. A game-theoretic memory mechanism for coevolution. In Cantú-Paz et al. [9], pages 286–297.

[25] D. B. Fogel. *Evolutionary Computation: The Fossil Record*. IEEE Press, 1998.

[26] D. Fudenberg and D. K. Levine. *The Theory of Learning in Games*. MIT Press, 1998.

[27] D. Fudenberg and J. Tirole. *Game Theory*. MIT Press, 1998.

[28] D. E. Goldberg. *Genetic Algorithms in Search, Optimization, and Machine Learning*. Addison-Wesley, 1989.

[29] D. Hillis. Co-evolving parasites improves simulated evolution as an optimization procedure. *Physica D*, 42:228–234, 1990.

[30] J. Hofbauer and K. Sigmund. *Evolutionary Games and Population Dynamics*. Cambridge University Press, 1998.

[31] G. Hornby and B. Mirtich. Diffuse versus true coevolution in a physics-based world. In W. Banzhaf et al., editors, *Proceedings of 1999 Genetic and Evolutionary Computation Conference*, pages 1305–1312. Morgan Kaufmann, 1999.

[32] H. Juillé. *Methods for Statistical Inference: Extending the Evolutionary Computation Paradigm*. PhD thesis, Brandeis University, 1999.

[33] H. Juillé and J. Pollack. Co-evolving intertwined spirals. In L. J. Fogel, P. J. Angeline, and T. Bäck, editors, *Proceedings of the Fifth Annual Conference on Evolutionary Programming*, pages 461–468. MIT Press, 1996.

[34] H. Juillé and J. Pollack. Dynamics of co-evolutionary learning. In P. Maes et al., editors, *Proceedings of the Fourth International Conference on Simulation of Adaptive Behavior*, pages 526–534. MIT Press, 1996.

[35] H. Juillé and J. B. Pollack. Coevolving the "ideal" trainer: Application to the discovery of cellular automata rules. In J. R. Koza et al., editors, *Proceedings of the Third Annual Genetic Programming Conference*, pages 519–527. Morgan Kaufmann, 1998.

[36] H. Juillé and J. B. Pollack. Coevolutionary learning and the design of complex systems. *Advances in Complex Systems*, 2(4):371–393, 2000.

[37] J. Knowles and D. Corne. Approximating the nondominated front using the Pareto archived evolution strategy. *Evolutionary Computation*, 8(2):149–172, 2000.

[38] K. J. Lang and M. J. Witbrock. Learning to tell two spirals apart. In D. Touretzky, G. Hinton, and T. Sejnowski, editors, *Proceedings of the 1988 Connectionist Models Summer School*, pages 52–59. Morgan Kaufmann, 1988.

[39] J. Maynard Smith. *Evolution and the Theory of Games*. Cambridge University Press, 1982.

[40] A. McLennan and J. Berg. Asymptotic expected number of Nash equilibria of two-player normal form games. *Games and Economic Behavior*, 51:264–295, 2005.

[41] J. Nash. Non-cooperative games. *The Annals of Mathematics*, 54(2):286–295, 1951. Second Series.

[42] J. Noble and R. A. Watson. Pareto coevolution: Using performance against coevolved opponents in a game as dimensions for Pareto selection. In Spector et al. [55], pages 493–500.

[43] S. Nolfi and D. Floreano. Co-evolving predator and prey robots: Do 'arm races' arise in artificial evolution? *Artificial Life*, 4(4):311–335, 1998.

[44] B. Olsson. *NK*-landscapes as test functions for evaluation of host-parasite algorithms. In Schoenauer et al. [53], pages 487–496.

[45] U.-M. O'Reilly et al., editors. *Proceedings of the 2005 Genetic and Evolutionary Computation Conference*, 2005. ACM Press.

[46] L. Pagie and P. Hogeweg. Evolutionary consequences of coevolving targets. *Evolutionary Computation*, 5(4):401–418, 1997.

[47] J. Paredis. Towards balanced coevolution. In Schoenauer et al. [53], pages 497–506.

[48] J. Reed, R. Toombs, and N. A. Barricelli. Simulation of biological evolution and machine learning: I. selection of self-reproducing numeric patterns by data processing machines, effects of hereditary control, mutation type and crossing. *Journal of Theoretical Biology*, 17:319–342, 1967.

[49] C. D. Rosin and R. Belew. Methods for competitive co-evolution: Finding opponents worth beating. In L. J. Eshelman, editor, *Proceedings of the Sixth International Conference on Genetic Algorithms*, pages 373–381. Morgan Kaufmann, 1995.

[50] C. D. Rosin and R. Belew. New methods for competitive co-evolution. *Evolutionary Computation*, 5(1):1–29, 1997.

[51] A. Samuel. Some studies in machine learning using the game of checkers. *IBM Journal of Research and Development*, 3(3):211–229, 1959.

[52] L. Samuelson. *Evolutionary Games and Equilibrium Selection*. MIT Press, 1997.

[53] M. Schoenauer et al., editors. *Parallel Problem Solving from Nature VI*, Lecture Notes in Computer Science 1917, 2000. Springer.

[54] K. Sims. Evolving 3d morphology and behavior by competition. In R. A. Brooks and P. Maes, editors, *Artificial Life IV*, pages 28–39. MIT Press, 1994.

[55] L. Spector et al., editors. *Proceedings of the 2001 Genetic and Evolutionary Computation Conference*, 2001.

[56] R. A. Watson. *Compositional Evolution: The Impact of Sex, Symbiosis, and Modularity on the Gradualist Framework of Evolution*. MIT Press, 2006.

[57] R. A. Watson and J. B. Pollack. Symbiotic combination as an alternative to sexual recombination in genetic algorithms. In Schoenauer et al. [53], pages 425–434.

[58] R. A. Watson and J. B. Pollack. Coevolutionary dynamics in a minimal substrate. In Spector et al. [55], pages 702–709.

[59] R. P. Wiegand. *An Analysis of Cooperative Coevolutionary Algorithms*. PhD thesis, George Mason University, 2003.

Constrained Optimization via Multiobjective Evolutionary Algorithms

Efrén Mezura-Montes[1] and Carlos A. Coello Coello[2*]

[1] Laboratorio Nacional de Informática Avanzada (LANIA A.C.)
Rébsamen 80, Centro, Xalapa, Veracruz, 91000 México
emezura@lania.mx
[2] CINVESTAV-IPN (Evolutionary Computation Group)
Departamento de Computación
Av. IPN No. 2508, Col. San Pedro Zacatenco, México D.F. 07360, México
ccoello@cs.cinvestav.mx

Summary. In this chapter, we present a survey of constraint-handling techniques based on evolutionary multiobjective optimization concepts. We present some basic definitions required to make this chapter self-contained, and then introduce the way in which a global (single-objective) nonlinear optimization problem is transformed into an unconstrained multiobjective optimization problem. A taxonomy of methods is also proposed and each of them is briefly described. Some interesting findings regarding common features of the approaches analyzed are also discussed.

1 Introduction

Nowadays, evolutionary algorithms (EAs) have become a popular choice to solve different types of optimization problems [20, 33, 45]. This chapter points out the application of some ideas originally designed to solve a specific type of optimization problem using EAs, which are now applied to solve a different type of problem. Despite being considered powerful search engines, EAs, in their original versions, lack a mechanism to incorporate constraints into the fitness function in order to solve constrained optimization problems. Hence, several approaches have been proposed to deal with this issue. Michalewicz and Schoenauer [36] and Coello Coello [7] have presented comprehensive surveys about constraint-handling techniques used with EAs. As indicated in such surveys, the most popular method adopted to handle constraints in EAs was taken from the mathematical programming literature: penalty functions (mostly exterior). Penalty functions were originally proposed by Courant in the 1940s [11] and later expanded by Carroll [4] and Fiacco and McCormick [17]. The idea of this method is to transform a constrained optimization problem into an unconstrained one by adding (or subtracting) a certain value to (or from) the objective function based on the degree of constraint violation present in

---

* Corresponding author

a certain solution. This aims to favour feasible solutions over infeasible ones during the selection process. The main advantage of the use of penalty functions is their simplicity. However, their main shortcoming is that penalty factors, which determine the severity of the punishment, must be set by the user and their values are problem dependent [50, 7]. This has motivated the design of alternative techniques like those based on special encodings and operators [34, 48] and on repair algorithms [35].

Unlike penalty functions, which combine the objective function and the constraint values into one fitness value, there are other approaches which handle these two values separately. The most representative approaches, which work based on this idea are: (1) the methods based on the superiority of feasible points [41, 15] and (2) the methods based on evolutionary multiobjective optimization concepts.

This chapter focuses on the latter type of technique (i.e., those based on multiobjective optimization concepts) and describes, tests and criticizes them.

In order to present our discussion of methods in a more organized way, we propose a simple taxonomy of techniques, based on the way they transform the nonlinear programming problem (NLP) into a multiobjective optimization problem (MOP):

1. Approaches which transform the NLP into an unconstrained bi-objective optimization problem (the original objective function and the sum of constraint violations).
2. Techniques which transform the NLP into an unconstrained MOP where the original objective function and each constraint of the NLP are treated as separate objectives. From this category, we observed two further subcategories:
   a) Methods which use non-Pareto concepts (mainly based on multiple populations) and
   b) Methods which use Pareto concepts (ranking and dominance) as their selection criteria.

The remainder of this chapter is organized as follows. In Section 2 we present the general NLP, and we recall some multiobjective optimization concepts used in this survey; we also show the transformation of an NLP into an MOP. After that, in Section 3 the approaches which solve the NLP as a bi-objective problem (using the original objective function and the sum of constraint violations) are presented. Section 4 shows techniques based on solving the problem by taking the original objective function and each of the constraints of the problem as different objectives, by using either Pareto or non-Pareto concepts. In Section 5, we provide some highlights of the methods previously discussed. A small comparative experiment using four representative approaches (from those previously discussed) is presented in Section 6. Finally, Section 7 presents our conclusions and some possible future paths of research in the area.

2 Problem Definition and Transformation

In the following definitions we will assume (without loss of generality) minimization. The general NLP is defined as to

$$\text{Find } \mathbf{X} \text{ which minimizes } f(\mathbf{X}) \tag{1}$$

subject to:

$$g_i(\mathbf{X}) \leq 0, \quad i = 1, \ldots, m \tag{2}$$

$$h_j(\mathbf{X}) = 0, \quad j = 1, \ldots, p \tag{3}$$

where $\mathbf{X} \in \mathbb{R}^n$ is the vector of solutions $\mathbf{X} = [x_1, x_2, \ldots, x_n]^T$, where each x_i, $i = 1, \ldots, n$ is bounded by lower and upper limits $L_i \leq x_i \leq U_i$ which define the search space $\mathcal{S}$; m is the number of inequality constraints; and p is the number of equality constraints (in both cases, constraints could be linear or nonlinear). The constraints define the feasible region $\mathcal{F} \subseteq \mathcal{S}$.

When solving NLPs with EAs, equality constraints are usually transformed into inequality constraints of the form

$$g_j(\mathbf{X}) = |h_j(\mathbf{X})| - \epsilon \leq 0, \quad j = m+1, m+2, \ldots, m+p \tag{4}$$

where ϵ is the tolerance allowed (a very small value). In the rest of the chapter we will refer only to inequality constraints because we will assume this transformation.

As discussed in the Introduction chapter of this volume, in a multiobjective problem, the optimum solution consists on a set of ("trade-off") solutions, rather than a single solution as in global optimization. This optimal set is known as the Pareto-optimal set.

Based on the review of the literature that we undertook, we found that researchers have adopted two different ways to transform the NLP into an MOP:

1. The first approach transforms the NLP into an unconstrained bi-objective problem. The first objective is the original objective function and the second objective is the sum of constraint violations as follows: optimize $\mathbf{F}(\mathbf{X}) = (f(\mathbf{X}), G(\mathbf{X}))$, where $G(\mathbf{X}) = \sum_{i=1}^{m+p} \max(0, g_i(\mathbf{X}))$, and each $g_i(\mathbf{X}), i = 1, \ldots, m+p$, must be normalized.
   Note, however, that when solving a transformed NLP, we are not looking for a set of solutions Instead, we seek a single solution, the global constrained optimum, where $f(\mathbf{X}) \leq f(\mathbf{Y})$ for all feasible $\mathbf{Y}$ and $G(\mathbf{X}) = 0$.
2. The second approach transforms the problem into an unconstrained MOP, in which we will have $k+1$ objectives, where k is the total number of constraints $(m+p)$ and the additional objective is the original NLP objective function. Then, we can apply a multiobjective evolutionary algorithm to the new vector $\mathbf{F}(\mathbf{X}) = (f(\mathbf{X}), g_1(\mathbf{X}), \ldots, g_{m+p}(\mathbf{X}))$, where $g_1(\mathbf{X}), \ldots, g_{m+p}(\mathbf{X})$ are the original constraints of the problem.
   As indicated before, we are looking again for the global constrained optimum instead of a set of trade-off solutions. Thus, we require the following: $g_i(\mathbf{X}) = 0$ for $1 \leq i \leq (m+p)$ and $f(\mathbf{X}) \leq f(\mathbf{Y})$ for all feasible $\mathbf{Y}$.

These apparently subtle differences in the way of stating an MOP point to changes in the way multiobjective concepts are applied (i.e., they influence the way in which MOEAs are actually used). In the following sections, we will describe the approaches reported in the specialized literature to deal with this special type of MOP that arises from a transformed NLP.

It is important to note that the use of multiobjective optimization concepts improves the solution procedure of a constrained problem with respect to a typical penalty function in two ways: (1) No penalty factors must be tuned and (2) the way to approach the feasible region becomes more robust because of the trade-offs among objective function and constraints of the problem. In contrast, in a typical penalty function this path to the constrained optimum is rather rigid and fixed. In fact, a penalty function forces the search to generate feasible solutions. On the other hand, by using a multiobjective approach we aim to decrease the violation of constraints, but at the same time we look for objective function improvement, and this behaviour may open up the potential of reaching the feasible region from different (and maybe promising) directions.

3 Transforming the NLP into a Bi-objective Problem

Surry and Radcliffe [51] proposed COMOGA (Constrained Optimization by Multiobjective Optimization Genetic Algorithms) where individuals are Pareto-ranked based on the sum of constraint violation. Then, solutions can be chosen using binary tournament selection based either on their rank or on their objective function value. This decision is based on a parameter called P_{cost} whose value is modified dynamically. The aim of the proposed approach to solve this bi-objective problem is based on reproducing solutions which are good in one of the two objectives with other competitive solutions in the other objective i.e., constraint violation (such as Schaffer's Vector Evaluated Genetic Algorithm (VEGA) promoted to solve MOPs [47]). COMOGA was tested on a gas network design problem, providing slightly better results than those obtained with a penalty function approach. Its main drawback is that it requires several extra parameters. Also, to the authors' best knowledge, this approach has not been used by other researchers.

Camponogara and Talukdar [3] proposed to solve the bi-objective problem in the following way. A set of Pareto fronts in the bi-objective space is generated by the EA. Two of them (S_i and S_j, where $i < j$) are selected. After that, two solutions $x_i \in S_i$ and $x_j \in S_j$ where x_i dominates x_j are chosen. Based in these two points, a search direction is generated as follows:

$$d = \frac{(x_i - x_j)}{|x_i - x_j|}. \tag{5}$$

A line search begins by projecting d over one variable axis on the decision variable space in order to find a new solution x which dominates both x_i and x_j. Other mechanisms of the approach allow, at pre-defined intervals, replacing of the worst half of the population with new random solutions to avoid premature convergence. This indicates some of the problems of the approach in maintaining diversity. Additionally, the use of line search within a GA adds some extra computational cost. Furthermore, it is not clear what the impact is of the segment chosen to search on the overall performance of the algorithm.

Zhou et al. [55] proposed a ranking procedure based on Pareto strength [56] for the bi-objective problem, i.e. counting the number of individuals which are dominated for a given solution. Ties are broken by the sum of constraint violations (second objective in the problem). The simplex crossover (SPX) operator is used to generate a set of offspring where the individual with the highest Pareto strength and the solution with the lowest sum of constraint violations are both selected to take part in the population for the next generation. The approach was tested on a subset of a well-known benchmark for evolutionary constrained optimization [30]. The results were competitive but the authors had to use different sets of parameters for different functions, which made evident the high sensitivity of the approach to the values of its parameters.

Wang and Cai [54] used a framework similar to the one proposed by Zhou et al. [55] because they also used the SPX with a set of parents to generate a set of offspring. However, instead of using just two individuals from the set of offspring, all nondominated solutions (in the bi-objective space) are used to replace the dominated solutions in the parent population. Furthermore, they use an external archive to store infeasible solutions with a low sum of constraint violations in order to replace some random solutions in the current population. The idea is to maintain infeasible solutions close to the boundaries of the feasible region in order to perform a better sampling of this region so as to find optimum solutions located there (i.e., when dealing with active constraints) [28]. The approach provided good results in 13 well-known test problems. However, different sets of values for the parameters were used, depending of the dimensionality of the problem.

Venkatraman and Yen [52] proposed a generic framework to solve the NLP. The approach is divided in two phases: The first one treats the NLP as a constraint satisfaction problem, i.e., the goal is to find at least one feasible solution. To achieve that, the population is ranked based only on the sum of constraint violations. The second phase starts when the first feasible solution has been found. Now both objectives (the original objective function and the sum of constraint violations) are taken into account and nondominated sorting [16] is used to rank the population (alternatively, the authors proposed a preference scheme based on feasibility rules [15], but in their experiments, they found that nondominated sorting provided better results). Also, to favour diversity, a niching scheme based on the distance of the nearest neighbours to each solution is applied. To decrease the effect of differences in values, all constraints are normalized before calculating the sum of those which are violated. The approach used a typical GA as a search engine with 10% elitism. The approach provided good quality results in 11 well-known benchmark problems and in some problems generated with the Test-Case Generator tool [32], but lacked consistency due to the fact that the way to approach the feasible region is mostly random because of the first phase, which only focuses on finding a feasible solution, regardless of the direction from which the feasible region is approached.

Wang et al. [53] also solved the bi-objective problem, but using selection criteria based on feasibility very similar to those proposed by Deb [15], where a feasible solution is preferred over an infeasible one; between two feasible solutions, the one with the best objective function value is selected and, finally, between two infeasible solutions, the one with the lowest sum of constraint violation is chosen. Furthermore, they proposed a new crossover operator based on uniform design methods [53]. This operator is able to explore regions closer to the parents. Finally, Gaussian noise is used as a mutation operator. The approach was tested on a subset of well-known benchmarks used to test evolutionary algorithms in constrained optimization [30]. No details are given by the authors about the influence of the extra parameters required to control the crossover operator (q) and the number of offspring generated (r).

4 Transforming the NLP into a Multiobjective Problem with Objective Function and Constraints as Separate Objectives

As indicated before, in this case we may use non-Pareto schemes or Pareto schemes. Each of these two subclasses of methods will be discussed next.

4.1 Techniques Based on Non-Pareto Schemes

Parmee and Purchase [40] used the idea proposed in VEGA [47] to guide the search of an evolutionary algorithm to the feasible region of an optimal gas turbine design problem with a heavily constrained search space. The aim of VEGA is to divide the population into subpopulations, and each subpopulation has then the goal of optimizing only one objective. In this case, the set of objectives are the constraints of the problem. Genetic operators are applied to all solutions regardless of the subpopulation of each solution. In Parmee and Purchase's approach, once the feasible region is reached, special operators are used to improve the feasible solutions found. The use of these special operators that preserve feasibility makes this approach highly specific to one application domain rather than providing a general methodology to handle constraints.

Schoenauer and Xanthakis [49] proposed a constraint-handling technique based on the notion of *behavioural memory* [13], which takes into account the information contained in the whole population after some genetic evolution. As it turns out, this approach consists of a form of "lexicographic ordering" [10]. The main idea is to satisfy each constraint of the problem in a sequential order. Once a certain number of solutions in the population satisfy the first constraint (based on a parameter of the approach), the second constraint is added in order to be also satisfied, but always enforcing that the solutions satisfy the first one. In this way, solutions which satisfy the second constraint but not the first one will be removed from the population (as in a death-penalty approach [7]). The success of the approach normally depends on the order in which constraints are processed. Besides, it may not be appropriate when solving problems with a large feasible region (with respect to the whole search space). However, this technique may be very effective to solve problems where constraints have a natural hierarchy to be evaluated.

Coello Coello [6] also used VEGA's idea [47] to solve NLPs. At each generation, the population was split into $m + 1$ subpopulations of equal (fixed) size, where m is the number of constraints of the problem. The additional subpopulation handled the objective function of the problem and the individuals contained within it were selected based on the unconstrained objective function value. The m remaining subpopulations took one constraint of the problem each as their fitness function. The aim was that each of the subpopulations tried to reach the feasible region corresponding to one constraint. By combining these different subpopulations, the approach would then reach the feasible region of the problem in terms of all of its constraints. The main drawback of the approach is that the number of subpopulations increases linearly with respect to the number of constraints.

This issue was further tackled by Liang and Suganthan [27], where a dynamic particle multiswarm optimization was proposed. They also used VEGA's idea to split the swarm into subswarms, and each subswarm optimized one objective. However, in this case, the subswarms are assigned dynamically. In this way, the number of subswarms depends on the complexity of the constraints to be satisfied instead of depending on the number of constraints. The authors also included a local search mechanism based on sequential quadratic programming to improve values of a set of randomly chosen *pbest* values. The approach provided competitive results in the extended version of a well-known benchmark adopted for evolutionary constrained optimization [27]. The main drawbacks of the approach are that it requires extra parameters to be tuned by the user and it has a strong dependency on the local search mechanism.

4.2 Techniques Based on Pareto Schemes

Carlos Fonseca was apparently the first to propose the idea of using the Pareto dominance relation to handle constraints [12, 19]. His proposal consisted of modifying the definition of Pareto dominance in order to incorporate constraints. It is worth noting that this proposal was really a small component of a multiobjective evolutionary algorithm (MOGA [18]), and was, therefore, mainly used to solve constrained multiobjective optimization problems. Because of this, Fonseca's proposal did not attract much interest from researchers working with constraint-handling techniques for single-objective optimization. Next, we will discuss several other constraint-handling techniques that directly incorporate Pareto-based schemes.

Jiménez et al. [24] proposed an approach that transforms the NLP (and also the constraint satisfaction and goal-programming problems) into an MOP by assigning priorities. Regarding the NLP, constraints are assigned a higher priority than the objective function. Then, a multiobjective algorithm based on a preselection scheme is applied. This algorithm generates from two parents a set of offspring which will be mutated to generate another set. The best individual from the first set of offspring (nonmutated) and the best one of the mutated ones will replace each of the two parents. The idea is to favour the generation of individuals close to their parents and to promote implicit niching. Comparisons among individuals are made by using Pareto dominance. A real-coded GA was used as a search engine with two types of crossover operators (uniform and arithmetic) and two mutation operators (uniform and nonuniform). The results on 11 problems taken from a well-known benchmark [30] were promising. The main drawback of the approach is the evident lack of knowledge about the effect of the parameter "q" related with the preselection scheme, which the authors do not discuss in their paper. Also, the authors do not provide any information regarding the number of evaluations performed by the approach.

Coello Coello [5] proposed a ranking procedure based on a counter which was incremented based on the number of individuals in the population which dominated a given solution based on some criteria (feasibility, sum of constraint violation and number of constraints violated). The approach was tested on a set of engineering design problems providing competitive results. An adaptive mechanism was also implemented in order to fine-tune the parameters of the approach. Its main drawbacks are the computational cost of the technique and its difficulties in handling equality constraints [29].

Ray et al. [42, 44] proposed the use of a Pareto ranking approach that operates on three spaces: objective space, constraint space and the combination of the two previous spaces. This approach also uses mating restrictions to ensure better constraint satisfaction in the offspring generated and a selection process that eliminates weaknesses in any of these spaces. To maintain diversity, a niche mechanism based on Euclidean distances is used. This approach can solve both constrained or unconstrained optimization problems with one or several objective functions. The mating restrictions used by this method are based on the information that each individual has about its own feasibility. Such a scheme is based on an idea proposed by Hinterding and Michalewicz [22]. The main advantage of this approach is that it requires a very low number of fitness function evaluations with respect to other state-of-the-art approaches. Its main drawback is that its implementation is considerably more complex than that of any of the other techniques previously discussed.

Ray extended his work to a simulation of social behaviour [1, 43], where a societies-civilization model is proposed. Each society has its leaders who influence its neighbours. Also, the leaders can migrate from one society to another, promoting exploration of new regions of the search space. Constraints are handled by a nondominated sorting mechanism [16] in the constraint space. A leader-centric operator is used to generate movements of the neighbours influenced by their leaders. The main drawback of the approach is its high computational cost derived from the nondominated sorting and a clustering technique required to generate the societies. Results reported on some engineering design problems are very competitive. However, to the authors' best knowledge, this technique has not been compared against state-of-the-art approaches adopting the same benchmark [30].

Coello Coello and Mezura-Montes [9] implemented a version of the Niched-Pareto Genetic Algorithm (NPGA) [23] to handle constraints in single-objective optimization problems. The NPGA is a multiobjective optimization approach in which individuals are selected through a tournament based on Pareto dominance. However, unlike the NPGA, Coello Coello and Mezura-Montes' approach does not require niches (or fitness sharing [14]) to maintain diversity in the population. Instead, it requires an additional parameter called S_r that controls the diversity of the population. S_r indicates the proportion of parents selected by four comparison criteria (based on Deb's proposal [15]), but when both solutions are infeasible, a dominance criterion in the constraint space is used to select the best solution. The remaining $1 - S_r$ parents are selected using a purely probabilistic approach. Results indicated that the approach was robust, efficient and effective. However, it was also found that the approach had scalability problems (its performance degraded as the number of decision variables increased).

The use of dominance to select between two infeasible solutions was taken to the differential evolution metaheuristic by Kukkonen and Lampinen [26]. In their approach, when the comparison between a parent vector and its child vector is performed, and both of them are infeasible, a dominance criterion is applied. The results on the extended version of the benchmark [26] were very competitive.

Angantyr et al. [2] proposed assigning a fitness value to solutions based on a two-ranking mechanism. The first rank is assigned according to the objective function value (regardless of feasibility). The second rank is assigned by using nondominated sorting [16] in the constraint space. These ranks have adaptive weights when defining the fitness value. The aim is to guide the search to the unconstrained optimum solution if there are many feasible solutions in the current population. If the rate of feasible solutions is low, the search will be biased to the feasible region. The goal is to promote an oscillation of the search between the feasible and infeasible regions of the search space. A typical GA with BLX crossover was used. The main advantage of this approach is that it does not add any extra parameters to the algorithm. However, it presented some problems when solving functions with equality constraints [2].

Hernandez et al. [21] proposed an approach named IS-PAES which is based on the Pareto Archived Evolution Strategy (PAES) originally proposed by Knowles and Corne [25]. IS-PAES uses an external memory to store the best set of solutions found. Furthermore, IS-PAES requires a shrinking mechanism to reduce the search space. The multiobjective concept is used in this case as a secondary criterion (Pareto dominance is used only to decide whether or not a new solution is inserted into the external memory). The authors acknowledge that the most important mechanisms of IS-PAES are its shrinking procedure and the information provided by the external memory which is used to determine the shrinking of the search space. Furthermore, despite its good performance as a global optimizer, IS-PAES is far from simple to implement.

Runarsson and Yao [46] presented a comparison of two versions of Pareto ranking applied in the constraint space: (1) considering the objective function value in the ranking process and (2) not considering it. These versions were compared against a typical overpenalized penalty function approach. The authors found in their work that using Pareto ranking leads to bias-free search, and thus concluded that it causes the search to spend most of its time searching in the infeasible region. Therefore, the approach is unable to find feasible solutions (or finds feasible solutions with a poor value of the objective function).

Oyama et al. [39] used a similar approach to the one proposed by Coello Coello and Mezura-Montes [9]. However, the authors proposed using a set of criteria based on feasibility to rank all the population (instead of using them in a tournament [9]). Moreover, this approach is designed to also solve constrained multiobjective optimization problems. A real-coded GA with BLX crossover was used as the search engine. This technique was used to solve one engineering design problem and a real-world NLP. No further experiments or comparisons were documented.

5 Remarks

Based on the features found in each of the methods previously discussed, we highlight the following findings:

- The transformation of the NLP into a multiobjective problem with constraints and objective function as separated objectives is a more popular approach than the transformation of the NLP to a bi-objective optimization problem.
- The use of subpopulations has been the least popular, although they may present certain advantages in some particular optimization problems (see, for example, [8]).
- There seems to be a certain trend towards using mean-centric crossover operators (BLX [2, 39], random mix [42, 44], SPX [54, 55]) over using parent-centric crossover (uniform design methods [53], leader centric operator [1, 43]) when adopting real-coded GAs. Furthermore, other authors used more than one crossover operator (uniform and arithmetic [24]). This choice seems to contradict the findings about competitive crossover operators that have been reported by other researchers when using other constraint-handling techniques such as GENOCOP and penalty functions [37, 38].

- The use of diversity mechanisms is found in most approaches, which is a clear indication of the loss of diversity experienced when adopting multiobjective optimization schemes for handling constraints [51, 3, 54, 52, 42, 44, 1, 43, 9, 2, 21].
- The use of explicit local search mechanisms is still scarce, despite the evident advantages that such mechanisms may bring into this area [27].
- The difficulty of using Pareto concepts when solving the NLP pointed out by Runarsson and Yao [46] has been confirmed by other researchers such as Mezura-Montes and Coello Coello [29]. However, the methods described in this survey provide several alternatives to deal with the inherent shortcoming for the lack of bias provided by Pareto ranking.

6 A Limited Comparative Study

Four techniques were selected from those discussed before to perform a small comparative study that aims to illustrate some practical issues of constraint-handling techniques based on multiobjective concepts. The techniques selected are the following: COMOGA [51] which transforms the constrained problem into a bi-objective problem; VEGA, as proposed in Coello Coello [6], which handles a problem of "$m+p+1$" objectives with the same number of subpopulations (where m is the number of inequality constraints and p is the number of equality constraints); the NPGA, as in [9], which calculates Pareto dominance in the constraint space (the number of objectives depends of the number of constraints of the problem); and the approach that uses Pareto ranking (called MOGA by us although it does not follow Fonseca's proposal [18] exactly) [5], where dominance is computed based on separated objectives (number of violated constraints and degree of constraint violation).

These techniques were chosen because all of them mainly modify the parent selection scheme of an EA and do not use additional mechanisms (specialized crossover or mutation operators, external memory, etc.). Therefore, and because of their simplicity they can be included inside a typical EA without further changes. In order to simplify our notation, the last three techniques previously indicated will be called CHVEGA, CHNPGA and CHMOGA, respectively (CH stands for constraint-handling).

To evaluate the performance of the techniques selected, we decided to use a well-known benchmark found in the specialized literature [36] and in three engineering design problems [31]. The Appendix at the end of the chapter includes the details of all the test functions adopted.

To get an estimate of how difficult is to generate feasible points through a purely random process, we computed the ρ metric (as suggested by Michalewicz and Schoenauer [36]) using the following expression:

$$\rho = |F|/|S| \tag{6}$$

where $|S|$ is the number of random solutions generated ($S = 1,000,000$ in our case), and $|F|$ is the number of feasible solutions found (out of the total $|S|$ solutions randomly generated).

The different values of ρ and the main features of each test function are shown in Table 1.

Problem	n	Type of function	ρ	LI	NI	LE	NE
1	5	quadratic	27.0079%	0	6	0	0
2	2	non linear	0.0057%	0	2	0	0
3	10	quadratic	0.0000%	3	5	0	0
4	7	non linear	0.5199%	0	4	0	0
5	8	linear	0.0020%	3	3	0	0
6	2	quadratic	0.0973%	0	0	0	1
7	4	quadratic	2.6859%	6	1	0	0
8	4	quadratic	39.6762%	3	1	0	0
9	3	quadratic	0.7537%	1	3	0	0

Table 1. Main features of the nine test problems used. n is the number of decision variables, LI is the number of linear inequalities, NI the number of nonlinear inequalities, LE is the number of linear equalities and NE is the number of nonlinear equalities

In our comparative study, we used a binary Gray-coded GA with two-point crossover and uniform mutation. Equality constraints were transformed into inequalities using a tolerance value of 0.001 (see [7] for details of this transformation). The number of fitness function evaluations is the same for all the approaches under study (80,000). The parameters adopted for each of the methods were the following:

- **COMOGA:**
  - Population size = 200
  - Crossover rate = 1.0
  - Mutation rate = 0.05
  - Desired proportion of feasible solutions = 10 %
  - $\epsilon = 0.01$
- **CHVEGA:**
  - Population size = 200
  - Number of generations = 400
  - Crossover rate = 0.6
  - Mutation rate = 0.05
  - Tournament size= 5
- **CHNPGA:**
  - Population size = 200
  - Number of generations = 400
  - Crossover rate = 0.6
  - Mutation rate = 0.05
  - Size of sample of the population = 10
  - Selection Ratio = 0.8
- **CHMOGA:**
  - Population size = 200
  - Number of generations = 400
  - Crossover rate = 0.6
  - Mutation rate = 0.05

A total of 100 runs per technique per problem were performed. A summary of all results is shown in Table 2, where P_i refers to the problem solved ($1 \le i \le 9$).

		Statistical results on nine test problems			
P.	**Approach**	**Optimal**	**Best**	**Mean**	**St. Dev.**	F_p
P_1	COMOGA	−30665.539	−30533.057	−30329.563	7.48E+1	0.24%
	CHVEGA		−30647.246	−30628.469	**7.88E+0**	41%
	CHNPGA		**-30661.033**	**-30630.883**	2.05E+1	35%
	CHMOGA		−30649.959	−30568.918	5.35E+1	3.5%
P_2	COMOGA	−6961.814	−6808.696	−5255.105	9.95E+2	0.20%
	CHVEGA		**−6942.747**	**−6762.048**	**1.02E+2**	4.3%
	CHNPGA		−6941.307	−6644.539	3.36E+2	2%
	CHMOGA		−6939.440	−6678.926	1.57E+2	2%
P_3	COMOGA*(8)	24.306	485.579	1567.294	9.24E+2	0.03%
	CHVEGA		28.492	34.558	2.93E+0	15%
	CHNPGA		**26.986**	**31.249**	**2.32E+0**	4.9%
	CHMOGA		29.578	45.589	1.52E+1	1.3%
P_4	COMOGA	680.63	733.00	983.63	1.16E+2	1.1%
	CHVEGA		693.64	739.31	2.51E+1	4.5%
	CHNPGA		**680.95**	**682.34**	**8.36E-1**	24%
	CHMOGA		681.71	692.97	1.09E+1	4.9%
P_5	COMOGA1*(71)	7049.25	10865.43	18924.58	3.85E+3	0.0001%
	CHVEGA*(63)		9842.45	17605.59	3.87E+3	0.005%
	CHNPGA*(29)		8183.30	13716.70	4.80E+3	0.05%
	CHMOGA		**7578.34**	**9504.36**	**1.50E+3**	2%
P_6	COMOGA	0.75	**0.75**	**0.75**	**4.95E-4**	0.029%
	CHVEGA		**0.75**	0.80	2.58E-2	1.1%
	CHNPGA		**0.75**	**0.75**	1.21E-2	2.6%
	CHMOGA		**0.75**	**0.75**	5.95E-4	1.7%
P_7	COMOGA	2.381	2.471158	2.726058	1.20E-1	0.03%
	CHVEGA		2.386833	**2.393504**	**3.80E-3**	35%
	CHNPGA		**2.382860**	2.420906	2.56E-2	20%
	CHMOGA		2.386333	2.504377	9.90E-2	5%
P_8	COMOGA	6059.946	6369.428	7795.412	7.01E+2	0.4%
	CHVEGA		6064.724	6259.964	1.70E+2	43%
	CHNPGA		**6059.926**	**6172.527**	**1.24E+2**	33%
	CHMOGA		6066.967	6629.064	3.85E+2	45%
P_9	COMOGA	0.012681	0.012929	0.014362	8.64E-4	2.11%
	CHVEGA		0.012688	0.012886	2.09E-4	25%
	CHNPGA		0.012683	**0.012752**	**6.20E-5**	10%
	CHMOGA		**0.012680**	0.012960	3.63E-4	4.8%

Table 2. Experimental results using the four approaches with the nine test problems. The symbol "*" and the number between parenthesis "(n)"mean that only in n runs feasible solutions were found; F_p is the average percentage of feasible solutions found during a single run (with respect to the full population)

6.1 Discussion of Results

Based on the results obtained, all of them summarized in Table 2, we will focus our discussion on the following topics.

- **Quality**: Which approach provides the "best" result overall (measured by the best result in column 4 in Table 2).
- **Consistency**: Which approach provides the "best" mean and standard deviation values (measured by the mean and standard deviation (Std. Dev.) results in columns 5 and 6, respectively, in Table 2).
- **Diversity**: To analyze the rate of feasible solutions of each approach during a single run.

Quality of the Results

CHNPGA provided the "best" best results in five problems (P1, P3, P4, P7 and P8) and slightly improved the best known solution in one of them (P8). CHVEGA obtained the "best" best result in problem P2 and CHMOGA in problems P5 and P9. All the four approaches reached the best solution in problem P6.

Consistency

CHNPGA provided the most consistent results in four problems (P3, P4, P8 and P9). In problem P1, CHNPGA showed a mean value closer to the optimal solution than that provided by CHVEGA; however, CHVEGA's standard deviation value was smaller than CHNPGA's. We consider the behaviour of CHNPGA to be more consistent because of its mean closeness value to the optimal result. CHVEGA presented the "best" mean and standard deviation values in two functions (P2 and P7). Finally, CHMOGA presented the best consistency in problem P5. It is important to note that, for function P5, only CHMOGA consistently found feasible solutions in each single run. The remaining techniques had problems reaching the feasible region in this problem. Again, problem P6 was easily solved with a similar performance by all four approaches.

Diversity

It is quite interesting to analyze the average number of feasible solutions that each algorithm maintains in a single run. The fact that the population size is the same for all four algorithms compared makes this a relatively fair point of comparison. For most of the problems, the approach which consistently reached the vicinity of the optimum, was able to achieve a rate of feasible solutions above the average rate of the four approaches. These approaches that reached near the optimum are CHNPGA in P1, P3 P4, P8 and P9, CHVEGA in P2, and P7, and CHMOGA in P5. As expected, this rate corresponds to the approximate size of the feasible region with respect to the whole search space (reported as ρ, the fourth column in Table 1), i.e., we have high rates on problems with larger feasible regions and low rates on problems with very small feasible regions. See, for example, test function P1 (27% of the search space is feasible and a 35% rate is maintained by CHNPGA) and P2 (0.0057% of the search space is feasible and a 4.3% rate is maintained by CHVEGA).

Summary of results

Based on the observations made on each aspect of our small set of experiments and analysis, we now summarize our main findings:

- The Pareto dominance tournament selection promoted by CHNPGA provided the most accurate and consistent results for the set of test problems used in the experiments (P1, P3, P4, P7, P8 and P9), regardless of the features of the problem to be solved (type of objective function and constraints, dimensionality, size of the feasible region with respect to the whole search space).
- The population-based mechanism used by CHVEGA was very effective in problems with a low dimensionality, small feasible regions and nonlinear objective function (P2 and P7).
- The Pareto ranking approach based on feasibility used by CHMOGA was very competitive in problems with average dimensionality, linear or quadratic objective function and very small feasible regions (P5 and P9). In fact, CHMOGA was the only approach that consistently found feasible solutions in problem P5.
- These three multiobjective-based constraint-handling mechanisms (CHNPGA, CHVEGA and CHMOGA) were able to maintain an appropriate rate of feasible solutions (with respect to the size of the feasible region of the problem) so as to reach the neighbourhood of the optimum.
- COMOGA was competitive only in problem P6, where all approaches were very competitive. This can be explained by the fact that COMOGA was explicitly designed to solve a specific type of problem rather than to be a general constraint-handling technique.

These findings are far from being conclusive, but provide some clues about the behaviour of these types of constraint-handling mechanisms.

7 Conclusions

We have presented in this chapter a survey of constraint-handling techniques based on multiobjective optimization concepts. A taxonomy of techniques based on the type of transformation made from the NLP to either a bi-objective (objective function and sum of constraint violation) or an MOP (with the objective function and each constraint considered as separate objectives) has been proposed. We have presented a discussion about the main features of each method (selection criteria, diversity handling mechanism, genetic operators, advantages and disadvantages, and validation). Furthermore, some interesting findings about all methods have been summarized and briefly discussed.

In the final part of the chapter, we included a small comparative experiment of four representative approaches. The aim of this study was to provide some basic guidelines of their use to those interested in adopting these techniques. In this study, emphasis was placed on relating each type of constraint-handling scheme to the type of problem being solved.

Based precisely on these preliminary results, we foresee several potential paths for future research in this area: (1) more intensive use of explicit local search mechanisms coupled with constraint-handling techniques, (2) in-depth studies of the influence of the genetic operators used in these types of methods, (3) novel and more effective proposals of diversity maintenance mechanisms, (4) the combination of multiobjective concepts (Pareto methods with population-based techniques) into one single constraint-handling approach.

Acknowledgments

The first author acknowledges support from CONACyT through project number 52048-Y. The second author acknowledges support from CONACyT through project number 42435-Y.

Appendix A

The details of each test function used in our experiments are presented below.

P1 Minimize $f(\mathbf{X}) = 5.3578547x_3^2 + 0.8356891x_1x_5 + 37.293239x_1 - 40792.141$

subject to
$$g_1(\mathbf{X}) = 85.334407 + 0.0056858x_2x_5 + 0.0006262x_1x_4$$
$$- 0.0022053x_3x_5 - 92 \leq 0$$
$$g_2(\mathbf{X}) = -85.334407 - 0.0056858x_2x_5 - 0.0006262x_1x_4$$
$$+ 0.0022053x_3x_5 \leq 0$$
$$g_3(\mathbf{X}) = 80.51249 + 0.0071317x_2x_5 + 0.0029955x_1x_2$$
$$+ 0.0021813x_3^2 - 110 \leq 0$$
$$g_4(\mathbf{X}) = -80.51249 - 0.0071317x_2x_5 - 0.0029955x_1x_2$$
$$- 0.0021813x_3^2 + 90 \leq 0$$
$$g_5(\mathbf{X}) = 9.300961 + 0.0047026x_3x_5 + 0.0012547x_1x_3$$
$$+ 0.0019085x_3x_4 - 25 \leq 0$$
$$g_6(\mathbf{X}) = -9.300961 - 0.0047026x_3x_5 - 0.0012547x_1x_3$$
$$- 0.0019085x_3x_4 + 20 \leq 0$$

where $78 \leq x_1 \leq 102$, $33 \leq x_2 \leq 45$, $27 \leq x_i \leq 45$ $(i = 3, 4, 5)$. The optimum solution is $\mathbf{X}^* = (78, 33, 29.995256025682, 45, 36.775812905788)$, where $f(\mathbf{X}^*) = -30665.539$. Constraints g_1 y g_6 are active.

P2 Minimize $f(\mathbf{X}) = (x_1 - 10)^3 + (x_2 - 20)^3$

subject to
$$g_1(\mathbf{X}) = -(x_1 - 5)^2 - (x_2 - 5)^2 + 100 \leq 0$$
$$g_2(\mathbf{X}) = (x_1 - 6)^2 + (x_2 - 5)^2 - 82.81 \leq 0$$

where $13 \leq x_1 \leq 100$ and $0 \leq x_2 \leq 100$. The optimum solution is $\mathbf{X}^* = (14.095, 0.84296)$, where $f(\mathbf{X}^*) = -6961.81388$. Both constraints are active.

P3 Minimize $f(\mathbf{X}) = x_1^2 + x_2^2 + x_1x_2 - 14x_1 - 16x_2 + (x_3 - 10)^2 + 4(x_4 - 5)^2 + (x_5 - 3)^2 + 2(x_6 - 1)^2 + 5x_7^2 + 7(x_8 - 11)^2 + 2(x_9 - 10)^2 + (x_{10} - 7)^2 + 45$

subject to
$$g_1(\mathbf{X}) = -105 + 4x_1 + 5x_2 - 3x_7 + 9x_8 \leq 0$$
$$g_2(\mathbf{X}) = 10x_1 - 8x_2 - 17x_7 + 2x_8 \leq 0$$
$$g_3(\mathbf{X}) = -8x_1 + 2x_2 + 5x_9 - 2x_{10} - 12 \leq 0$$
$$g_4(\mathbf{X}) = 3(x_1 - 2)^2 + 4(x_2 - 3)^2 + 2x_3^2 - 7x_4 - 120 \leq 0$$
$$g_5(\mathbf{X}) = 5x_1^2 + 8x_2 + (x_3 - 6)^2 - 2x_4 - 40 \leq 0$$
$$g_6(\mathbf{X}) = x_1^2 + 2(x_2 - 2)^2 - 2x_1 x_2 + 14x_5 - 6x_6 \leq 0$$
$$g_7(\mathbf{X}) = 0.5(x_1 - 8)^2 + 2(x_2 - 4)^2 + 3x_5^2 - x_6 - 30 \leq 0$$
$$g_8(\mathbf{X}) = -3x_1 + 6x_2 + 12(x_9 - 8)^2 - 7x_{10} \leq 0$$

where $-10 \leq x_i \leq 10$ ($i = 1, \ldots, 10$). The global optimum is $\mathbf{X}^* = (2.171996, 2.363683, 8.773926, 5.095984, 0.9906548, 1.430574, 1.321644, 9.828726, 8.280092, 8.375927)$, where $f(\mathbf{X}^*) = 24.3062091$. Constraints g_1, g_2, g_3, g_4, g_5 and g_6 are active.

P4 Minimize $f(\mathbf{X}) = (x_1 - 10)^2 + 5(x_2 - 12)^2 + x_3^4 + 3(x_4 - 11)^2 + 10x_5^6 + 7x_6^2 + x_7^4 - 4x_6 x_7 - 10x_6 - 8x_7$

subject to
$$g_1(\mathbf{X}) = -127 + 2x_1^2 + 3x_2^4 + x_3 + 4x_4^2 + 5x_5 \leq 0$$
$$g_2(\mathbf{X}) = -282 + 7x_1 + 3x_2 + 10x_3^2 + x_4 - x_5 \leq 0$$
$$g_3(\mathbf{X}) = -196 + 23x_1 + x_2^2 + 6x_6^2 - 8x_7 \leq 0$$
$$g_4(\mathbf{X}) = 4x_1^2 + x_2^2 - 3x_1 x_2 + 2x_3^2 + 5x_6 - 11x_7 \leq 0$$

where $-10 \leq x_i \leq 10$ ($i = 1, \ldots, 7$). The global optimum is $\mathbf{X}^* = (2.330499, 1.951372, -0.4775414, 4.365726, -0.6244870, 1.038131, 1.594227)$, where $f(\mathbf{X}^*) = 680.6300573$. Two constraints are active (g_1 and g_4).

P5 Minimize $f(\mathbf{X}) = x_1 + x_2 + x_3$

subject to
$$g_1(\mathbf{X}) = -1 + 0.0025(x_4 + x_6) \leq 0$$
$$g_2(\mathbf{X}) = -1 + 0.0025(x_5 + x_7 - x_4) \leq 0$$
$$g_3(\mathbf{X}) = -1 + 0.01(x_8 - x_5) \leq 0$$
$$g_4(\mathbf{X}) = -x_1 x_6 + 833.33252x_4 + 100x_1 - 83333.333 \leq 0$$
$$g_5(\mathbf{X}) = -x_2 x_7 + 1250x_5 + x_2 x_4 - 1250x_4 \leq 0$$
$$g_6(\mathbf{X}) = -x_3 x_8 + 1250000 + x_3 x_5 - 2500x_5 \leq 0$$

where $100 \leq x_1 \leq 10000$, $1000 \leq x_i \leq 10000$, ($i = 2, 3$), $10 \leq x_i \leq 1000$, ($i = 4, \ldots, 8$). The global optimum is: $\mathbf{X}^* = (579.19, 1360.13, 5109.92, 182.0174, 295.5985, 217.9799, 286.40, 395.5979)$, where $f(\mathbf{X}^*) = 7049.248$. g_1, g_2 and g_3 are active.

P6 Minimize $f(\mathbf{X}) = x_1^2 + (x_2 - 1)^2$

subject to
$$h(\mathbf{X}) = x_2 - x_1^2 = 0$$

where $-1 \leq x_1 \leq 1$, $-1 \leq x_2 \leq 1$. The optimum solution is $\mathbf{X}^* = (\pm 1/\sqrt{2}, 1/2)$, where $f(\mathbf{X}^*) = 0.75$.

P7 Design of a welded beam A welded beam is designed for minimum cost subject to constraints on shear stress (τ), bending stress in the beam (σ), buckling load on the bar (P_c), end deflection of the beam (δ), and side constraints. There

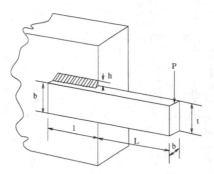

Fig. 1. Welded beam.

are four design variables as shown in Figure 1: h (x_1), l (x_2), t (x_3) and b (x_4). The problem can be stated as follows.

Minimize $f(\mathbf{X}) = 1.10471x_1^2x_2 + 0.04811x_3x_4(14.0 + x_2)$

subject to
$$g_1(\mathbf{X}) = \tau(\mathbf{X}) - \tau_{max} \leq 0$$
$$g_2(\mathbf{X}) = \sigma(\mathbf{X}) - \sigma_{max} \leq 0$$
$$g_3(\mathbf{X}) = x_1 - x_4 \leq 0$$
$$g_4(\mathbf{X}) = 0.10471x_1^2 + 0.04811x_3x_4(14.0 + x_2) - 5.0 \leq 0$$
$$g_5(\mathbf{X}) = 0.125 - x_1 \leq 0$$
$$g_6(\mathbf{X}) = \delta(\mathbf{X}) - \delta_{max} \leq 0$$
$$g_7(\mathbf{X}) = P - P_c(\mathbf{X}) \leq 0$$

where $\tau(\mathbf{X}) = \sqrt{(\tau')^2 + 2\tau'\tau''\frac{x_2}{2R} + (\tau'')^2}$ $\tau' = \frac{P}{\sqrt{2}x_1x_2}, \tau'' = \frac{MR}{J}$,

$M = P\left(L + \frac{x_2}{2}\right)$, $R = \sqrt{\frac{x_2^2}{4} + \left(\frac{x_1+x_3}{2}\right)^2}$

$J = 2\left\{\frac{x_1x_2}{\sqrt{2}}\left[\frac{x_2^2}{12} + \left(\frac{x_1+x_3}{2}\right)^2\right]\right\}$ $\sigma(\mathbf{X}) = \frac{6PL}{x_4x_3^2}, \delta(\mathbf{X}) = \frac{4PL^3}{Ex_3^3x_4}$

$P_c(\mathbf{X}) = \frac{4.013\sqrt{\frac{EGx_3^2x_4^6}{36}}}{L^2}\left(1 - \frac{x_3}{2L}\sqrt{\frac{E}{4G}}\right)$

$P = 6000$ lb, $L = 14$ in, $E = 30 \times 10^6$ psi, $G = 12 \times 10^6$ psi $\tau_{max} = 13,600$ psi, $\sigma_{max} = 30,000$ psi, $\delta_{max} = 0.25$ in, where $0.1 \leq x_1 \leq 2.0$, $0.1 \leq x_2 \leq 10.0$, $0.1 \leq x_3 \leq 10.0$ and $0.1 \leq x_4 \leq 2.0$.

P8 Design of a pressure vessel A cylindrical vessel is capped at both ends by hemispherical heads as shown in Figure 2. The objective is to minimize the total cost, including the cost of the material, forming and welding. There are four design variables: T_s (thickness of the shell), T_h (thickness of the head), R (inner radius) and L (length of the cylindrical section of the vessel, not including the head). T_s and T_h are integer multiples of 0.0625 inch, which are the available thicknesses of rolled steel plates, and R and L are continuous.

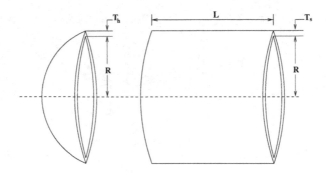

Fig. 2. Pressure vessel

The problem can be stated as follows.

Minimize $f(\mathbf{X}) = 0.6224x_1x_3x_4 + 1.7781x_2x_3^2 + 3.1661x_1^2x_4 + 19.84x_1^2x_3$

subject to
$g_1(\mathbf{X}) = -x_1 + 0.0193x_3 \leq 0$
$g_2(\mathbf{X}) = -x_2 + 0.00954x_3 \leq 0$
$g_3(\mathbf{X}) = -\pi x_3^2 x_4 - \frac{4}{3}\pi x_3^3 + 1,296,000 \leq 0$
$g_4(\mathbf{X}) = x_4 - 240 \leq 0$

where $1 \leq x_1 \leq 99$, $1 \leq x_2 \leq 99$, $10 \leq x_3 \leq 200$ and $10 \leq x_4 \leq 200$.

P9 Minimization of the weight of a tension/compression spring This problem consists of minimizing the weight of a tension/compression spring (see Figure 3) subject to constraints on minimum deflection, shear stress, surge frequency, outside diameter and design variables. The design variables are the mean coil diameter D (x_2), the wire diameter d (x_1) and the number of active coils N (x_3). Formally, the problem can be expressed as

Minimize $(N + 2)Dd^2$

subject to
$g_1(\mathbf{X}) = 1 - \frac{D^3 N}{71785d^4} \leq 0$
$g_2(\mathbf{X}) = \frac{4D^2 - dD}{12566(Dd^3 - d^4)} + \frac{1}{5108d^2} - 1 \leq 0$
$g_3(\mathbf{X}) = 1 - \frac{140.45d}{D^2 N} \leq 0$

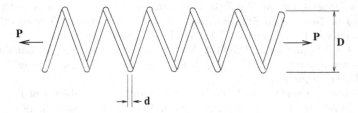

Fig. 3. Tension-compression spring

$g_4(\mathbf{X}) = \frac{D+d}{1.5} - 1 \leq 0$

where $0.05 \leq x_1 \leq 2$, $0.25 \leq x_2 \leq 1.3$ and $2 \leq x_3 \leq 15$.

References

[1] Shamim Akhtar, Kang Tai, and Tapabrata Ray. A Socio-Behavioural Simulation Model for Engineering Design Optimization. *Engineering Optimization*, 34(4):341–354, 2002.

[2] Anders Angantyr, Johan Andersson, and Jan-Olov Aidanpaa. Constrained Optimization based on a Multiobjective Evolutionary Algorithms. In *Proceedings of the Congress on Evolutionary Computation 2003 (CEC'2003)*, volume 3, pages 1560–1567, Piscataway, New Jersey, December 2003. Canberra, Australia, IEEE Service Center.

[3] Eduardo Camponogara and Sarosh N. Talukdar. A Genetic Algorithm for Constrained and Multiobjective Optimization. In Jarmo T. Alander, editor, *3rd Nordic Workshop on Genetic Algorithms and Their Applications (3NWGA)*, pages 49–62, Vaasa, Finland, August 1997. University of Vaasa.

[4] C. W. Carroll. The created response surface technique for optimizing nonlinear restrained systems. *Operations Research*, 9:169–184, 1961.

[5] Carlos A. Coello Coello. Constraint-handling using an evolutionary multiobjective optimization technique. *Civil Engineering and Environmental Systems*, 17:319–346, 2000.

[6] Carlos A. Coello Coello. Treating Constraints as Objectives for Single-Objective Evolutionary Optimization. *Engineering Optimization*, 32(3):275–308, 2000.

[7] Carlos A. Coello Coello. Theoretical and Numerical Constraint Handling Techniques used with Evolutionary Algorithms: A Survey of the State of the Art. *Computer Methods in Applied Mechanics and Engineering*, 191(11-12):1245–1287, January 2002.

[8] Carlos A. Coello Coello and Arturo Hernández Aguirre. Design of Combinational Logic Circuits through an Evolutionary Multiobjective Optimization Approach. *Artificial Intelligence for Engineering, Design, Analysis and Manufacture*, 16(1):39–53, 2002.

[9] Carlos A. Coello Coello and Efrén Mezura-Montes. Handling Constraints in Genetic Algorithms Using Dominance-Based Tournaments. In I.C. Parmee,

editor, *Proceedings of the Fifth International Conference on Adaptive Computing in Design and Manufacture (ACDM'2002)*, volume 5, pages 273–284, University of Exeter, Devon, UK, April 2002. Springer-Verlag.

[10] Carlos A. Coello Coello, David A. Van Veldhuizen, and Gary B. Lamont. *Evolutionary Algorithms for Solving Multi-Objective Problems.* Kluwer Academic Publishers, June 2002.

[11] R. Courant. Variational Methods for the Solution of Problems of Equilibrium and Vibrations. *Bulletin of the American Mathematical Society*, 49:1–23, 1943.

[12] Carlos Manuel Mira de Fonseca. *Multiobjective Genetic Algorithms with Applications to Control Engineering Problems.* PhD thesis, Department of Automatic Control and Systems Engineering, University of Sheffield, Sheffield, UK, September 1995.

[13] Hugo de Garis. Genetic Programming: Building Artificial Nervous Systems using Genetically Programmed Neural Networks Models. In R. Porter and B. Mooney, editors, *Proceedings of the 7th International Conference on Machine Learning*, pages 132–139. Morgan Kaufmann Publishers, 1990.

[14] Kalyanmoy Deb. Genetic algorithms in multimodal function optimization. Master's thesis, University of Alabama, Alabama, USA, 1989.

[15] Kalyanmoy Deb. An Efficient Constraint Handling Method for Genetic Algorithms. *Computer Methods in Applied Mechanics and Engineering*, 186(2/4):311–338, 2000.

[16] Kalyanmoy Deb, Amrit Pratap, Sameer Agarwal, and T. Meyarivan. A Fast and Elitist Multiobjective Genetic Algorithm: NSGA–II. *IEEE Transactions on Evolutionary Computation*, 6(2):182–197, April 2002.

[17] A. V. Fiacco and G. P. McCormick. Extensions of SUMT for nonlinear programming: equality constraints and extrapolation. *Management Science*, 12(11):816–828, 1968.

[18] Carlos M. Fonseca and Peter J. Fleming. Genetic Algorithms for Multiobjective Optimization: Formulation, Discussion and Generali zation. In Stephanie Forrest, editor, *Proceedings of the Fifth International Conference on Genetic Algorithms*, pages 416–423, San Mateo, California, 1993. Morgan Kauffman Publishers.

[19] Carlos M. Fonseca and Peter J. Fleming. Multiobjective optimization and multiple constraint handling with evolutionary algorithms—Part I: A unified formulation. *IEEE Transactions on Systems, Man, and Cybernetics, Part A: Systems and Humans*, 28(1):26–37, 1998.

[20] David Goldberg. *The Design of Innovation.* Kluwer Academic Publishers, New York, June 2002. ISBN 1-4020-7098-5.

[21] Arturo Hernández-Aguirre, Salvador Botello-Rionda, Carlos A. Coello Coello, Giovanni Lizárraga-Lizárraga, and Efrén Mezura-Montes. Handling Constraints using Multiobjective Optimization Concepts. *International Journal for Numerical Methods in Engineering*, 59(15):1989–2017, April 2004.

[22] Robert Hinterding and Zbigniew Michalewicz. Your Brains and My Beauty: Parent Matching for Constrained Optimisation. In *Proceedings of the 5th International Conference on Evolutionary Computation*, pages 810–815, Anchorage, Alaska, May 1998.

[23] Jeffrey Horn, Nicholas Nafpliotis, and David E. Goldberg. A Niched Pareto Genetic Algorithm for Multiobjective Optimization. In *Proceedings of the First IEEE Conference on Evolutionary Computation, IEEE World Congress*

on Computational Intelligence, volume 1, pages 82–87, Piscataway, New Jersey, June 1994. IEEE Service Center.

[24] F. Jiménez, A.F. Gómez-Skarmeta, and G. Sánchez. How Evolutionary Multi-objective Optimization can be used for Goals and Priorities based Optimization. In *Primer Congreso Español de Algoritmos Evolutivos y Bioinspirados (AEB'02)*, pages 460–465. Mérida España, 2002.

[25] Joshua D. Knowles and David W. Corne. Approximating the Nondominated Front Using the Pareto Archived Evolution Strategy. *Evolutionary Computation*, 8(2):149–172, 2000.

[26] Saku Kukkonen and Jouni Lampinen. Constrained Real-Parameter Optimization with Generalized Differential Evolution. In *2006 IEEE Congress on Evolutionary Computation (CEC'2006)*, pages 911–918, Vancouver, Canada, July 2006. IEEE Press.

[27] J.J. Liang and P.N. Suganthan. Dynamic Multi-Swarm Particle Swarm Optimizer with a Novel Constraint-Handling Mechanism. In *2006 IEEE Congress on Evolutionary Computation (CEC'2006)*, pages 316–323, Vancouver, Canada, July 2006. IEEE Press.

[28] Efrén Mezura-Montes and Carlos A. Coello Coello. Adding a Diversity Mechanism to a Simple Evolution Strategy to Solve Constrained Optimization Problems. In *Proceedings of the Congress on Evolutionary Computation 2003 (CEC'2003)*, volume 1, pages 6–13, Piscataway, New Jersey, December 2003. Canberra, Australia, IEEE Service Center.

[29] Efrén Mezura-Montes and Carlos A. Coello Coello. Multiobjective-Based Concepts to Handle Constraints in Evolutionary Algorithms. In Edgar Chávez, Jesús Favela, Marcelo Mejía, and Alberto Oliart, editors, *Proceedings of the Fourth Mexican International Conference on Computer Science (ENC'2003)*, pages 192–199, Los Alamitos, CA, September 2003. Apizaco, Tlaxcala, México, IEEE Computer Society.

[30] Efrén Mezura-Montes and Carlos A. Coello Coello. A Simple Multimembered Evolution Strategy to Solve Constrained Optimization Problems. *IEEE Transactions on Evolutionary Computation*, 9(1):1–17, February 2005.

[31] Efrén Mezura-Montes, Carlos A. Coello Coello, and Ricardo Landa-Becerra. Engineering Optimization Using a Simple Evolutionary Algorithm. In *Proceedings of the Fiftheenth International Conference on Tools with Artificial Intelligence (ICTAI'2003)*, pages 149–156, Los Alamitos, CA, November 2003. Sacramento, California, IEEE Computer Society.

[32] Zbigniew Michalewicz, Kalyanmoy Deb, Martin Schmidt, and Thomas Stidsen. Test-Case Generator for Nonlinear Continuous Parameter Optimization Techniques. *IEEE Transactions on Evolutionary Computation*, 4(3):197–215, September 2000.

[33] Zbigniew Michalewicz and David B. Fogel. *How to Solve It: Modern Heuristics*. Springer, Germany, 2nd edition, 2004.

[34] Zbigniew Michalewicz and Cezary Z. Janikow. Handling Constraints in Genetic Algorithms. In R. K. Belew and L. B. Booker, editors, *Proceedings of the Fourth International Conference on Genetic Algorithms (ICGA-91)*, pages 151–157, San Mateo, California, 1991. University of California, San Diego, Morgan Kaufmann Publishers.

[35] Zbigniew Michalewicz and G. Nazhiyath. Genocop III: A co-evolutionary algorithm for numerical optimization with nonlinear constraints. In David B. Fogel,

editor, *Proceedings of the Second IEEE International Conference on Evolutionary Computation*, pages 647–651, Piscataway, New Jersey, 1995. IEEE Press.

[36] Zbigniew Michalewicz and Marc Schoenauer. Evolutionary Algorithms for Constrained Parameter Optimization Problems. *Evolutionary Computation*, 4(1):1–32, 1996.

[37] D. Ortiz-Boyer, C. Hervás-Martínez, and N. García-Pedrajas. Crossover Operator Effect in Functions Optimization with Constraints. In J. J. Merelo-Guervós and et al., editors, *Proceedings of the 7th Parallel Problem Solving from Nature (PPSN VII)*, pages 184–193, Heidelberg, Germany, September 2002. Granada, Spain, Springer-Verlag. Lecture Notes in Computer Science Vol. 2439.

[38] Domingo Ortiz-Boyer, Rafael Del-Castillo-Gomariz, Nicolas Garcia-Pedrajas, and Cesar Hervas-Martinez. Crossover effect over penalty methods in function optimization with constraints. In *2005 IEEE Congress on Evolutionary Computation (CEC'2005)*, volume 2, pages 1127–1134, Edinburgh, Scotland, September 2005. IEEE Service Center.

[39] Akira Oyama, Koji Shimoyama, and Kozo Fujii. New Constraint-Handling Method for Multi-Objective Multi-Constraint Evolutionary Optimization and Its Application to Space Plane Design. In R. Schilling, W. Haase, J. Periaux, H. Baier, and G. Bugeda, editors, *Evolutionary and Deterministic Methods for Design, Optimization and Control with Applications to Industrial and Societal Problems (EUROGEN 2005)*, Munich, Germany, 2005.

[40] I. C. Parmee and G. Purchase. The development of a directed genetic search technique for heavily constrained design spaces. In I. C. Parmee, editor, *Adaptive Computing in Engineering Design and Control-'94*, pages 97–102, Plymouth, UK, 1994. University of Plymouth.

[41] David Powell and Michael M. Skolnick. Using genetic algorithms in engineering design optimization with non-linear constraints. In Stephanie Forrest, editor, *Proceedings of the Fifth International Conference on Genetic Algorithms (ICGA-93)*, pages 424–431, San Mateo, California, July 1993. University of Illinois at Urbana-Champaign, Morgan Kaufmann Publishers.

[42] Tapabrata Ray, Tai Kang, and Seow Kian Chye. An Evolutionary Algorithm for Constrained Optimization. In Darrell Whitley, David Goldberg, Erick Cantú-Paz, Lee Spector, Ian Parmee, and Hans-Georg Beyer, editors, *Proceedings of the Genetic and Evolutionary Computation Conference (GECCO'2000)*, pages 771–777, San Francisco, California, July 2000. Morgan Kaufmann.

[43] Tapabrata Ray and K.M. Liew. Society and Civilization: An Optimization Algorithm Based on the Simulation of Social Behavior. *IEEE Transactions on Evolutionary Computation*, 7(4):386–396, August 2003.

[44] Tapabrata Ray and Kang Tai. An Evolutionary Algorithm with a Multilevel Pairing Strategy for Single and Multiobjective Optimization. *Foundations of Computing and Decision Sciences*, 26:75–98, 2001.

[45] Colin B. Reeves. *Modern Heuristic Techniques for Combinatorial Problems*. John Wiley and Sons, Great Britain, 1993.

[46] Thomas Philip Runarsson and Xin Yao. Search biases in constrained evolutionary optimization. *IEEE Transactions on Systems, Man, and Cybernetics Part C—Applications and Reviews*, 35(2):233–243, May 2005.

[47] J. David Schaffer. Multiple Objective Optimization with Vector Evaluated Genetic Algorithms. In *Genetic Algorithms and their Applications: Proceedings*

of the First International Conference on Genetic Algorithms, pages 93–100. Lawrence Erlbaum, 1985.

[48] Marc Schoenauer and Zbigniew Michalewicz. Evolutionary Computation at the Edge of Feasibility. In H.-M. Voigt, W. Ebeling, I. Rechenberg, and H.-P. Schwefel, editors, *Proceedings of the Fourth Conference on Parallel Problem Solving from Nature (PPSN IV)*, pages 245–254, Heidelberg, Germany, September 1996. Berlin, Germany, Springer-Verlag.

[49] Marc Schoenauer and Spyros Xanthakis. Constrained GA Optimization. In Stephanie Forrest, editor, *Proceedings of the Fifth International Conference on Genetic Algorithms (ICGA-93)*, pages 573–580, San Mateo, California, July 1993. University of Illinois at Urbana-Champaign, Morgan Kauffman Publishers.

[50] Alice E. Smith and David W. Coit. Constraint Handling Techniques—Penalty Functions. In Thomas Bäck, David B. Fogel, and Zbigniew Michalewicz, editors, *Handbook of Evolutionary Computation*, chapter C 5.2. Oxford University Press and Institute of Physics Publishing, 1997.

[51] Patrick D. Surry and Nicholas J. Radcliffe. The COMOGA Method: Constrained Optimisation by Multiobjective Genetic Algorithms. *Control and Cybernetics*, 26(3):391–412, 1997.

[52] Sangameswar Venkatraman and Gary G. Yen. A Generic Framework for Constrained Optimization Using Genetic Algorithms. *IEEE Transactions on Evolutionary Computation*, 9(4), August 2005.

[53] Yuping Wang, Dalian Liu, and Yiu-Ming Cheung. Preference bi-objective evolutionary algorithm for constrained optimization. In Yue Hao et al., editor, *Computational Intelligence and Security. International Conference, CIS 2005*, volume 3801, pages 184–191, Xi'an, China, December 2005. Springer-Verlag. Lecture Notes in Artificial Intelligence.

[54] Wang Yong and Cai Zixing. A Constrained Optimization Evolutionary Algorithm Based on Multiobjective Optimization Techniques. In *2005 IEEE Congress on Evolutionary Computation (CEC'2005)*, volume 2, pages 1081–1087, Edinburgh, Scotland, September 2005. IEEE Service Center.

[55] Yuren Zhou, Yuanxiang Li, Jun He, and Lishan Kang. Multi-objective and MGG Evolutionary Algorithm for Constrained Optimization. In *Proceedings of the Congress on Evolutionary Computation 2003 (CEC'2003)*, volume 1, pages 1–5, Piscataway, New Jersey, December 2003. Canberra, Australia, IEEE Service Center.

[56] E. Zitzler and L. Thiele. Multiobjective evolutionary algorithms: a comparative case study and the strength Pareto approach. *IEEE Transactions on Evolutionary Computation*, 3(4):257–271, 1999.

Tackling Dynamic Problems with Multiobjective Evolutionary Algorithms

Lam T. Bui[1], Minh-Ha Nguyen[1], Jürgen Branke[2], and Hussein A. Abbass[1]

[1] The Artificial Life and Adaptive Robotics Laboratory, School of ITEE, ADFA,
University of New South Wales, Canberra, Australia
{l.bui, m.nguyen, h.abbass}@adfa.edu.au
[2] Institute AIFB, University of Karlsruhe, 76128 Karlsruhe, Germany
{branke@aifb.uni-karlsruhe.de}

Summary. In this chapter, we discuss the use of multiobjective evolutionary algorithms (MOEAs) for solving single-objective optimization problems in dynamic environments. Specifically, we investigate the consideration of a second (artificial) objective, with the aim of maintaining greater population diversity and adaptability. The paper suggests and compares a number of alternative ways to express this second objective. An empirical comparison shows that the best alternatives are competitive with other evolutionary algorithm variants designed for handling dynamic environments.

1 Introduction

Many real-world optimization problems are dynamic. Changes in the environment can take various forms such as changes in the parameters, objective functions, or problem constraints. For example, portfolio decisions need to be re-optimized every now and then to reflect changes in the stock market. In scheduling, new jobs have to be inserted into the schedule. Even the optimization of a flight path needs to adapt to changes on the fly; especially under the new free-flight control arrangements.

In such dynamic environments the optimum changes, and an optimization algorithm has to track the changing optimum over time. One difficulty when using evolutionary algorithms (EAs) in dynamic environments is that the population loses its genetic diversity, thereby also losing its ability to adapt. A variety of methods has been proposed in the literature to address the diversity loss, e.g., by maintaining a separate memory to store the best solutions found in each generation, or by using multi-populations to simultaneously track several areas in a changing landscape.

In this chapter, we investigate the use of multiobjective evolutionary algorithms (MOEAs), which usually do not fully converge but naturally maintain a certain diversity of the population along the Pareto front. The idea is to supplement the dynamically changing original objective by adding a second (artificial) objective, aiming at maintaining diversity. We suggest a number of alternatives for this second objective, compare them empirically, and discuss their computational complexity.

The moving peaks benchmark (MPB) [22] is used as a standard testbed for our experiments. NSGA2 [9], a widely used MOEA variant, is employed as the evolutionary multiobjective technique. Our proposed approaches are empirically compared against each other, and against a traditional EA and two classical variants for dynamic environments: the random immigrants [12] and the hypermutation algorithm [6] on different types of changes.

In the remainder of the paper, we first review some related work on changing environments, and the use of MOEAs for solving single-objective optimization problems. In Section 3, we introduce the different alternatives used as the second objective. The empirical comparison and a discussion of the results can be found in Section 4. The paper concludes with a summary and some ideas for future work.

2 Related Work

2.1 Changing Environments

A number of criteria have been proposed in the literature to classify dynamic environments, such as frequency, severity, and predictability [4, 24, 27].

The frequency of change determines how often the environment changes. As the frequency increases, the time left for adaptation gets shorter and tracking the optima gets harder. The severity of a change indicates the degree of a change, e.g., the distance the optimum moves, or how strongly the heights of local optima change. The predictability of the change defines the pattern of the change, such as linearity, randomness, or circularity. In the latter, the cycle length defines the amount of time needed before the changes repeat themselves.

Dynamic environments have been studied extensively in the EA literature. Some detailed reviews can be found in [4, 15]. Generally speaking, there are three main approaches to date: diversity control, memory-based and multi-population approaches.

Diversity control is a common topic in EA in general. To control diversity in a dynamic problem, one can either increase diversity whenever a change is detected — such as with the hypermutation method [6] and the variable local search technique [26] — or maintain high diversity throughout the evolutionary run as in random immigrants [12], the Thermodynamical Genetic Algorithm [18, 19], or sentinels [20].

Memory-based techniques employ an extra memory that implicitly or explicitly stores useful information to guide future search. Implicit memory usually uses redundant representations [8, 11, 13, 17] and leaves it up to the evolutionary algorithm how to make use of this available memory. In explicit memories [3, 29, 30], specific information (e.g., the best solution at certain time intervals) is stored and retrieved when needed by the evolutionary mechanism.

The last approach uses subpopulations to simultaneously track several optima in different areas of the search space. Basically, it maintains a self-adaptive memory of several promising areas of the search space. Examples in this group are the self-organizing scouts method [4] and the multinational GA [25].

A number of authors have suggested different benchmark problems including the moving peaks problem [3, 4], a close variant thereof [21], the XOR-DOP generator [31], and a class of dynamic trap functions [2]. In our later analysis, we use the moving peaks problem.

2.2 Using MOEAs to Solve Single-Objective Optimization Problems

Many real-world problems involve several, usually conflicting objectives. In those cases, there is usually no single optimal solution, but a set of equally good alternatives with different trade-offs. Evolutionary algorithms are particularly suitable to solve such problems, since they maintain a population of solutions and can search for a large number of alternatives simultaneously. Most work on MOEAs is based on the concept of dominance. A solution x is said to dominate a solution y if and only if solution x is at least as good as y in all objectives, and strictly better in at least one objective. This concept is then used during selection by favouring nondominated solutions. For comprehensive books on the topic, the reader is referred to [7, 9].

There are several approaches applying MOEAs to solve single-objective problems. Knowles, Watson, and Corne [16] hypothesised that adding objectives may reduce the number of local optima in a problem, and showed good performance on benchmark TSP problems. Also, in [14] Jensen thought that adding objectives to flow-shop scheduling problems might lead to good building blocks for better solutions, and so chose to use additional objectives that rewarded parts of a good solution (flow times of individual jobs). Abbass and Deb [1] proposed adding an artificial objective to promote diversity. Three different artificial objectives were discussed in their work. The first is based on a time-stamp for each chromosome using the generation number. The second is by generating the second fitness at random. The third is by reversing the optimization of the first objective (i.e., maximizing the function if the original problem is minimization and vice versa). In comparison to single-objective algorithms, this approach offered a better convergence rate while maintaining good diversity.

Toffolo and Benini [23] used diversity explicitly as an additional objective for multiobjective evolutionary algorithms. They used the sum of the Euclidean distances between an individual and all other individuals in the population as a measure on how much that individual contributes to the diversity. As an alternative measure, they also suggested the distance to the closest individual. The approach was tested against several state-of-the-art MOEAs. The results showed that the approach is very effective at converging to the POF and at distributing the solutions along it.

In all of the previous work, environments were assumed to be stationary. With the purpose of solving optimization problems in dynamic environments, Yamasaki [28] introduced a technique which records all evaluations an individual ever received, and uses all this information to decide which individuals to keep. In this process, the concept of Pareto dominance is also used.

In this chapter, we look at the use of MOEAs for dynamic environments. We propose and compare a number of different choices for the second artificial objective, including ideas from Abbass and Deb [1] and Toffolo and Benini [23]. Some of these objectives explicitly address diversity, while others do so only implicitly. The chapter is an extended version of [5], with an analysis on the complexity of the method as well as a more detailed empirical evaluation.

3 Methodology

3.1 Designing the Artificial Objective

One of the main challenges in using MOEAs to solve single objective optimization problems is defining the supplementary objectives. In this work, we propose to transform the original single-objective optimization problem into a bi-objective problem. The first objective is always taken as the objective of the original single optimization problem, while the second objective is an artificial one. While in principle it would be possible to consider also the case of more than two objectives, we assume only one artificial objective. There are a number of ways to define the artificial objective. Here, we classify them into two categories, namely by whether they address diversity explicitly or implicitly. For those using implicit diversity, the second objective is built based on some information that may help indirectly support diversity of the population. The artificial objectives considered are as follows:

- **Time stamp:** The first artificial objective is a time stamp at the time when an individual is generated. As in [1], we stamp each individual in the initial population with a different time stamp represented by a counter that gets incremented whenever a new individual is created. From the second population on, all individuals in the population get the same time stamp, set to the population size plus the generation index. The time stamp then serves as a second objective to be minimized (i.e., old individuals are favoured in the selection).
- **Random:** The second artificial objective is to minimize a random value assigned to each individual. Some bad individuals may be assigned small random values and get a chance to survive, which may be useful when the environment changes.
- **Inversion:** The third approach inverts the original objective function by minimizing it if it was a maximization problem, and vice versa, as proposed in [1].

The objectives explicitly addressing diversity will rely on the Euclidean distance between solutions. We propose three different options as follows, and they are all to be maximized:

- **Distance from the closest neighbour (DCN):** The artificial objective of a solution is the distance from the solution to its closest neighbour. So, a pair of very similar solutions will have a rather poor second objective value, and a certain spread of solutions is encouraged.
- **Average distance from all individuals (ADI):** This objective takes into account a solution's average distance from all other solutions in the population. Again, it encourages the spread of the population over the search space, favouring solutions at the edge of the population.
- **Distance from the best individual of the population (DBI):** This objective will help the algorithm to avoid any possible trap caused from local optima. The best individual will be located first; then the distance from the solution to the best individual will be calculated.

3.2 Complexity Analysis

Obviously, the three artificial objectives with implicit diversity preservation can be calculated very quickly (in constant time). For the others, the computational complexity is as follows:

- **DCN**: We need to calculate the second objective value for all solutions in the population using Euclidean distance. Further, for each solution, we need to search for its closest neighbour. Therefore, its computational complexity is $O(nN^2)$ where N is the population size and n is the number of variables.
- **ADI**: Similarly, this case requires $O(nN^2)$ operations since, for each solution, we need to count all the distances from the solution to all others.
- **DBI**: Since the best solution is used for calculation of the objective value for all solutions, complexity is only linear in N, that is, $O(nN)$, much smaller than for DCN and ADI.

However, in the following, we will compare the approaches solely on the basis of function evaluations, assuming that the time to calculate distances is negligible compared to the time to evaluate an individual.

4 Experimental Studies

4.1 Selected MOEA

NSGA2 [9, 10] is one of today's most successful and most widely used MOEAs. It is based on two principles, convergence and diversity. Convergence to the Pareto-optimal front is ensured by nondominated sorting. This method ranks individuals by iteratively determining the nondominated solutions in the population (nondominated front), assigning those individuals the next best rank, and removing them from the population. Diversity within one rank is maintained by favouring individuals with a large crowding distance, which is defined as the sum of distances between a solution's neighbours on either side in each dimension of the objective space. Furthermore, NSGA2 is an elitist algorithm, i.e., it keeps as many nondominated solutions as possible (up to the size of the population).

4.2 Moving Peaks Benchmark

We test our approach on the Moving Peaks Benchmark [3, 22]. This is a dynamic benchmark problem with a number of peaks changing over time in location, width, and height. Formally, the benchmark function can be formulated as follows:

$$F(\overrightarrow{x}, t) = max[B(\overrightarrow{x}), max_{i=1,..,m}P(\overrightarrow{x}, h_i(t), w_i(t), \overrightarrow{p}_i(t))], \tag{1}$$

where B is a time-invariant basis landscape, P is the function defining the shape of the individual peaks with peak-height h, peak-width w, and peak-location p, and i ranges from 1 to m, where m is the number of peaks.

The benchmark exhibits three different types of change:

- Moving the peaks within the search space does not create new optima. It requires a tracking behaviour of the EA, which has to follow a moving optimum. Naturally, the faster the peaks move, the more difficult the tracking, up to the point when the population is unable to follow.
- Changing the peak heights leads to a "jumping" behaviour of the optimum, when the current optimum is reduced in height and another, previously local optimum becomes the new global optimum.

- Changing the width makes the basin of attraction narrower or wider over time. Because narrower peaks mean that a small peak is less likely to be "covered" by a higher peak, narrower peaks generally mean a rougher landscape, slowing down the tracking process.

The benchmark generator has a number of parameters. For a basic scenario, we used the parameters specified in Table 1.

Table 1. Parameters for the benchmark problem

Parameters	Values
Number of peaks	50
Number of dimensions	5
Change frequency	25 generations
Peak function	cone
Peak heights	20..70
Peak width	1..12
Correlation coefficient	0.0

We run three sets of experiments. In the first set of experiments, we have a small shift length of 1.0 for moving the peaks, and consider all combinations of height severities 7 and 15 and width severities 1 and 3.

In the second set of experiments, we fix the severities for height and width to 7 and 1, respectively, and consider different shift lengths: 1, 3, 5, 7, and 10.

The last set of experiments uses the same settings as in [4], which allows comparison with results from the literature. Details can be found in Table 2.

Table 2. Parameter settings for the third set of experiments

Parameters	Values
Number of peaks	10
Change frequency	50 generations
Shift severity	1

4.3 Parameter Settings and Performance Assessment

We will compare our MOEAs with some other approaches from the literature, namely

- a traditional EA,
- random immigrants [12], which, in each generation, replaces a certain percentage of the individuals with random individuals, and
- hypermutation, which drastically increases the mutation rate for one generation after a change.

All MOEAs use NSGA2 with binary representation (with 30 bits for each variable), binary tournament selection, single-point crossover, and bit-flip mutation. The random immigrants algorithm replaces 20% of the population at each generation with new individuals, as recommended in [12]. The mutation rate of the hypermutation algorithm is set to 0.5 [6]. In order to have a fair comparison, we also employ elitism for all three algorithms, where the best solution of the previous generation is always included in the current one.

The behaviour of the evolutionary methods largely depends on the crossover and mutation rates used. Therefore, it is necessary to examine the different methods with a wide range of parameter values to identify a good setting. The crossover rate p_c is varied between 0.5 and 1 with a step size of 0.05 and the mutation rate p_m is varied between 0 and 0.2 with a step size of 0.01. For each pair of p_c and p_m, 30 runs are performed with different random seeds, a population size of 100, and 1,000 the number of generations.

We report on two performance measures: For the *average generation error* (AGEr), we record the best individual in each generation as measured on the original single objective function. The difference between the objective value of this individual and the current global optima is known as the *generation error*. The average generation error (AGEr) is then the average over all generation errors just before a new change occurs. The *offline error* [4] also takes into account how quickly the new best solution is located. It is defined as the average of all current errors, i.e., the average deviation of the best individual evaluated since the last change from the optimum. Finally, the diversity of the population is also recorded over time. It is calculated as the average Euclidean distance between all pairs of individuals in the population in phenotype space.

4.4 Parameter Tuning and Effect of Changing the Peak Width and Height

Firstly, we try to find the best performance each approach achieved over a wide range of crossover and mutation rates, and for different change severities of height and width (our first scenario). These results are summarized in Tables 3 and 4. The result associated with each approach in the table corresponds to the minimum value of the AGEr and its associated pair of p_c and p_m. As can be seen, the height severity has a larger influence than the width severity.

In general, the traditional EA has the worst performance. This indicates the need for introducing diversity in the population during the optimization process, especially after the change, such that the population has enough diversity to adapt to the new area of the optimum. Of the traditional approaches, hypermutation performs best. In fact, it is even the best of all tested approaches for the test case with large height severity and small width severity. The random immigrants approach didn't differ much from the baseline traditional EA. The best overall performance is achieved by the MOEA with DCN or ADI as artificial objective (note that the difference in results between DCN and ADI is not statistically significant).

The small mutation rate for the random immigrants and hypermutation approaches may contribute to the fact that diversity is already maintained through the introduction of new offspring at random for the random immigrants and the higher mutation rate for the hypermutation algorithm. An overall high mutation rate is not helpful.

Table 3. The best p_c and p_m for each approach and the AGEr $\pm$ the standard error with height change severity of 7.0 and width change severity of 1.0 and 3.0

Width severity	1.0		3.0	
The artificial objective function	p_c (p_m)	AGEr	p_c (p_m)	AGEr
Traditional EA	0.55 (0.04)	11.48 $\pm$ 0.60	0.55 (0.04)	13.69 $\pm$ 0.75
Random Immigrants	1.00 (0.01)	11.47 $\pm$ 0.56	0.65 (0.01)	13.51 $\pm$ 1.06
hypermutation	0.75 (0.01)	11.95 $\pm$ 0.59	0.85 (0.01)	13.78 $\pm$ 0.63
Time-based objective	0.60 (0.11)	12.06 $\pm$ 0.64	0.50 (0.11)	12.96 $\pm$ 0.81
Random objective	0.60 (0.10)	11.29 $\pm$ 0.55	0.50 (0.08)	12.30 $\pm$ 0.96
Inverse objective	0.55 (0.06)	12.37 $\pm$ 0.87	0.50 (0.09)	13.96 $\pm$ 0.87
DCN	0.75 (0.05)	**9.52 $\pm$ 0.45**	0.65 (0.04)	10.42 $\pm$ 0.71
ADI	0.70 (0.06)	9.74 $\pm$ 0.35	0.55 (0.04)	**9.31 $\pm$ 0.51**
DBI	0.50 (0.09)	12.24 $\pm$ 0.55	0.80 (0.10)	11.79 $\pm$ 0.71

Table 4. The best p_c and p_m for each approach and the AGE $\pm$ the standard error with height change severity of 15.0 and width change severity of 1.0 and 3.0

Width severity	1.0		3.0	
The artificial objective function	p_c (p_m)	AGEr	p_c (p_m)	AGEr
Traditional EA	0.55 (0.07)	16.14 $\pm$ 0.71	0.50 (0.04)	14.79 $\pm$ 0.85
Random Immigrants	0.55 (0.03)	15.38 $\pm$ 0.81	0.80 (0.02)	14.67 $\pm$ 0.70
hypermutation	0.70 (0.02)	**11.96 $\pm$ 0.80**	0.80 (0.02)	12.70 $\pm$ 0.66
Time-based objective	1.00 (0.10)	12.06 $\pm$ 0.80	0.55 (0.09)	15.06 $\pm$ 1.00
Random objective	0.55 (0.10)	14.79 $\pm$ 0.66	0.65 (0.09)	14.20 $\pm$ 0.83
Inverse objective	0.50 (0.07)	15.98 $\pm$ 0.89	0.60 (0.07)	15.28 $\pm$ 0.88
DCN	0.50 (0.07)	12.68 $\pm$ 0.60	0.50 (0.06)	**12.56 $\pm$ 0.62**
ADI	0.65 (0.06)	13.18 $\pm$ 0.52	0.50 (0.05)	13.00 $\pm$ 0.63
DBI	0.60 (0.06)	14.05 $\pm$ 0.61	0.60 (0.07)	13.96 $\pm$ 0.74

Figures 1 and 2 visualize the values of AGEr for different values of crossover and mutation rates for the two best MOEA representatives and traditional EA approaches: ADI and hypermutation. Obviously, the absence of mutation deteriorates the quality of solutions. This effect is stronger for ADI than for hypermutation. Hypermutation can still introduce new genetic material when triggered. With ADI, genetic material, once lost, can never be regenerated. Over all considered change severities, the surface for ADI is smoother than that for hypermutation, indicating that the approach produces more reliable results. A good performance was achieved for MOEAs within an area centred around $p_c = 0.6, p_m = 0.07$. In this area, the DCN and ADI approaches produced the best results. The traditional EA and random immigrants approaches are very sensitive to the mutation rate (not shown).

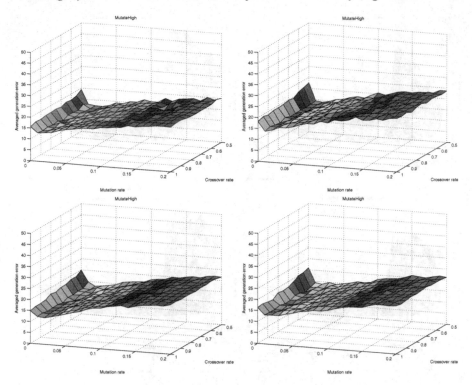

Fig. 1. The AGEr achieved from the hypermutation over different ranges of p_c and p_m in four different cases of change severities: (7,1),(7,3),(15,1),(15,3)

4.5 Analysis of the Effect of Moving Peaks

In the second scenario, we investigate the performance of the algorithms under the effect of moving peaks with different shift lengths. Height change severity and width change severity were fixed at 7.0 and 1.0, respectively, and the best algorithm parameters found for this scenario in the previous subsection were used. We again measure the value of AGEr for all algorithms and report them in Table 5.

The results still show the best performance for DCN and ADI. The results also demonstrate the inefficiency of the three proposed objectives using implicit diversity (time stamp, random, and inverse). They seem not to make any improvement in comparison with other single-objective methods (EA, hypermutation, and random immigrants).

4.6 Comparison with Self-organizing Scouts (SOS)

In order to compare our approaches with more complex state-of-the-art approaches, the third problem set uses the same parameter settings as used in [4] for testing the

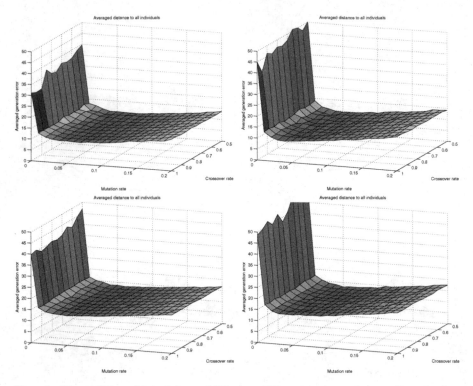

Fig. 2. The AGEr achieved from ADI over different ranges of p_c and p_m in four different cases of change severities: (7,1),(7,3),(15,1),(15,3)

Table 5. The obtained AGEr and the standard error of all algorithms with different shift severities: 1, 3, 5, 7, 10

Algs	1	3	5	7	10
Traditional EA	11.48 ±0.6	12.31 ±0.90	12.73 ±0.89	13.41 ±0.88	14.86 ±0.92
Rand. Imm.	11.47 ±0.56	12.57 ±0.90	12.79 ±0.96	13.19 ±0.93	13.98 ±0.98
Hypermutation	11.95 ±0.59	11.71 ±0.88	12.35 ±0.92	12.86 ±0.84	12.97 ±0.92
Time objective	12.06 ±0.64	13.09 ±0.87	14.17 ±0.88	14.96 ±0.87	15.21 ±0.93
Rand. obj.	11.29 ±0.55	12.52 ±0.85	13.36 ±0.87	13.91 ±0.87	14.11 ±0.92
Inv. obj.	12.37 ±0.87	13.37 ±0.85	13.59 ±0.86	14.03 ±0.86	14.67 ±0.93
DCN	**9.52 ±0.45**	**10.64 ±0.80**	**11.10 ±0.79**	**11.77 ±0.80**	**12.36 ±0.86**
ADI	9.74 ±0.35	11.20 ±0.97	12.84 ±1.05	13.13 ±0.93	13.67 ±0.89
DBI	12.24 ±0.55	13.44 ±0.94	14.27 ±0.98	14.59 ±0.91	15.04 ±0.95

self-organizing scouts (see Table 2). The offline errors of all tested algorithms are reported in Table 6.

Note in particular that the change frequency is lower than in the previous experiments. This allows algorithms more time to recover from the change. However, MOEAs with implicit diversity objective showed a slow ability of recovering in comparison to the other algorithms. Their values of offline error are the worst. Meanwhile, the algorithms using explicit diversity (especially DCN and ADI) are still the best, with the smallest values of offline error.

Since self-organizing scouts used real-valued representation, for consistency we also tested the approaches with real-valued representation. For this, the SBX crossover and polynomial mutation [9] were used. The offline error was also recorded for each approach and reported in Table 6. Although the trend does not change regarding the performance of EMO and the traditional methods, it is interesting to see that the approaches with real-valued representation generally obtained better results in comparison with those using binary representation.

For self-organizing scouts, Branke [4] reported an offline error of 4.01, which is a bit lower than the 4.60 we found for our MOEA with DCN as artificial objective. Still, we believe this is a promising result, as our approach is much simpler than self-organizing scouts and basically uses standard MOEA software.

Table 6. The off-line error and its standard error obtained with the problem instance from the literature

Algorithms	Binary Rep.	Real-valued Rep.
Traditional EA	6.63±0.158	5.72±0.11
Random Immigrants	6.80±0.132	5.82±0.109
Hypermutation	6.11±0.206	5.88±0.082
Time objective	8.04±0.299	6.38±0.179
Random objective	7.52±0.211	5.67±0.119
Inverse objective	8.45±0.244	8.08±0.215
DCN	5.84±0.131	**4.60±0.085**
ADI	**5.35±0.162**	5.25±0.161
DBI	6.21±0.155	6.40±0.235

4.7 Diversity Measurement

We now investigate the underlying reason for the superior performance of DCN and ADI. We hypothesised that diversity plays a key role in solving the dynamic optimization problem. We implemented both explicit and implicit diversity as the second objective function to make the single-objective dynamic problem a bi-objective problem. Especially for DCN and ADI, they directly impose the geometrical spread of the population and discourage crowding consistently over time. Further, NSGA2 with crowding distance also provides a good mechanism to preserve diversity in the population.

To further support our hypothesis, we measure the phenotypic diversity of the population for each approach over time as the average Euclidean distance between all pairs of individuals. A typical example is shown in Figure 3. Clearly, ADI maintains the highest level of diversity throughout the run. On average, DCN, hypermutation, and random immigrants have a similar level of diversity. But while hypermutation is strongly oscillating, random immigrants and DCN maintain a more or less constant level of diversity. The fact that DCN performs much better than random immigrants in terms of AGEr is probably due to a more intelligent strategy to maintain this diversity. Instead of introducing just random (and mostly useless) individuals, diversity is maintained by keeping good but different individuals.

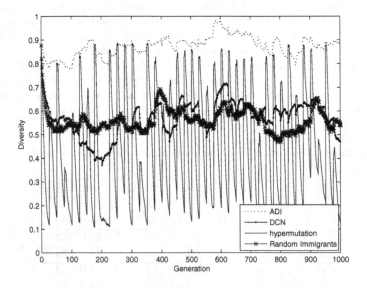

Fig. 3. The diversity of the population over time with change severities of height = 7.0, width = 1.0, and with shift = 5.0 (using binary representation)

In summary, the above results show that integrating diversity into EMO offers a promising solution to dynamic optimization problems. The diversity-based objective helps the EMO approach outperform the random immigrant and hypermutation algorithms.

5 Conclusion

This chapter has proposed a new way to maintain diversity in the population in order to improve the EA's ability to continuously adapt to a dynamic environment. The main idea is to add an artificial objective with the goal of encouraging diversity, and then to use a standard MOEA technique to solve the (now two-objective) optimization problem.

We looked into a number of alternatives for the artificial objective and compared these alternatives empirically with each other and with some other results from the literature. Of the six tested artificial objectives, only those which explicitly take into account diversity seem to perform well in all scenarios. The best ones clearly outperform the standard EA and classical diversity control techniques, such as random immigrants or hypermutation. When compared to self-organizing scouts, a state-of-the-art evolutionary algorithm specifically designed for dynamic environments, our approach is slightly inferior, but it is much simpler and easier to implement.

In future, we intend to investigate other types of diversity as the second objective such as using the niche count (as in fitness sharing) instead of the average distance to all other individuals.

Acknowledgments

This work is supported by the University of New South Wales grant PS04411 and the Australian Research Council (ARC) Centre on Complex Systems grant number CEO0348249.

References

[1] H. A. Abbass and K. Deb. Searching under multi-evolutionary pressures. In Zitzler et al., editor, *Proceedings of the Fourth Conference on Evolutionary Multi-Criterion Optimization*, Spain, 2003.

[2] H. A. Abbass, K. Satry, and D. Goldberg. Oiling the wheels of change: The role of adaptive automatic problem decomposition in non-stationary environments. Technical report, IlliGAL, University of Illinois at Urbana-Champaign, 2004.

[3] J. Branke. Memory enhanced evolutionary algorithms for changing optimisation problems. In *In Congress on Evolutionary Computation CEC99*, pages 1875–1882. IEEE, 1999.

[4] J. Branke. *Evolutionary optimization in dynamic environments*. Kluwer Academic Publishers, Massachusetts USA, 2002.

[5] L. T. Bui, J. Branke, and H. Abbass. Multiobjective optimization for dynamic environments. In *Congress on Evolutionary Computation*, pages 2349–2356. IEEE Press, 2005.

[6] H.G. Cobb. An investigation into the use of hypermutation as an adaptive operator in genetic algorithms having continuous, time-dependent nonstationary environments. Technical Report AIC-90-001, Naval Research Laboratory, 1990.

[7] C. A. C. Coello, D. A. V. Veldhuizen, and G. B. Lamont. *Evolutionary Algorithms for Solving Multi-Objective Problems*. Kluwer Academic Publishers, New York USA, 2002.

[8] D. Dasgupta. Incorporating redundancy and gene activation mechanisms in genetic search. In L. Chambers, editor, *Practical Handbook of Genetic Algorithms*, pages 303–316. CRC Press, 1995.

[9] K. Deb. *Multiobjective Optimization using Evolutionary Algorithms*. John Wiley and Son Ltd, New York, 2001.

[10] K. Deb, A. Pratap, S. Agarwal, and T. Meyarivan. A fast and elitist multiobjective genetic algorithm: NSGA-II. *IEEE Transactions on Evolutionary Computation*, 6(2):182–197, 2002.

[11] D.E. Goldberg and R.E. Smith. Nonstationary function optimisation using genetic algorithms with dominance and diploidy. In J.J. Grefenstette, editor, *Second International Conference on Genetic Algorithms*, pages 59–68. Lawrence Erlbaum Associates, 1987.

[12] J.J. Grefenstette. Genetic algorithms for changing environments. In R. Männer and B. Manderick, editors, *Parallel Problem Solving from Nature*, pages 137–144. Elsevier Science Publisher, 1992.

[13] B.S. Hadad and C.F. Eick. Supporting polyploidy in genetic algorithms using dominance vectors. In *International Conference on Evolutionary Programming*, volume 1213 of *Lecture Notes in Computer Science*, pages 223–234, 1997.

[14] M.T. Jensen. Helper-objectives: Using multiobjective evolutionary algorithms for single-objective optimization. *Journal of Mathematical Modelling and Algorithms*, 1(25), 2004.

[15] Y. Jin and J. Branke. Evolutionary optimization in uncertain environments – a survey. *IEEE Transactions on Evolutionary Computation*, to appear.

[16] J. Knowles, R. A. Watson, and D. Corne. Reducing local optima in single-objective problems by multi-objectivization. In Zitzler et al., editor, *Proceedings of the First Conference on Evolutionary Multi-Criterion Optimization*, pages 269–283, Zurich Switzerland, 2001.

[17] J. Lewis, E. Hart, and G. Ritchie. A comparison of dominance mechanisms and simple mutation on non-stationary problems. In A. E. Eiben, T. Bäck, M. Schoenauer, and H.-P. Schwefel, editors, *Parallel Problem Solving from Nature*, number 1498 in LNCS, pages 139–148. Springer, 1998.

[18] N. Mori, S. Imanishia, H. Kita, and Y. Nishikawa. Adaptation to changing environments by means of the memory based thermodynamical genetic algorithms. In T. Bäck, editor, *Seventh International Conference on Genetic Algorithms*, pages 299–306. Morgan Kaufmann, 1997.

[19] N. Mori, H. Kita, and Y. Nishikawa. Adaptation to changing environments by means of the thermodynamical genetic algorithms. In H.-M. Voigt, editor, *Parallel Problem Solving from Nature*, volume 1411 of *Lecture Notes in Computer Science*, pages 513–522, Berlin, 1996. Elsevier Science Publisher.

[20] R. W. Morrison. *Designing Evolutionary Algorithms for Dynamic Environments*. Springer, 2004.

[21] R.W. Morrison and K. A. DeJong. A test problem generator for non-stationary environments. In *Proceedings of 1999 Congress on Evolutionary Computation*, 1999.

[22] The moving peaks benchmark. http://www.aifb.uni-karlsruhe.de/~jbr/MovPeaks

[23] A. Toffolo and E. Benini. Genetic diversity as an objective in multi-objective evolutionary algorithms. *Evolutionary Computation*, 11(2):151–168, 2003.

[24] K. Trojanowski and Z. Michalewicz. Evolutionary algorithms for non-stationary environments. In *Proc. of 8th Workshop: Intelligent Information systems*, pages 229–240, Porland, 1999. ICS PAS Press.

[25] R.K. Ursem. Multinational GAs: multimodal optimization techniques in dynamic environments. In *Proceedings of the Genetic and Evolutionary Computation Conference (GECCO-2000)*, pages 19–26. San Francisco, CA: Morgan Kaufmann, 2000.

[26] F. Vavak, K. Jukes, and T.C. Fogarty. Learning the local search range for genetic optimisation in nonstationary environments. In *IEEE International Conference on Evolutionary Computation*, pages 355–360. IEEE Publishing, 1997.

[27] K. Weicker. *Evolutionary Algorithms and Dynamic Optimization Problems*. Der Andere Verlag, 2003.

[28] K. Yamasaki. Dynamic Pareto optimum GA against the changing environments. In J. Branke and T. Bäck, editors, *Evolutionary Algorithms for Dynamic Optimization Problems*, pages 47–50, San Francisco, California, USA, 7 2001.

[29] S. Yang. Memory-based immigrants for genetic algorithms in dynamic environments. In Hans-Georg Beyer et al., editors, *Genetic and Evolutionary Computation Conference*, pages 1115–1122. ACM, 2005.

[30] S. Yang. Associative memory scheme for genetic algorithms in dynamic environments. In F. Rothlauf et al., editors, *Applications of Evolutionary Computing*, volume 3907 of *LNCS*, pages 788–799. Springer, 2006.

[31] S. Yang and X. Yao. Population-based incremental learning algorithms for dynamic optimization problems. *Soft Computing*, 9(11):815–834, 2005.

Computational Studies of Peptide and Protein Structure Prediction Problems via Multiobjective Evolutionary Algorithms

Vincenzo Cutello[1], Giuseppe Narzisi[1,2], and Giuseppe Nicosia[1]

[1] Department of Mathematics and Computer Science
University of Catania
Viale A. Doria 6, 95125 Catania, Italy {vctl, nicosia}@dmi.unict.it
[2] Courant Institute of Mathematical Sciences
New York University
715 Broadway #1010, New York, NY 10003, USA narzisi@nyu.edu

Summary. Finding the native structure of a protein starting from its amino acid sequence remains one of the most challenging open problems in bioinformatics and molecular biology. The Protein Structure Prediction (PSP) problem has been tackled from many different directions. The common approach is to cast it in the form of a global single-objective optimization problem using energy functions to evaluate the physical state of the conformations. In this work we reformulate the PSP as a multiobjective optimization problem motivated by the fact that the folded state of a protein is a small *ensemble* of conformational structures. A 2-objective decomposition of the CHARMM energy function is proposed based on local and nonlocal interactions between atoms, supported by experimental evidence that these objectives are in fact conflicting. A new MOEA algorithm is used to search for (observed) Pareto-optimal sets of conformations with respect to the 2-objective formulation and tested on a large set of medium-size proteins (26-70 residues), with results demonstrating the effectiveness of this approach and providing different measures of protein complexity. Results also point to instances in which the CHARMM energy model suffers from low accuracy owing to the required trade-off between differing objectives in finding "good" conformations.

1 Introduction

Central to the field of protein structural biology is a set of observations, hypotheses and paradoxes. The *thermodynamic hypothesis* postulates that the native state of a protein is the state of lowest free energy of the protein system under physiological conditions [2]. *Anfinsen's hypothesis* claims that the information determining the three-dimensional structure of a protein is a consequence of both its amino acid sequence and the solvent environment [1]. The function of proteins is directly related to the three-dimensional conformations assumed by the proteins. From the protein

structure it is theoretically possible to infer the protein function. Hence, sequence →
structure → function is another well-known paradigm in molecular biology. According to *Levinthal's paradox*, it would take the present age of the universe for a protein
to explore all possible configurations and locate the minimum energy configuration
[18].

The free energy of a protein can be modeled as a function of the different interactions within the protein. These interactions (local, nonlocal, hydrophobic, entropic
effects, hydrogen bonding) depend on the positions of the atoms of the protein. The
set of atomic coordinates providing the minimum possible value of free energy corresponds to the native conformation of the protein. Since the interactions comprising
the energy function are highly nonconvex, the protein structure prediction problem
must be tackled as a global optimization problem.

For the past fifty years, the protein structure prediction problem has been defined
as a *large single-objective optimization problem*, with researchers employing Molecular Dynamics, Monte Carlo methods and Evolutionary Algorithms [3, 27, 23, 6, 12].
In this chapter, we reason by computational experiments that it would be more suitable to model the protein structure prediction problem as a *multiobjective optimization problem*. The goal of our research is to find a set of *equivalent* three-dimensional
folded conformations, relying on the observation that the folded state is one of only
a small *ensemble* of all possible conformations [24]. We adopt a multiobjective approach in order to obtain "good" nondominated compact solutions near or inside
the folded state. At any stage the protein exists in an ensemble of conformations
which comprise an *approximated Pareto front*. As described later in the chapter, we
consider *two conflicting forces/objectives*: *local and nonlocal interactions*. A more
thorough comparison of our approach with other state-of-the-art PSP methods may
be found in [8, 7]. Here we explore the advantages of this approach and attempt to
validate it through detailed computational studies on a large test-bed of proteins.

To our knowledge there are only two previous works that study PSP as a multiobjective optimization problem. In a recent article [10], Lamont and coauthors
reformulated the PSP problem as a MOP using a multiobjective evolutionary algorithm (MOfmGA), but limited their study to two small protein sequences: [Met]-
enkephelin (five residues) and polyalanine (14 residues). Schulze-Kremer has also
used multiobjective optimization.[3] In particular he used $RMSD$ as an additional
objective with the rationale that "more . . . information about genuine protein conformations should improve the fitness function to guide the genetic algorithm towards
native-like conformations". This approach is problematic in that it assumes advance
knowledge of the native conformation, which of course is not the case for newly
discovered protein sequences.

The chapter is organized as follows. We begin by explaining the energy model
used for the PSP problem. After discussing how conformations are represented, we
introduce the multiobjective formulation and present experimental evidence supporting the notion that various objectives are in fact conflicting. Section 4 describes
the designed multiobjective evolutionary algorithm and the mutation operators used.
Next, a simple angle-based decision making method is presented to select solutions
from the computed Pareto sets as a post-processing phase. Section 5 reports the
results of the computational studies on the proposed test-bed of proteins. The first
set of results presents an exhaustive study of six proteins extracted from the lit-

---

[3] http://www.techfak.uni-bielefeld.de/bcd/Curric/ProtEn

erature. The dynamics of the Pareto fronts are presented, coupled with analysis of the structural quality of the conformations during the evolutionary process guided by the MOEA. Additional analysis at the level of correlation between the energy function and the metrics ($RMSD$ and DME) used is reported. The second set of results evaluates the performance of our approach on another seven proteins extracted from the literature. We conclude with a summary of results and possible future investigations.

2 Multiobjective Optimization

Many optimization problems involve *multiple, conflicting, and noncommensurate objective functions having strong nonlinear interdependence* and *several constraints*; in general, it is desirable *to optimize all objective functions simultaneously*. The primary goal of such optimization problems is to obtain a class of equivalent solutions, usually called *efficient* or Pareto optimal.

A conformation c^{po} is Pareto optimal if there exists no feasible point c which would decrease some criterion without causing a *simultaneous* increase in at least one other criterion. A Pareto optimal set that truly meets this definition is called a true Pareto-optimal set, $\mathcal{P}^*_{true}$. In contrast, a Pareto-optimal set that is obtained by means of an optimization method is referred to as an observed Pareto-optimal set, P^*_{obs}. In reality, an observed Pareto optimal set is an *estimate* of a true Pareto-optimal set. Identifying a good estimate P^*_{obs} is the key factor for the decision maker's selection of a compromise solution, which satisfies the objectives as much as possible. We denote the observed Pareto-optimal set obtained at time-step t using an optimization method by $P^{*,t}_{obs}$ (or the current observed Pareto-optimal set). Moreover, we have

$$P^{*,t}_{obs} = \{\mathbf{x}^t_1, \ldots, \mathbf{x}^t_{np}\} \tag{1}$$

where $np = \mid P^{*,t}_{obs} \mid$ is the total number of observed Pareto solutions at time step t. Consequently, the major problem a decision maker faces is to find "the best"

$$\mathbf{x} \in P^*_{obs}.$$

Definition 3 *For a given MOP $\mathbf{f}(\mathbf{x})$ and Pareto-optimal set P^*, the Pareto front, PF^*, is defined as:*

$$PF^* = \{\mathbf{u} = \mathbf{f} = (f_1(\mathbf{x}), \ldots, f_k(\mathbf{x})) \mid \mathbf{x} \in \mathcal{P}^*\}. \tag{2}$$

As for the Pareto optimal set, we can define the *observed Pareto front* at time step t by an optimization method:

$$PF^{*,t}_{obs} = \{\mathbf{u}^t_1, \mathbf{u}^t_2, \ldots, \mathbf{u}^t_N\}, \tag{3}$$

where $N = \mid PF^{*,t}_{obs} \mid$ is the total number of observed Pareto front solutions at time step t.

The goals of our research work are to find a PF^*_{obs} as close as possible to PF^* and to find a PF^*_{obs} as diverse as possible by the application of multiobjective optimization algorithms for PSP. Identifying a good estimate of $PF^{*,t}_{obs}$ is crucial for the decision maker's selection of a good protein conformation in terms of given metrics. In sum, an MOP consists of the following procedure: finding the optimal (or the observed) Pareto front and choosing one of the candidate solutions in the Pareto front using some higher-level information. The design, application, and evaluation of procedures for determining the most preferred alternative from a set of nondominated alternatives is currently a topic of interest dealt with in the field of MCDM (see the introduction chapter, page 14).

3 Energy Modeling

3.1 Representation Models

A nontrivial task that preempts use of any search procedure to attack the PSP problem is the selection of a good *representation* for the conformations. Few conformation representations are commonly used: all-atom 3D coordinates, all-heavy-atom coordinates, backbone atom coordinates + sidechain centroids, C_α coordinates, and backbone and sidechain torsion angles. Some algorithms use multiple representations simultaneously for different purposes. The relative advantages of different representations is in and of itself a topic of great interest in evolutionary computation, and in optimization in general.

In the current work, we use an *internal coordinates representation* (torsion angles), which is currently the most widely used representation model for real proteins. Each residue type requires a fixed number of torsion angles to fix the 3D coordinates of all atoms. Bond lengths and angles are fixed at their ideal values. All the ω torsion angles are fixed at their ideal value at $180°$. So, the degrees of freedom in this representation are the backbone and sidechain torsion angles (ϕ, ψ and χ_i). The number of χ angles depends on the residue type.

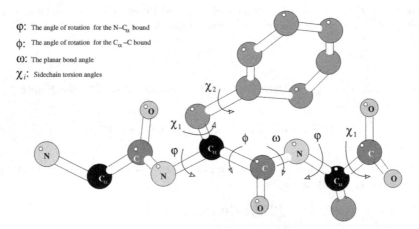

φ: The angle of rotation for the N–C_α bound

ϕ: The angle of rotation for the C_α–C bound

ω: The planar bond angle

χ_i: Sidechain torsion angles

Fig. 1. Internal coordinates representation (torsion angles)

Figure 1 illustrates the representation used and the meaning of the different torsion angles

3.2 Multiobjective Formulation

In order to evaluate the structure of a molecule, we need to use some cost functions. Sometimes called potential energy functions or force fields, these functions return a value for the energy based on the conformation of the molecule. As such, they provide information on which conformations of the molecule are better or worse. A lower the energy value represents a better conformation.

The literature on cost functions is enormous [19, 13, 5]. In this work we use the CHARMM (version 27) energy function. CHARMM (Chemistry at HARvard Macromolecular Mechanics) is a popular all-atom force field used mainly for studying macromolecules [20]. It is a composite sum of several molecular mechanics equations grouped into two major types: *bonded* (stretching, bending, torsion, Urey-Bradley, impropers) and *non-bonded* (van der Waals, electrostatics) interactions.

Following the structure of the energy function, we can think of a protein as a collection of atoms linked by chemical bonds. With the symbol $a_i \leftrightarrow a_j$ we represent a chemical bond between the two atoms a_i and a_j. Using this notation we can divide all the atoms into two categories: *bonded atoms* and *non-bonded atoms*:

$$A_{bond} = \{< a_i, \ldots, a_{i+k} > | \exists a_i \leftrightarrow a_{i+1}\}, \forall i = 1 \ldots k, 1 \leq k \leq 3, \qquad (4)$$

$$A_{\neg bond} = \{< a_i, \ldots, a_j > | \neg \exists a_i \leftrightarrow a_j\}, \forall i, j. \qquad (5)$$

The bond set A_{bond} represents the set of all atom chains with a maximum length of four. In this way we consider only bonds, angles and torsion interactions between atoms, i.e., *local interaction*. The $A_{\neg bond}$ set represents all the atoms not connected by a chemical bond, which are atoms separated by at least three or more covalent bonds, and whose interactions are thus *nonlocal*. This division reflects the decomposition of the CHARMM energy function into two partial sums, bond and non-bond atom energies:

$$f_1 = E_{bond}(A_{bond}, \mathbf{C}_{bond}), \quad f_2 = E_{\neg bond}(A_{non-bond}, \mathbf{C}_{non-bond}), \qquad (6)$$

where symbols $\mathbf{C}_{bond}$ and $\mathbf{C}_{\neg bond}$ are, respectively, the force constants involved for bonded and non-bonded atoms (for a detailed description of these constants see [20]).

The bond energy characterizes the interactions between residues that are neighbours along the primary sequence. The non-bond term represents the interaction between residues that are separated in the primary sequence by at least two intervening residues (one to four interactions).

We reduce the size of the conformational space bounding the backbone torsion angles by limiting it to regions that satisfy secondary and supersecondary constraints. Each torsion angle is constrained to lie in a range which reflects the secondary structure prediction for the residue. Table 1 reports the different ranges for each class. The native classes are based on the full DSSP 8-class classification. The predicted classes are based on the artificial neural network method presented in [28]. Side-chain torsion angles are constrained in regions derived from the backbone-independent rotamer library of Roland L. Dunbrack [26]. Side-chain constraint regions are of the form $[\mu - \sigma, \mu + \sigma]$, where μ and σ are the mean and the standard deviation for each side-chain torsion angle computed from the rotamer library. It is important to note that under these constraints the conformation is still highly

Table 1. Corresponding regions of the Secondary and Supersecondary Structure Constraints both for the native (a) and predicted (b) case

	(a) Native			(b) Predicted	
Structures	ϕ	ψ	Structures	ϕ	ψ
H (α-helix)	$[-67°, -47°]$	$[-57°, -37°]$	H (α-helix)	$[-75°, -55°]$	$[-50°, -30°]$
B (β-bridge)	$[-130°, -110°]$	$[110°, 130°]$	E (β-strand)	$[-130°, -110°]$	$[110°, 130°]$
E (β-strand)	$[-130°, -110°]$	$[110°, 130°]$	a	$[-150°, -30°]$	$[-100°, 50°]$
G ($3 - 10$-helix)	$[-59°, -39°]$	$[-36°, 16°]$	b	$[-230°, -30°]$	$[100°, 200°]$
I (pi-helix)	$[-67°, -47°]$	$[-80°, -60°]$	e	$[30°, 130°]$	$[130°, 260°]$
T (turn)	$[-180°, 180°]$	$[-180°, 180°]$	l	$[30°, 150°]$	$[-60°, 90°]$
S (bend)	$[-180°, 180°]$	$[-180°, 180°]$	t	$[-160°, -50°]$	$[50°, 100°]$
U (undefined)	$[-180°, 180°]$	$[-180°, 180°]$	U (undefined)	$[-180°, 180°]$	$[-180°, 180°]$

flexible and the structure can take on various shapes that are vastly different from the native shape. We use both the native and predicted secondary structures to compare the quality of the obtained conformations.

Using this representation, the multiobjective formulation is the following: the two functions f_1 and f_2 represent our minimization *objectives*, the torsion angles of the protein are the *decision variables* of the multiobjective problem, and the constraint regions derived by the secondary structures are the *variable bounds*.

3.3 Bond Energy vs. Non-bond Energy

Before we analyze the quality of the obtained results, we offer experimental validation for this particular multiobjective approach. It is based on the fact that local interaction (bond energy) and nonlocal interaction (non-bond energy) among atoms are in conflict, this being the typical characteristic of a multiobjective optimization problem. The literature on energy functions is vast, and most of the major energy functions are based on the combined use of bond and non-bond energies. While there is no *formal proof* about the conflict between them, we present an intuitive argument of the conflicting nature of bond and non-bond interactions and then show how it is possible to verify it experimentally.

Mentioned above is the *thermodynamic hypothesis*, which states that the native structure of a protein is that conformation which sits at the global minimum of the thermodynamical potential (free energy) of the protein [1]. This is a valid principle that governs the protein conformational search even if it is a simplified version of the real-world scenario. Along the pathway to reach the native structure, the protein is forced "to decide" what to do next. It is quite clear that the protein moves in such a way as to decrease the bond energy of the system locally, not globally. For example, the electrostatic interactions or the short distance van der Waals interaction of atoms nearby in space but far in primary sequence could be penalized although an improvement is reached in bond, angles or torsion energies. This example represents the conflict between minimizing local and nonlocal interactions.

This simple intuition is demonstrated experimentally by the plot in Figure 2. In this figure we show the typical bond and non-bond energy course during the iterations of the multiobjective evolutionary algorithm (see Section 4). It is clear that the two functions are in conflict. If one interaction decreases, the other always increases. At the same time, there are compensatory effects that result in the minimization of the total energy.

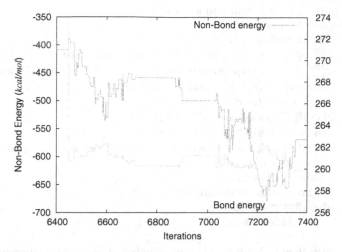

Fig. 2. Landscape illustrating the conflict between bond and non-bond energies. Left y axis range is for non-bond energy, right y axis range is for bond energy

4 The Multiobjective Evolutionary Algorithm

The PAES (Pareto Archived Evolutionary Strategy) algorithm was proposed for the first time by Knowles and Corne in 1999 [16]. PAES is a multiobjective optimizer which uses a simple (1+1) local search evolution strategy. Nonetheless, it is capable of finding diverse solutions in the Pareto-optimal set because it maintains an archive of nondominated solutions which it exploits to accurately estimate the quality of new candidate solutions. At each iteration t, a candidate solution c_t and a mutated solution m_t are compared for dominance. In the case that either one dominates the other, the dominating solution becomes the current solution of the next iteration. If neither solution dominates the other, the new candidate solution is compared with a reference population of previously archived nondominated solutions. If the comparison fails to favour one solution over the other, the chosen solution is the one which resides in the least crowded region of the space. A maximum size of the archive is always maintained. The crowding procedure is based on recursively dividing up each dimension of the objective space into 2^d equal-sized hypercubes, where d is a user defined 'depth' parameter. The algorithm continues until a given, fixed number of iterations is reached.

PAES by itself has proved to be a useful MOEA with successful application in several different fields. However, when applied to the PSP problem, we have observed poor performance both in terms of energy function and final structure obtained (see Table 2). The complexity of the funnel landscape of the PSP problem, which is

I-PAES(*depth, archive_size, objectives*)
1. $t := 0$;
2. Initialize(c); /*Generate initial random solution*/
3. Evaluate(c); /*Evaluation of initial solution*/
4. AddToArchive(c); /*Add c to archive*/
5. **while**(not(Termination()))
          /*Start Immune phase*/
6. $(c_1^{clo}, c_2^{clo}) :=$ Cloning(c); /*Clonal expansion phase*/
7. $(c_1^{hyp}, c_2^{hyp}) :=$ Hypermutation(c_1^{clo}, c_2^{clo}); /*Affinity maturation*/
8.   Evaluate(c_1^{hyp}, c_2^{hyp}); /*Evaluation phase*/
9.   **if**(c_1^{hyp} dominates c_2^{hyp}) $m := c_1^{hyp}$;
10.   **else if**(c_2^{hyp} dominates c_1^{hyp}) $m := c_2^{hyp}$;
11.   **else** $m :=$ Best(c_1^{hyp}, c_2^{hyp}); /*min E_{charmm} selection*/
12.   AddToArchive(Worst(c_1^{hyp}, c_2^{hyp})); /*max E_{charmm} selection*/
          /*End Immune phase*/
          /*Start (1+1)-PAES*/
13.   **if**(c dominates m) discard m;
14.   **else if**(m dominates c)
15.   AddToArchive(m);
16. $c := m$;
17.   **else if**(m is dominated by any member of the archive) discard m;
18.   **else** test($c, m, archive\_size, depth$);
19. $t := t + 1$;
20. **endwhile**

Fig. 3. Pseudocode of I-PAES

characterized by a huge number of local minima, coupled with the goal of producing a "good" conformation from a structural point of view ($RMSD$ and DME), clearly poses many problems (e.g., premature convergence, trapping in local minima, etc). These kinds of results are reported in Section 5.1, and they have motivated us to develop a more sophisticated algorithm.

I-PAES [8, 7] is a modified version of PAES [16, 17] with a different solution representation (polypeptide chain) and immune-inspired (*cloning* and *hypermutation*) operators [22]. The algorithm starts by initializing a random conformation. The torsion angles (ϕ, ψ, χ_i) are generated randomly from the constraint regions. Next, the energy of the conformation (a point in the landscape) is evaluated. The protein structure in internal coordinates (torsion angles) is transformed into Cartesian coordinates. The CHARMM energy potential of the structure is then computed using routines from the TINKER Molecular Modeling Package.[4]

At this point, we have the main loop of the algorithm. From the current solution, two clones will be generated, producing the solutions (c_1^{clo}, c_2^{clo}) which will be mutated into (c_1^{hyp}, c_2^{hyp}). After evaluation, the better clone (min E_{charmm}) of c_1^{hyp} and c_2^{hyp} is selected as the new mutated solution m, while the other one, if possible, is

---

[4] http://dasher.wustl.edu/tinker/

added to the archive following the standard method of PAES to update the archive. From this moment on, the algorithm proceeds following the standard structure of PAES. Figure 3 shows the pseudocode of the algorithm.

4.1 Mutation Operators

Two kinds of mutation operators are used in the affinity maturation phase (line 7 of I-PAES). The first clone is mutated using the first mutation operator and the second clone using the second mutation operator. The first mutation operator, M_1, may change the conformation dramatically. When this operator acts on a peptide chain, all the values of the backbone and sidechain torsion angles of a randomly chosen residue are reselected from their corresponding constrained regions. The probability for the application of this operator is regulated by the following law:

$$M_1(ffe) = e^{\left(\frac{-2 \times (ffe)}{T_{max}}\right)}, \tag{7}$$

where T_{max} is the maximum number of evaluations allowed and ffe is the number of fitness function evaluations performed. The probability of mutation decreases as the search method proceeds. The second mutation operator, M_2, performs a local search of the conformational space. It will perturb some torsion angles (ϕ, ψ, χ_i) of a randomly chosen residue with the law $\theta' = \theta + N(0, 1)$, where θ is the generic torsion angle, and $N(0, 1)$ is a real number generated by a Gaussian distribution of mean $\mu = 0$ and standard deviation $\sigma = 1$. The mutation rate used is similar to that in the scheme presented in [6]. The number of mutations decreases as the search method proceeds following the law

$$M_2(ffe) = 1 + \left(\frac{L}{k}\right) \times e^{\left(\frac{-2 \times (ffe)}{T_{max}}\right)}, \tag{8}$$

where ffe and T_{max} are defined as before, L is the number of residues and k is a constant set to 4.

The simple idea behind these equations operators is to maintain a high level of diversity in the PAES archive while preventing premature convergence. As we will see later in the results section, these operators will play an important role, improving the performance of the algorithm in contrast to the use of a static mutation rate.

4.2 Decision-making Phase

After a Pareto front is found, one has to choose a solution (or a class of solutions) in this front using some "higher-level information". Such a *decision making* phase can be difficult to accomplish, in particular when the number of objectives and solutions is large. Although there is no universally accepted method, in general the most interesting solutions of the observed Pareto front are characterized by the fact that a small improvement in one objective will cause a large deterioration in at least one other objective. These solutions are called *knees* [4, 11]. A simple and effective algorithm for finding the knees in the Pareto front is based on an angle-based method which uses the four closest neighbours. Given a conformation point P, if we denote by A_1 and A_2 the two closest points from the left, and by B_1 and B_2 the two closest points from the right, we can form four angles: $\widehat{A_1PB_1}$, $\widehat{A_1PB_2}$,

$\widehat{A_2 P B_1}$, and $\widehat{A_2 P B_2}$. The greatest of these four angles is then assigned to P. The knees of the Pareto front are the angles greater than a given threshold. Using this idea, we can adopt the following decision-making scheme:

1. detect the solutions which lie in the knees of the observed Pareto front, using the angle-based method with four neighbours described above; and
2. select the solution with the lowest energy function value from these samples.

This simple method is able to select solutions with a good trade-off between energy and metrics values (DME and $RMSD$).

This is just one possible approach for the decision-making phase. It is possible to use other types of higher-level information to select solutions from the Pareto front, using for instance structure stability, compactness, or hydrophobic score.

5 Computational Studies

5.1 First Set of Results and Discussions

In the first set of experiments, we apply the approach to six protein sequences, five extracted from reference [6] and one from [9]: 1ZDD, 1ROP, 1CRN, 1UTG, 1R69 and 1CTF. 1ZDD (*Disulfide-Stabilized Mini Protein A Domain*) is a two-helix peptide of 34 residues. For this protein the secondary structure constraints were predicted by the SCRATCH[5] prediction server [25]. 1ROP (*repressor of primer*) is a 4-helix bundle protein that is composed of two identical monomers. Each monomer has 56 residues and forms an α-turn-α structure. 1CRN (*crambin*) is a 46-residue protein with two α-helices and a pair of β-strands. It has three disulphide bonds, whose constraints we do not use. 1UTG (*uteroglobin*) is a 4-helix protein that has 70 residues. 1R69 (*Amino-Terminal domain of phage 434 Repressor*) is 63-residue protein with five short α-helices; two of these form a helix-turn-helix motif. 1CTF (*C-terminal domain of the ribosomal protein L7/L12*) is 68-residue protein. It has six secondary structures (three α-helix and three β-strands). For these proteins we set the maximum number of energy function evaluations to 2.5×10^5. For the proteins 1ROP, 1CRN, 1R69 and 1CTF the supersecondary structure constraints were predicted by a well-known artificial neural network method [28].

The discussion is as follows. First, we compare the performance of different versions of the PAES and I-PAES algorithms on the first protein set. Then we study the stability of the approach with respect to the native and predicted secondary structure constraints. Finally, we show specific results for each protein in terms of the obtained observed Pareto-optimal sets at different time steps, $P_{obs}^{*,t}$, and various dynamics of the algorithm during the evolution.

A Class of PAES Algorithms for the PSP

We start our computational studies with a comparison of different versions of the PAES evolution strategy. Four versions of the PAES algorithm are presented: I-PAES$_m$ and I-PAES$_s$ are instances of the modified version of PAES proposed in [8] featuring, respectively, dynamic (exponential decay) and static (single mutation)

---

[5] http://www.ics.uci.edu/%7Ebaldig/scratch/

rates. (1+1)-PAES$_1$ and (1+1)-PAES$_2$ are two instances of the standard PAES algorithm where the first version uses only the local mutation operator, M_2, while the second version uses the combination of both operators, global (M_1) and local (M_2). Note that for both versions of (1+1)-PAES a residue is selected according to the standard PAES mutation strategy, with probability $1/l$, where l is the length of the protein (number of residues).

In Table 2 we show the comparative results on ten independent runs between the four different versions of the PAES algorithm. The first interesting result is connected to the use of the global (M_1) mutation operators, which clearly plays an important role in the exploration of the conformational space. In fact, in terms of energy, lower energy values are obtained using both the local and the global mutation operators. This is seen by the fact that the (1+1)-PAES$_1$ algorithm, which uses only the local mutation operator, is unable to efficiently optimize the energy function value. On the other hand, if we inspect the results at the algorithm level comparing (1+1)-PAES with I-PAES, we observe a better overall behaviour of the I-PAES algorithm. This is related to the different selection scheme used by I-PAES with respect to (1+1)-PAES.

Table 2. Comparative results between I-PAES$_s$, I-PAES$_m$, (1+1)-PAES$_1$ and (1+1)-PAES$_2$. For each protein we report the Protein Data Bank (PDB) identifier, the length (number of residues), the approximate class (α-helix, β-sheet), and the energy values of the native structures. The last three columns show the best results obtained for each protein on ten independent runs. The DME and $RMSD$ values are measured on C_α atoms from the native structure. Energy values are calculated using the ANALYZE routine from TINKER

Protein	Algorithm	DME_{min} (Å)	$RMSD_{min}$ (Å)	Min energy ($kcal/mol$)
1ROP(56 aa)	I-PAES$_s$	2.01	4.11	−661.48
class: α	I-PAES$_m$	**1.684**	**3.70**	**−902.36**
energy: *-667.05 kcal/mol*	(1+1)-PAES$_1$	4.91	6.31	2640.77
	(1+1)-PAES$_2$	5.99	8.665	−409.95
1UTG(70 aa)	I-PAES$_s$	4.49	5.11	**282.24**
class: α	I-PAES$_m$	**3.79**	**4.60**	573.89
energy: *-142.46 kcal/mol*	(1+1)-PAES$_1$	4.71	6.04	7563.07
	(1+1)-PAES$_2$	4.82	5.56	397.12
1CRN(46 aa)	I-PAES$_s$	4.13	4.73	**232.29**
class: $\alpha + \beta$	I-PAES$_m$	**3.72**	**4.31**	509.09
energy: *202.73 kcal/mol*	(1+1)-PAES$_1$	4.67	6.18	1653.93
	(1+1)-PAES$_2$	6.05	7.89	509.52
1R69(63 aa)	I-PAES$_s$	5.93	8.42	**211.26**
class: α	I-PAES$_m$	**4.91**	**5.05**	264.56
energy: *-676.53 kcal/mol*	(1+1)-PAES$_1$	5.16	7.59	9037.89
	(1+1)-PAES$_2$	6.88	8.52	659.49
1CTF(68 aa)	I-PAES$_s$	8.08	10.69	**71.55**
class: $\alpha + \beta$	I-PAES$_m$	**6.82**	**10.12**	218.99
energy: *230.08 kcal/mol*	(1+1)-PAES$_1$	9.61	12.09	1424.33
	(1+1)-PAES$_2$	8.84	10.21	617.69

Table 3. Comparative results using Native and Predicted Secondary Structure constraints with I-PAES. For each protein we report the Protein Data Bank (PDB) identifier, the length (number of residues), and the approximate class (α-helix, β-sheet). The last three columns show the best results obtained for each protein on ten independent runs. The DME and $RMSD$ values are measured on C_α atoms from the native structure

Protein	N_{res}	class	Best DME_{min} (Å)		Best $RMSD_{min}$ (Å)		Best energy ($kcal/mol$)	
			native	predicted	native	predicted	native	predicted
1ZDD	34	α	2.01	**1.54**	3.26	**2.27**	**−1500.31**	−1052.09
1ROP	56	α	**1.52**	1.62	**3.47**	3.70	**−992.77**	−902.36
1CRN	46	$\alpha + \beta$	4.75	**3.72**	3.86	4.43	**126.76**	509.09
1UTG	70	α	7.01	**3.79**	6.92	**4.60**	**−667.91**	573.89
1R69	63	α	6.51	**4.09**	5.93	**5.05**	**−648.72**	264.56
1CTF	68	$\alpha + \beta$	9.39	**6.82**	12.94	**10.12**	**−409.13**	218.98

Native Versus Predicted Constraints

In this section we study the impact of the use of the native secondary structure information with respect to the predicted one. Table 3 shows the results for each protein.

One would expect an improvement in the quality of the computed conformations when the native (known) secondary structures are used. Evident from inspection of the results in Table 3 is that while the imposed native constraints allow one to find better energies for all the proteins, the obtained conformational structures in fact do not take advantage of the native secondary structure except for proteins with "simple" native 3D conformations. I-PAES continues to show a good performance on 1ZDD and 1ROP proteins while for 1UTG, 1CTF and 1R69 the results get worse. A different behaviour is observed for the 1CRN protein which shows results somehow intermediate between the simple and the more complex proteins. A possible explanation of this phenomena is the following: the conformational constraints imposed by the native secondary structure do not alter the flexibility of the residues along the connecting peptides (turns, twists, etc.). In fact, while the constraint regions of the torsion angles are still relatively large, the secondary structural elements (α-helix, β-strands) now have a more stable and fixed structure which probably does not help the protein folding generated by the evolutionary algorithm.

Conformational Energy and Structure Dynamics

In order to better understand the exploration of the conformational space carried out by the proposed evolutionary algorithm, we save, for each run at each iteration, statistical information for the solutions in the archive: energy values, archive size, metrics ($RMSD$ and DME). Figure 4 shows the final best Pareto fronts obtained by I-PAES for each studied protein. It is evident how approximate Pareto fronts of the nondominated solutions are effectively obtained for each protein. The distribution of points along the fronts shows a higher number of points for the non-bond objective compared to the bond objective. The algorithm samples more points along the non-bond objective relative to the bond one, which means that the non-bond energy has more variability and a higher weight during the evolution of the algorithm. This

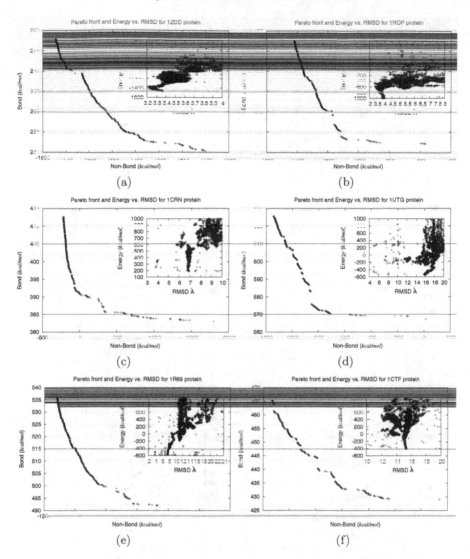

Fig. 4. Final best Pareto front with the characteristic energy function vs. $RMSD$ values plot (inset figure) for (a) 1ZDD, (b) 1ROP, (c) 1CRN, (d) 1UTG, (e) 1R69, and (f) 1CTF

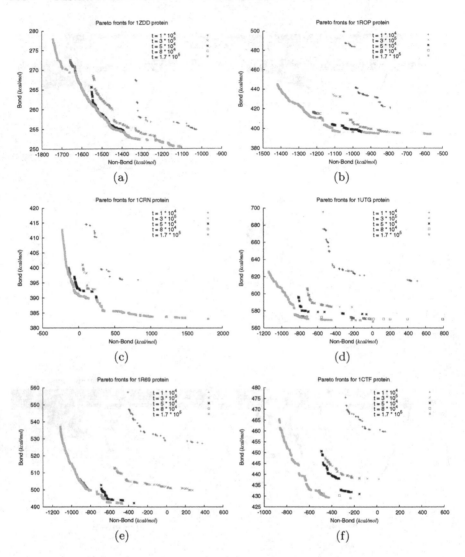

Fig. 5. Observed Pareto fronts at different iterations for (a) 1ZDD, (b) 1ROP, (c) 1CRN, (d) 1UTG, (e) 1R69, and (f) 1CTF

reflects experimentally the theory that the non-bond energy term, which includes van der Waals and electrostatic atom energies, is one of the driving forces during the protein folding.

The inset plot in each figure shows the total energy value and the $RMSD$ value for each conformation that has been sampled by the algorithm. This type of plot illustrates the level of correlation between the energy function and the $RMSD$ values and is an index of the accuracy of the energy function for the specific protein, suggesting that minimizing the energy by varying the conformation will tend to drive the conformation towards the native structure. The proteins 1ZDD, 1ROP, 1CRN and 1R69 show a higher correlation with respect to 1UTG and 1CTF. In particular, 1UTG has the lowest level of correlation, meaning that CHARMM is not a good energy function for exploring its conformational space. On the other hand, even if the 1CTF protein shows an acceptable level of correlation, the complexity of the structure, in terms of secondary structural elements and native 3D conformation, strongly reduces the performance of the algorithm. Moreover, by inspecting the plots, it is clear how the algorithm is able to produce a high sampling of the search space. Generally, more than $1,000$ conformations lie in a range of 1 Å.

Figure 5 presents snapshots of the dynamics of observed Pareto fronts at different iteration times. As a first observation we note that the size of solutions in the front increases as the search process proceeds. Almost all the proteins reach (or get very close to) the maximum allowed size of $1,000$ solutions used in all the experiments (see Figure 8). This behaviour satisfies one of the requirements of good convergence in multiobjective optimization, where a wide range of values for the objective should be produced. Moreover, as we will see later, the detailed dynamics of the archive size will give more additional information on the level of complexity of the protein. The number of knees and the discontinuous regions are other measures of the level of complexity of the protein. Simple proteins, with few secondary structure elements and simple native conformation, tend to present a continuous front with few knees (e.g., 1ZDD and 2MLT, discussed in the next section).

It is also possible to select solutions from the observed Pareto fronts in figure 5 and reconstruct the folding pathway generated by the algorithm for each protein. Figure 6 shows the reconstruction for the 1CRN protein. The conformation at time 0 is the first randomly generated structure; then each of the other conformations is representative of the observed Pareto front at time step t (iteration).

Another measure of complexity is given by the analysis of the dynamics of the $RMSD$ values in the archive (see figure 7). Simple proteins (e.g., 1ZDD and 1ROP) show quick convergence towards native-like structures, while complex proteins (e.g., 1UTG and 1CTF) demonstrate lower convergence rates and greater fluctuations in the value of the best $RMSD$ in the archive. The distance between the two curves is an indicator of the quality of the final archive. A smaller distance indicates a larger size of the ensemble of native-like structures generated.

Finally, the analysis of the archive size at each iteration provides an experimental proof for the previous comments. Figure 8 shows the dynamics of the archive size at each iteration for each protein. As observed before for all the proteins except 1CTF, the algorithm is able to fill up the archive; but, again, simple protein sequences show a faster growth rate while complex sequences need more time before a good conformation is generated. For example, 1CTF protein shows less than 100 conformations in the archive for the first $10,000$ iterations; moreover, 1CTF is the only protein for which the final size of the archive does not reach the allowed maximum. The

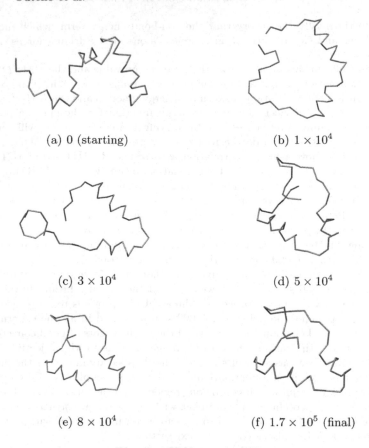

(a) 0 (starting)   (b) 1×10^4

(c) 3×10^4 (d) 5×10^4

(e) 8×10^4 (f) 1.7×10^5 (final)

Fig. 6. 1CRN protein folding

strong oscillations in the curves reflect the update scheme of PAES and I-PAES: if a new nondominated solution is generated it will be added to the archive, and all the solutions which are dominated by the new one are removed.

5.2 Second Set of Results and Discussions

The second set of experiments presents further results using additional protein sequences in the literature. The protein set is composed of seven proteins extracted from [15, 21, 9]. The PDB IDs are 2MLT, 1VII, 1G2H, 1EOM, 2GP8, 1BW6, 1ED0 and 1NKL.

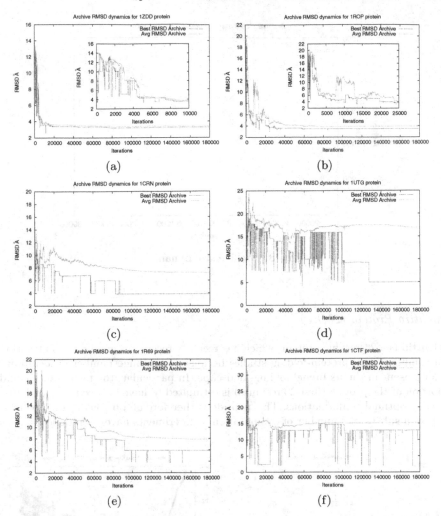

Fig. 7. Archive $RMSD$ dynamics with respect to the crystal structure for 1ZDD (a), 1ROP (b), 1CRN (c), 1UTG (d), 1R69 (e) and 1CTF (f)

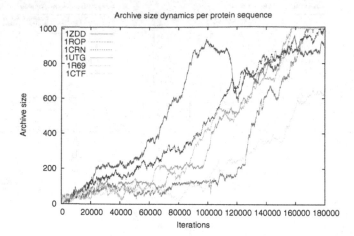

Fig. 8. Archive size dynamics

Melittin Protein (2MLT)

Melittin is a 26-residue proteins which has recently received a good deal of attention in computational protein-folding studies because of the huge number of local minima present in the its funnel folding-landscape. In particular, the membrane-bound portion of the protein (first 20 residues) is estimated to have between 10^{34} and 10^{54} locally optimal conformations. This peptide is therefore an obvious test subject for which a substantial number of computational experiments have been done [15].

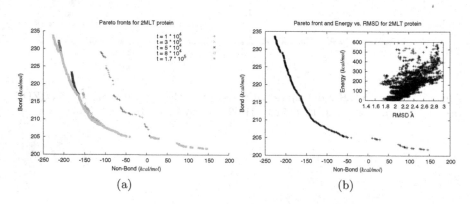

Fig. 9. 2MLT: (a) observed Pareto fronts at different iterations; (b) final best Pareto front with the characteristic energy function vs. $RMSD$ values plot (inset plot)

As reported in Table 4, the best conformation obtained with I-PAES has $DME = 0.77$Å and $RMSD = 1.92$Å (see figure 10). Inspecting the conformations

Fig. 10. Native (left plot) and predicted (right plot) for 2MLT protein ($DME = 0.77$Å, $RMSD = 1.92$Å)

in the final archive we found that they all present good characteristics in terms of both metrics and energy. This is also evident from the results reported in Table 4, where the $RMSD$ and DME values for the conformation with best energy are still competitive. This is an index of a good accuracy level of CHARMM energy function for this peptide. Good performance of I-PAES on the 2MLT protein is also evident from the the plots in Figure 9, where the Pareto front dynamics and the correlation between the energy function and the $RMSD$ values are shown. As in the work of Floudas [15], we have also studied only the membrane-bound portion of the proteins (20 residues), and, in this case, the best predicted structure matches the crystal structure with $DME = 0.66$Å and $RMSD = 1.36$Å.

Second Protein Set

Finally, we have tested the multiobjective approach with additional protein sequences which have been recently studied using constraint logic programming [9] and Monte Carlo methods [21]. Table 4 reports the best results obtained with the I-PAES algorithm in terms of DME, $RMSD$ and the energy value for each protein. The results continue to show the low correlation between the energy function CHARMM and the metrics: the values of the metrics for the conformation with lowest energy are always the worst with respect to the best conformation. Moreover, we would like to highlight the fact that the protein conformation that has the minimum energy in the knees is often better than the one obtained from the whole Pareto front generated. Thus, as mentioned above, the energy landscape produced by the CHARMM energy function does not seem to fit the real landscape well. Despite this, the proposed multiobjective approach is able to generate good ensembles of native-like conformations. Finally, the fact that solutions in the Pareto front that are not minimum energy solutions are better (in terms of $RMSD$ and DME) clearly justifies a multiobjective approach, rather than one which focuses on a single objective (i.e., energy).

6 Conclusion

As reported by Plotkin and Onuchic in 2002 [24], "the folded state is a small ensemble of conformational structures compared to the conformational entropy present in the unfolded ensemble". This claim inspired our research goal of finding a set of

Table 4. Simulation results using I-PAES. For each protein we report the Protein Data Bank (PDB) identifier, the length (number of residues), and the approximate class (num. of α-helix and/or β-sheet). The other columns list, respectively, the best $RMSD$, DME, and energy results obtained for each protein on ten independent runs. The last column shows the values of DME and $RMSD$ for the conformation with minimum energy (sixth column). The DME and $RMSD$ values are measured on C_α atoms from the native structure. Energy values are calculated using the ANALYZE routine from TINKER

Protein	N_{res}	class α, β	Best DME_{min} Å	Best $RMSD_{min}$ Å	Best energy $kcal/mol$	$(DME, RMSD)$ (Å,Å)
2MLT	26	3α	0.77	1.92	-18.25	$(0.88, 2.01)$
1VII	36	3α	4.71	6.39	-407.25	$(4.81, 8.25)$
1G2H	36	4α	7.61	8.51	-193.77	$(8.78, 8.68)$
1E0M	37	3β	5.46	7.27	-108.78	$(5.52, 7.39)$
2GP8	40	2α	5.27	6.84	-705.25	$(5.45, 7.26)$
1BW6	43	5α	3.88	6.98	-1403.17	$(5.38, 7.36)$
1ED0	46	2α	7.24	8.79	112.49	$(10.38, 11.22)$
1NKL	70	6α	9.92	9.87	-1064.14	$(10.92, 12.21)$

equivalent three-dimensional conformations inside the folded state. To reach this goal we adopted a modified version of PAES multiobjective evolutionary algorithm in order to obtain a set of nondominated compact solutions close to the folded state. We considered the bond and non-bond interactions as main forces to direct the folding towards the native state. Our model is based on the fact that local interaction (bond energy) and nonlocal interaction (non-bond energy) between atoms are in conflict.

Namely, it is clear that although it is possible to make movements that are able to locally decrease the bond energy of the protein conformation, this is not possible globally. Moreover, the electrostatic interactions or the short distance van der Waals interaction of atoms nearby in space but far in the protein primary sequence could be penalized even though an improvement is obtained in bond, angles or torsion energies. If one interaction decreases, the other always increases. At the same time, there are compensatory effects that will bring about minimization of the total energy. We have provided experimental evidence for the conflict between the two types of interactions, which in turn allows for casting the PSP problem in the form of a multiobjective optimization problem.

The analysis of the I-PAES final archives points to the low-accuracy problem inherent in the CHARMM energy function (we would not be surprised if a similar behaviour is observed with other available energy function models). Specifically, low energy usually does not reflect good (native-like) structure, especially for long/complex proteins. A trade-off is required between these objectives in order to find good conformations. The multiobjective approach has shown to be useful in order to overcome this problem.

We proposed a modified version of the algorithm PAES, I-PAES, that uses immune-inspired principles (clonal expansion and hypermutation operators) as a new search method for PSP. Computational studies for large peptides and proteins, employing I-PAES, demonstrate the effectiveness of the evolutionary multiobjective approach, and the results are comparable in terms of $RMSD$ and DME to other approaches in the literature. The I-PAES algorithm is highly efficient, though the scheme lacks a theoretical guarantees of convergence.

The proposed method is the starting point for many other possible future lines of investigation. A straightforward approach consists of increasing the number of objectives in order to include additional driving forces relevant to protein folding, for example, the hydrophobic fitness score proposed in [14]. In theory, while any number of objectives can be used at the same time, there arises the problem of correctly handling objectives in different units. It would be interesting to scale up this approach to very long protein sequences (1,000+ residues) by, for example, cutting the protein into smaller subsequences and applying the multiobjective approach to each of them. The predicted structure of the full protein will be then assembled using the prediction for each of the subsequences.

References

[1] C. Anfinsen. Principles that govern the folding of protein chains. *Science*, 181: 223–230, 1973.

[2] C. Anfinsen, E. Haber, M. Sela, and J. F. H. White. The kinetics of formation of native ribonuclease during oxidation of the reduced polypeptide chain. *Proc. Natl. Acad. Sci. USA*, 47:1309–1314, 1961.

[3] J. U. Bowie and D. Eisemberg. An evolutionary approach to folding small alpha-helical proteins that uses sequence information and an empirical guiding fitness function. *Proc. Natl Acad Sci USA*, 91:4436–4440, 1994.

[4] J. Branke, K. Deb, H. Dierolf, and M. Osswald. Finding knees in multi-objective optimization. In *PPSN*, pages 722–731, 2004.

[5] W. D. Cornell, P. Cieplak, C. I. Bayly, I. R. Gould, K. M. Merz, D. M. Ferguson, D. C. Spellmeyer, T. Fox, J. W. Caldwell, , and P. A. Kollman. A second generation force field for the simulation of proteins, nucleic acids, and organic molecules. *J. Am. Chem. Soc.*, 117(19):5179–5197, 1995.

[6] Y. Cui, R. S. Chen, and W. H. Wong. Protein folding simulation using genetic algorithm and supersecondary structure constraints. *Proteins: Structure, Function and Genetics*, 31(3):247–257, 1998.

[7] V. Cutello, G. Narzisi, and G. Nicosia. A class of pareto archived evolution strategy algorithms using immune inspired operators for ab-initio protein structure prediction. In *EvoWorkshops*, pages 54–63, 2005.

[8] V. Cutello, G. Narzisi, and G. Nicosia. A multi-objective evolutionary approach to the protein structure prediction problem. *Journal of Royal Society Interface*, 3(6):139–151, Feb. 2006.

[9] A. Dal Palu, A. Dovier, and F. Fogolari. Constraint logic programming approach to protein structure prediction. *BMC Bioinformatics*, 5(1):186, 2004. URL http://www.biomedcentral.com/1471-2105/5/186.

[10] R. O. Day, J. B. Zydallis, and G. B. Lamont. Solving the protein structure prediction problem through a multiobjective genetic algorithm. *ICNN*, 2:31–35, 2001.

[11] J. Handl and J. D. Knowles. Exploiting the trade-off - the benefits of multiple objectives in data clustering. In *EMO*, pages 547–560, 2005.

[12] U. H. Hansmann and Y. Okamoto. Numerical comparisons of three recently proposed algorithms in the protein folding problem. *J Comput Chem*, 18:920–933, 1997.

[13] J. Hermans, H. J. C. Berendsen, W. F. V. Gunsteren, and J. P. M. Postma. A consistent empirical potential for water-protein interactions. *Biopolymers*, 23 (8):1513–1518, 1984.

[14] E. Huang, S. Subbiah, and M. Levitt. Recognizing native folds by the arrangement of hydrophobic and polar residues. *J. Mol. Biol.*, 252:709–720, 1995.

[15] J. L. Klepeis, M. J. Pieja, and C. A. Floudas. Hybrid Global Optimization Algorithms for Protein Structure Prediction: Alternating Hybrids. *Biophys. J.*, 84(2):869–882, 2003.

[16] J. D. Knowles and D. W. Corne. The pareto archived evolution strategy : A new baseline algorithm for pareto multiobjective optimisation. *In Proceedings of the 1999 Congress on Evolutionary Computation (CEC'99)*, 10:98–105, 1999.

[17] J. D. Knowles and D. W. Corne. Approximating the nondominated front using the pareto archived evolution strategy. *Evolutionary Computing*, 8(2):149–172, 2000.

[18] C. Levinthal. Are there pathways to protein folding ? *J. Chem. Phys.*, 65 (44-45), 1968.

[19] F. A. Momany, R. F. McGuire, A. W. Burgess, , and H. A. Scheraga. Energy parameters in polypeptides. vii. geometric parameters, partial atomic charges, nonbonded interactions, hydrogen bond interactions, and intrinsic torsional potentials for the naturally occurring amino acids. *J. Phys. Chem.*, 79(22):2361–2381, 1975.

[20] N. F. N and J. A. D. MacKerell A. D. All-atom empirical force field for nucleic acids: I. parameter optimization based on small molecule and condensed phase macromolecular target data. *J Comput Chem*, 21:86–104, 2000.

[21] M. Nanias, M. Chinchio, J. Pillardy, D. R. Ripoll, and H. A. Scheraga. Packing helices in proteins by global optimization of a potential energy function. *PNAS*, 100(4):1706–1710, 2003.

[22] G. Nicosia. *Immune Algorithms for Optimization and Protein Structure Prediction*. PhD thesis, Department of Mathematics and Computer Science, University of Catania, 2004.

[23] J. T. Pendersen and J. Moult. Protein folding simulations with genetic algorithms and a detailed molecular description. *J Mol Biol*, 169:240–259, 1997.

[24] S. S. Plotkin and J. N. Onuchic. Understanding protein folding with energy landscape theory. *Quarterly Reviews of Biophysics*, 35(2):111–167, 2002.

[25] G. Pollastri, D. Przybylski, B. Rost, and P. Baldi. Improving the prediction of protein secondary structure in three and eight classes using recurrent neural networks and profiles. *Proteins: Structure, Function, and Genetics*, 47(2):228–235, 2002.

[26] J. R. L. Dunbrack and F. E. Cohen. Bayesian statistical analysis of protein sidechain rotamer preferences. *Protein Science*, 6:1661–1681, 1997.

[27] K. T. Simons, C. Kooperberg, E. Huang, and D. Baker. Assembly of of protein tertiary structures from fragments with similar local sequences using simulated annealing and bayesian scoring function. *J Mol Biol*, 306:1191–1199, 1997.

[28] Z. Sun, X. Rao, L. Peng, and D. Xu. Prediction of protein supersecondary structures based on the artificial neural network method. *Protein Eng.*, 10: 763–769, 1997.

Can Single-Objective Optimization Profit from Multiobjective Optimization?

Frank Neumann[1] and Ingo Wegener[2*]

[1] Algorithms and Complexity, Max-Planck-Institut für Informatik,
  Stuhlsatzenhausweg 85, 66123 Saarbrücken, Germany
[2] FB Informatik, LS 2, Univ. Dortmund, 44221 Dortmund, Germany

Summary. Many real-world problems are multiobjective optimization problems, and evolutionary algorithms are quite successful on such problems. Since the task is to compute or approximate the Pareto front, multiobjective optimization problems are considered as more difficult than single-objective problems. One should not forget that the fitness vector with respect to more than one objective contains more information that in principle can direct the search of evolutionary algorithms. Therefore, it is possible that a single-objective problem can be solved more efficiently via a generalized multiobjective model of the problem. That this is indeed the case is proved by investigating the single-source shortest paths problem and the computation of minimum spanning trees.

1 Introduction

Typical textbooks on optimization problems focus on single-objective optimization problems; see, e. g., Cormen, Leiserson, Rivest, and Stein (2001). The function f to be optimized is defined on a search space S and takes real values, i. e., $f \colon S \to \mathbb{R}$. For minimization problems on discrete search spaces S there may be many optimal search points $s \in S$ such that $f(s) \leq f(s')$ for all $s' \in S$ but only one optimal value $f_{\min} := \min\{f(s) \mid s \in S\}$. One is interested in the optimal value $f_{\min}$ and one optimal search point s.

In the case of multiobjective optimization problems the objective function f is vector-valued, i. e., $f \colon S \to \mathbb{R}^k$. Since there is no canonical complete order on $\mathbb{R}^k$, one compares the quality of search points with respect to the canonical partial order on $\mathbb{R}^k$, namely $f(s) \leq f(s')$ iff $f_i(s) \leq f_i(s')$ for all $i \in \{1, \ldots, k\}$. A Pareto-optimal search point s is a search point such that (in the case of minimization problems)

* This work was supported by the Deutsche Forschungsgemeinschaft (DFG) as part of the Collaborative Research Center "Computational Intelligence" (SFB 531) and by the German-Israeli Foundation (GIF) in the project "Robustness Aspects of Algorithms".

$f(s)$ is minimal with respect to this partial order and all $f(s'), s' \in S$. Again, there can be many Pareto-optimal search points but they do not necessarily have the same objective vector. The Pareto front consists of all objective vectors $y = (y_1, \ldots, y_k)$ such that there exists a search point s where $f(s) = y$ and $f(s') \leq f(s)$ implies $f(s') = f(s)$. The problem is to compute the Pareto front, and for each element y of the Pareto front, one search point s such that $f(s) = y$. As for any optimization problem, one may be satisfied with approximate solutions. This can be formalized as follows. For each element y of the Pareto front we have to compute a solution s such that $f(s)$ is close enough to y. Close enough is measured by an appropriate metric and an approximation parameter. In the single-objective case one switches to the approximation variant if exact optimization is too difficult. The same reason may hold in the multiobjective case. There may be another reason. The size of the Pareto front may be too large for exact optimization.

Sometimes, people try to turn multiobjective problems into single-objective ones, e. g., by optimizing a weighted sum of the objective values of the single criterion. This may be useful in some applications but, in general, we do not obtain the information contained in the Pareto front and the corresponding search points.

Multiobjective optimization has been an issue in operations research for a long time. Due to the typically high computational complexity of multiobjective problems the application of randomized search heuristics is a way to obtain satisfactory solutions. Many variants of evolutionary algorithms specialized to multiobjective optimization problems have been developed and applied; for a survey see the monographs of Deb (2001) and Coello Coello et al. (2002).

A conclusion from this discussion is that "multiobjective optimization is more (at least as) difficult than (as) single-objective optimization". This is true at least if the objective values for the different criteria are "somehow independent". Without such an assumption there is no reason to believe in the conclusion above.

Only a few publications point out that re-formulating a problem in terms of more objective functions can reduce the run time of the optimization process. Jensen (2004) successfully used additional "helper-objectives" to guide the search of evolutionary algorithms in high-dimensional spaces. A similar approach was proposed by Knowles, et al. (2001) where single-objective problems are "multiobjectivized", i.e., decomposed into multiobjective problems which are easier to solve than the original problems.

The aim of this chapter is to show by rigorous analyses that formulating some well-known single-objective optimization problems as multiobjective ones might speed up the computation process. More precisely, we discuss the following scenario. The considered problem is a single-objective problem. It is possible to add some further criteria such that the Pareto front of the newly created multiobjective optimization problem is not too large and such that the solution of the multiobjective problem includes the solution of the single-objective problem. Solving the multiobjective problem instead of the single-objective problem implies computing the Pareto front instead of a single optimal value. Each considered search point contains more information than in the single-objective case since it contains also the objective values for the additional criteria. At least in principle it is possible that this additional information improves the search behaviour of evolutionary algorithms. This would imply that for solving difficult single-objective optimization problems one should also think about the possibility of reformulating them as generalized multiobjective optimization problems.

The purpose of this chapter is to prove that the considered scenario is not a fiction. We investigate not artificial problems to support this claim but two combinatorial optimization problems contained in any textbook on algorithms, namely, the single-source shortest paths problem and the computation of minimum spanning trees. (Nobody should expect that evolutionary algorithms beat the well-known problem-specific algorithms.)

In Section 2, we consider the single-source shortest paths problem and show how a well-known evolutionary algorithm can solve this problem efficiently using a multiobjective fitness function. Section 3 is devoted to the minimum spanning tree problem. We compare a simple multiobjective EA with a single-objective one and point out the advantages of the multiobjective approach by a rigorous run time analysis as well as by experiments on randomly chosen instances.

2 The Single-Source Shortest Paths Problem

The single-source shortest paths problem (SSSP) is a fundamental combinatorial optimization problem. The usual description is the following one. The problem instance is described by a distance matrix $D = (d_{ij})_{1 \leq i,j \leq n}$ where $d_{ij} \in \mathbb{N} \cup \{\infty\}$ is the length of the direct connection from place i to place j. The problem is to compute for the source $s := n$ and for each place i a shortest path from s to i. The naive description of all shortest paths may need a storage space of $\Theta(n^2)$. Dijkstra's famous algorithm has a computation time of $\Theta(n^2)$ and computes a description of all shortest paths which needs only storage space $\Theta(n)$. For each place i the place v_i is the direct predecessor on a shortest path from s to i.

In order to consider EAs for the SSSP we use the following model of the problem. The search space consists of all $v = (v_1, \ldots, v_{n-1}) \in \{1, \ldots, n\}^{n-1}$ where $v_i \neq i$. Place v_i is considered as the direct predecessor of place i. Hence, each search point v describes a directed graph on $V = \{1, \ldots, n\}$ where $s = n$ has indegree 0 and all other nodes have indegree 1. However, there are invalid graphs which are not trees rooted at s. Figure 1 shows a valid tree and an invalid graph.

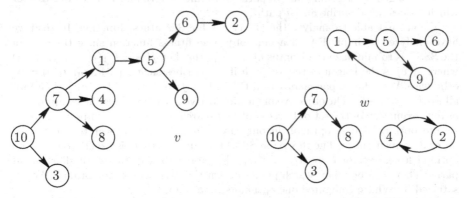

Fig. 1. Illustration of the search points $v = (7, 6, 10, 7, 1, 5, 10, 7, 5)$ leading to a tree of s-i paths and $w = (9, 4, 10, 2, 1, 5, 10, 7, 5)$ leading to an invalid graph

As mentioned before, we want to analyze an EA which is not designed for SSSP. Hence, we transform the (1+1) EA working on the boolean search space $\{0,1\}^n$ to work on the search space of all permutations on $\{1,\ldots,n\}$. The local operation of the (1+1) EA on $\{0,1\}^n$ is the flip of one bit. Moreover, the (1+1) EA does not accept worsenings. In order not to get stuck forever in a local optimum, the (1+1) EA flips each bit independently from the others with probability $1/n$. Hence, the number of local operations is asymptotically Poisson distributed with parameter $\lambda = 1$.

A local operation for SSSP is to replace the predecessor v_i of some place $i \leq n-1$ with another predecessor $v_i' \in \{1,\ldots,n\} - \{i,v_i\}$. This operation changes the considered paths for all places in the subtree of place i. The number of different local operations equals $(n-1)(n-2)$ and a flip is a randomly chosen local operation. For a mutation step, we choose S according to a Poisson distribution with parameter $\lambda = 1$ and perform sequentially $S+1$ flips.

Finally, we have to describe an appropriate fitness function f. The first idea is to define $f(v) = \infty$ for all invalid v and $f(v)$ as the sum of the lengths of the s-i paths in the tree $T(v)$ described by v. However, this leads to a difficult problem for all randomized search heuristics, at least for certain problem instances. Let $d_{i,i-1} < \infty$ and $d_{ij} = \infty$ if $j \neq i-1$. Then the search point $v^* = (2,3,\ldots,n-2,n-1,n)$ is optimal, and it is the only search point where $f(v^*) < \infty$. Hence, this optimization problem is equivalent to the well-known scenario referred to as "needle in a haystack". There is a unique global optimum, and all other search points have the same fitness. Then, nothing is better than random search, which takes exponential time with overwhelming probability.

We can hope for better results of randomized search heuristics only if the fitness function provides more information. We may restrict the possible problem instances by considering only distance matrices where $d_{ij} \in \{1,\ldots,d^*\} \cup \{\infty\}$ for some parameter d^* (possibly depending on n). If a search point v describes for j places paths of finite length, then $f(v)$ is defined as the sum of the sum of the lengths of these paths and $(n-1-j)nd^*$. Here "non-paths" and paths of infinite length contribute nd^* to the fitness, which is more than the maximal length of a path of finite length. However, we cannot distinguish between non-paths and paths of infinite length. This can be changed by assigning nd^* to paths of infinite length and n^2d^* to places i for which v does not describe an s-i path.

We are not able to analyze the (1+1) EA for this fitness function. Instead we have analyzed a simple EA on a multiobjective fitness function since the core of the SSSP is to minimize the lengths of $n-1$ paths. Let $f(v) = (f_1(v),\ldots,f_{n-1}(v))$ where $f_i(v)$ is the length of the s-i path if v describes such a path and $f_i(v) = \infty$ otherwise. We define a partial order on $\mathbb{R}^{n-1}$. It is $f(v) \leq f(v')$ iff $f_i(v) \leq f_i(v')$ for all $i \in \{1,\ldots,n-1\}$. The objective in multiobjective optimization is the computation or approximation of the set of Pareto optimal search points. A search point is called Pareto optimal if it is optimal, in our case minimal, with respect to the partial order described above. The theory on SSSP tells us that there is exactly one Pareto optimal fitness vector $l^* = (l_1^*,\ldots,l_{n-1}^*)$ describing the lengths of all shortest s-i paths. There can be many search points v such that $f(v)$ is Pareto optimal. We are satisfied if we have computed one optimal search point.

Now, we have a vector-valued fitness function and a partial order on the fitness vectors. The multiobjective (1+1) EA chooses a search point v uniformly at random.

Then it applies the mutation operator described above and accepts v' iff $f(v') \leq f(v)$.

There are SSSP instances with a unique optimal search point (this seems to be a typical case in applications). For these instances we can prove an $\Omega(n^2)$ bound on the expected optimization time of the multiobjective (1+1) EA.

Theorem 1. *The expected optimization time of the multiobjective (1+1) EA on SSSP is bounded below by $\Omega(n^2)$.*

Proof. The search space consists of $(n-1)^{(n-1)}$ search points. For instances with a unique optimal search point, the probability of choosing the optimal one in the initialization step is $(n-1)^{-(n-1)}$. Otherwise, we investigate the final step producing the optimal individual. For each non-optimal individual there is at most one operation to change it into the optimal one since the predecessor for one place has to be altered to the right one. Therefore, the probability of a success is bounded above by $\frac{1}{(n-1)(n-2)}$ and the expected waiting time is bounded below by $(n-2)(n-1) = \Omega(n^2)$. □

Theorem 2. *The expected optimization time of the multiobjective (1+1) EA on SSSP is bounded above by $O(n^3)$.*

We prove a more sophisticated bound. Let t_i be the smallest number of edges on a shortest s-i path, $m_j := \#\{i \mid t_i = j\}$, and $T = \max\{j \mid m_j > 0\}$. Then we prove that the upper bound is

$$en^2 \sum_{1 \leq j \leq T} (\ln m_j + 1).$$

This bound has its maximal value $\Theta(n^3)$ for $m_1 = \cdots = m_{n-1} = 1$. We also obtain the bound $O(n^2 T \log n)$ which is much better than $O(n^3)$ in the typical case where T is small.

Proof. The proof is based on the following simple observation. Whenever $f_i(v) = l_i^*$, we only accept search points v' where $f_i(v') = l_i^*$. Hence, we do not forget the length of shortest paths we have found (although we may switch to another shortest path). Now we assume that we have a search point v where $f_i(v) = l_i^*$ for all i where $t_i < t$. Then we wait until this property holds for all i where $t_i \leq t$. For each place i where $t_i = t$ and $f_i(v) > l_i^*$ there exists a place j such that $t_j = t-1$, j is the predecessor of i on a shortest s-i path using t edges, and $f_j(v) = l_j^*$. Then a mutation flipping only v_i into j is accepted and leads to a search point v' where $f_i(v') = l_i^*$. The probability of such a mutation equals $1/(e(n-1)(n-2))$ ($1/e$ is the probability of flipping exactly one position, $1/(n-1)$ is the probability of flipping the correct position, and $1/(n-2)$ is the probability of flipping it to the correct value). If we have r such places, the success probability is at least $r/(en^2)$ and the expected waiting time is bounded above by en^2/r. The largest value for r is m_t and we have to consider each of the values $m_t, \ldots, 1$ at most once. Hence, the total expected time of this phase is bounded above by $en^2(1 + \frac{1}{2} + \cdots + \frac{1}{m_t}) \leq en^2(\ln m_t + 1)$. Since t can take the values $1, \ldots, T$ we have proved the claimed bound. □

The upper bound of Theorem 2 holds even in the case where we allow infinite distance values. Let us consider the special case where $d_{i,i-1} = 1$ and $d_{ij} = \infty$ otherwise. This is the needle-in-a-haystack scenario for the single-objective optimization problem. Theorem 7 implies an $O(n^3)$ bound of the (1+1) EAin the multiobjective optimization problem. This bound is tight for this problem instance. As long as $v_{n-1} \neq n$ we have $f(v) = (\infty, \ldots, \infty)$. The probability of starting with $v_{n-1} = n$ equals $1/(n-1)$. In the negative case, we have to wait for a mutation where v_{n-1} is mutated into n. The probability that a local operation does this change is $1/(n-1)(n-2)$. The expected number of local changes per step equals 2. Hence, the expected time until $v_{n-1} = n$ equals $\Theta(n^2)$. Until $v_{n-1} = n$, the value of v_{n-2} does not influence the fitness vector. Therefore, we can repeat the arguments for $v_{n-2}, \ldots, v_1$ and obtain an expected optimization time of $\Theta(n^3)$.

Altogether, the multiobjective (1+1) EAon SSSP has an expected optimization time of $O(n^3)$, and $\Omega(n^2)$ if the solution is unique. For typical problem instances, the more sophisticated bound which follows from the proof of Theorem 7 is "much closer" to n^2 than to n^3. Hence, the multiobjective (1+1) EAis an efficient heuristic to solve SSSP (without beating Dijkstra's algorithm). Our results on SSSP also show that multiobjective problems should not be transformed artificially into single-objective problems.

3 Minimum Spanning Trees

3.1 Simple Evolutionary Algorithms for Multiobjective Optimization

The rigorous analysis of the expected optimization time of evolutionary algorithms is not easy. Most of such results are on simple evolutionary algorithms like the (1+1) EA(Droste, Jansen, and Wegener (2002)). This is even more true for multiobjective optimization. Therefore, we investigate and analyze the algorithm called SEMO (Simple Evolutionary Multiobjective Optimizer) due to Laumanns et al. (2002). The algorithm starts with an initial solution $s \in \{0,1\}^n$. All nondominated solutions are stored in the population P. In each step a search point from P is chosen uniformly at random and one bit is flipped to obtain a new search point s'. The new population contains for each nondominated objective vector $f(s)$, $s \in P \cup \{s'\}$, one corresponding search point, and in the case where $f(s')$ is not dominated, s' is chosen.

In applications, we need a stopping criterion. Here we are interested in the expected number of rounds until $f(P) := \{f(s)|s \in P\}$ equals the Pareto front. This is called the expected optimization time. Note that the described algorithm differs from the original version of SEMO, by replacing an individual s'' of P by s' if $f(s'') = f(s')$ holds. Applying our version of SEMO to a single-objective optimization problem, we obtain the algorithm known as RLS (randomized local search). All our results also hold for the original version of SEMO but it seems to be more typical for search heuristics to replace search points by other ones with the same quality (e.g., simulated annealing works this way). If SEMO starts with a search point s which is a local optimum, then $P = \{s\}$ forever. The use of this local mutation operator was motivated by the fact that this choice simplifies the analysis. Giel

SEMO

1. Choose an initial solution s.
2. Determine $f(s)$ and initialize $P := \{s\}$.
3. Repeat
   - choose $s \in P$ uniformly at random,
   - choose $i \in \{1, \ldots, n\}$ uniformly at random,
   - define $s' = (s'_1, \ldots, s'_n)$ by $s'_j = s_j$, if $j \neq i$, and $s'_i = 1 - s_i$,
   - determine $f(s')$,
   - let P unchanged, if there is an $s'' \in P$ such that $f(s'') \leq f(s')$ and $f(s'') \neq f(s')$
   - otherwise, exclude all s'' where $f(s') \leq f(s'')$ from P and add s' to P.

(2003) has generalized the investigations of Laumanns et al. (2002) and Zitzler et al. (2003) by considering the usual mutation operator of evolutionary algorithms.

GSEMO (Global SEMO)

GSEMO works like SEMO, but s' is defined in a different way. For each i, $s'_i = 1 - s_i$ with probability $1/n$ and $s'_i = s_i$ otherwise.

Note that GSEMO applied to single-objective optimization problems equals the well-known (1+1) EA. Hence, we compare SEMO and GSEMO with RLS and (1+1) EA.

3.2 A Two-Objective Model of the Minimum Spanning Tree Problem

An instance of the minimum spanning tree problem consists of an undirected graph $G = (V, E)$ with n vertices and m edges and a positive integer weight $w(e)$ for each edge. The problem is to find an edge set E' connecting all vertices of V with minimal total weight.

Neumann and Wegener (2004) have analyzed RLS (with 1-bit flips and 2-bit flips) and the (1+1) EA for the minimum spanning tree problem. They have used the following model of the problem. The search space is $S = \{0, 1\}^m$ and $s \in S$ describes the edge set of all edges e_i where $s_i = 1$. Raidl and Julstrom (2003) have shown that edge sets are appropriate for the minimum spanning tree problem. Neumann and Wegener (2004) have penalized edge sets which do not describe connected graphs (and in one model have additionally penalized edge sets containing cycles). They were able to prove the following results:

- The expected optimization time of RLS and the (1+1) EA is bounded by $O(m^2(\log n + \log w_{max}))$ where w_{max} is the largest weight of the considered graph.
- There are graphs with $m = \Theta(n^2)$ and $w_{max} = \Theta(n^2)$ such that the expected optimization time of RLS and the (1+1) EA equals $\Theta(m^2 \log n)$.

This is one of the first rigorous analyses of the expected optimization time of evolutionary algorithms on combinatorial optimization problems contained in textbooks. Previous results considered the computation of shortest paths (Scharnow, Tinnefeld, and Wegener (2002)) and maximum matchings (Giel and Wegener (2003)).

We discuss the reason for the expected optimization time of RLS and the (1+1) EA. If a search point describes a non-minimum spanning tree, one-bit flips are not accepted. The new search point describes either an unconnected graph or a connected graph with a larger weight. We have to wait until a mutation step includes an edge and excludes a heavier one from the newly created cycle. The expected waiting time for a specified 2-bit flip equals $\Theta(m^2)$.

As already mentioned, the considered algorithms penalize the number of connected components. This motivates the following two-objective optimization model of the minimum spanning tree problem.

- The search space S equals $\{0,1\}^m$ for graphs on m edges and the search point s describes an edge set.
- The fitness function $f : S \rightarrow \mathbb{R}^2$ is defined by $f(s) = (c(s), w(s))$ where $c(s)$ is the number of connected components of the graph described by s and $w(s)$ is the total weight of all chosen edges.
- Both objectives have to be minimized.

We discuss some simple properties of this problem.

- The parameter $c(s)$ is an integer from $\{1, \ldots, n\}$.
- The first property implies that the populations of SEMO and GSEMO contain at most n search points and the Pareto front contains exactly n elements.
- The parameter $w(s)$ is an integer.

We have to be careful when discussing this problem. There exists another type of multiobjective minimum spanning tree problem. Each edge has k different types of weights, i.e., $w(e) = (w_1(e), \ldots, w_k(e))$. Unconnected graphs are penalized, and the aim is to minimize $f(s)$ where s is not legal if s does not describe a connected graph, and $f(s)$ is the sum of all $w(e_i)$ corresponding to $s_i = 1$, otherwise. Similarly to other optimization problems this multiobjective variant of a polynomially solvable problem is NP-hard (Ehrgott (2000)). This problem has been attacked in different ways, e.g., by Hamacher and Ruhe (1994). Zhou and Gen (1999) present experimental results for evolutionary algorithms, and Neumann (2004) has analyzed which parts of the Pareto front can be obtained in expected pseudopolynomial time.

3.3 The Analysis of the Expected Optimization Time

Our results hold for SEMO as well as for GSEMO. The essential steps are 1-bit flips. In the definition of SEMO and GSEMO we have not specified how to choose the first search point. We discuss two possibilities.

- The first search point is chosen uniformly at random. This is the typical choice for evolutionary algorithms.
- The first search point is $s = 0^m$ describing the empty edge set. This is quite typical, e.g., for simulated annealing.

Our analysis is simplified by knowing that P contains 0^m. Note that $f(0^m) = (n, 0)$ belongs to the Pareto front and 0^m is the only search point s with $c(s) = n$. First, we investigate the expected time until the population contains the empty edge set.

Theorem 3. *Starting with an arbitrary search point the expected time until the population of SEMO or GSEMO contains the empty edge set is bounded above by $O(mn(\log n + \log w_{max}))$.*

Proof. One might expect that we have to wait only until all edges of the initial search point s have been excluded. This is not true. In this way, it is possible that we accept the inclusion of edges since this decreases the number of connected components (although it increases the total weight). Later, we may exclude edges of the new search point s' without increasing the number of connected components. It is possible to construct a search point s'' which dominates s. Then s is eliminated and all search points in the population (perhaps only one) have more edges than s.

Hence, the situation is more complicated. Instead of the minimal number of edges of all search points in P we analyze the minimal weight of all search points in P. The search point s^* with minimal weight has the largest number of connected components (otherwise, the search point s^{**} with $c(s^{**}) > c(s^*)$ is dominated by s^* and will be excluded from P). We analyze $w(s^*)$. We have reached the aim of our investigations if $w(s^*) = 0$, since this implies $s^* = 0^m$. After initialization, $w(s^*) \leq W := w_1 + \cdots + w_m \leq m \cdot w_{max}$.

We only investigate steps where s^* is chosen for mutation. The probability of such a step is always at least $1/n$, since $|P| \leq n$. Hence, the expected time is larger only by a factor of at most n than the expected number of steps where s^* is chosen.

By renumbering, we may assume that s^* has chosen the first k edges. We investigate only steps flipping exactly one bit. This has probability 1 for SEMO and probability at least e^{-1} for GSEMO, where $e = 2.71\ldots$. These steps are accepted if they flip one of the first k edges. If the edge i is flipped, we obtain a search point whose weight is $w(s^*) - w_i$ and the minimal weight has been decreased by a factor of $1 - \frac{w_i}{w(s^*)}$. The average factor of the weight decrease equals

$$\frac{1}{m}\left(\sum_{1 \leq i \leq k}(1 - \frac{w_i}{w(s^*)}) + \sum_{k+1 \leq i \leq m} 1\right) \geq 1 - \frac{1}{m}$$

if the choice of a nonexistent edge is considered as a weight decrease by a factor of 1. The bound $1 - \frac{1}{m}$ does not depend on the population. After $M := \lceil (\ln 2) \cdot m \cdot (\log W + 1)\rceil$ steps choosing the current s^*, the expected weight of the new s^* is bounded above by $(1 - 1/m)^M \cdot W \leq \frac{1}{2}$. Applying Markoff's inequality, the probability that $w(s^*) \geq 1$ is bounded above by $1/2$. Hence, $w(s^*) < 1$ holds with probability at least $1/2$. Since weights are integers, $w(s^*) < 1$ implies $w(s^*) = 0$. The expected number of phases of length M until $w(s^*) = 0$ is at most 2. Hence, altogether the expected waiting time for $s^* = 0^m$ is bounded above by $2 \cdot n \cdot M = O(mn(\log n + \log w_{max}))$ for SEMO. The corresponding value for GSEMO is larger at most by a factor of 3. □

One may expect that this upper bound is an overestimate for many graphs and starting points.

Theorem 4. *Starting with a population containing the empty edge set, the expected optimization time of SEMO or GSEMO is bounded by $O(mn^2)$.*

Proof. As long as the algorithm has not reached its goal, we consider the smallest i such that the population contains, for each j, $i \leq j \leq n$, a Pareto-optimal search point s_j with $f(s_j) = (j, \cdot)$. This implies that the graph described by s_j consists of j connected components and has the minimal possible weight among all possible search points describing graphs with j connected components. After initialization, the population includes 0^m which has the smallest weight among all search points representing graphs with n connected components. Hence, i is well defined. The search point s_j is only excluded from the population if a search point s'_j with $f(s'_j) = f(s_j)$ is included in the population. Hence, the crucial parameter i can only decrease and the search is successful if $i = 1$.

Finally, we investigate the probability of decreasing i. It is well known that a solution with $i - 1$ components and minimal weight can be constructed from a solution with i components and minimal weight by introducing a lightest edge that does not create a cycle. Therefore, it is sufficient to choose s_i for mutation (probability at least $1/n$) and to flip exactly one bit concerning a lightest edge connecting two components in the graph described by s_i (probability at least $1/m$ for SEMO and at least $1/(em)$ for GSEMO). Hence, the expected waiting time to decrease the parameter i is bounded above by $O(nm)$. After at most $n - 1$ of such events the search is successful. □

Corollary 1. *If the weights are bounded above by 2^n, SEMO and GSEMO find the Pareto front in the two-objective variant of the minimum spanning tree problem in an expected number of $O(mn^2)$ rounds independently of the choice of the first search point.*

For dense graphs, this bound beats the bound $O(m^2 \cdot \log n)$ for the application of RLS and the (1+1) EA to the single-objective variant of the minimum spanning tree problem.

3.4 Discussion

The most interesting case is the one of polynomially bounded weights. The expected optimization time of the (1+1) EA is then $O(m^2 \log n)$ (see Section 3.2), and this bound is best possible as shown by the analysis of special input graphs G_n. The reason for this bound is the following. If the considered search point is a spanning tree, we obtain a better spanning tree by eliminating i edges of the tree and inserting i edges producing a tree of smaller weight. For G_n, steps for $i = 1$ are much more likely than steps for larger i. The expected waiting time for a special 2-bit flip is $O(m^2)$. It is likely to need $\Theta(n)$ of such 2-bit flips. One might think that the expected optimization time is $\Theta(nm^2)$. However, this is not the case since in the beginning there are several good 2-bit flips.

In Section 2, we have obtained an upper bound of $O(mn^2)$ on the expected optimization time of both SEMO variants. This is only an upper bound, and it would be better to have an upper bound which is asymptotically best possible. It is an open question whether there exist graphs G_n with $m = \Theta(n^2)$ where SEMO has an expected optimization time of $\Theta(n^4)$. The best lower bound we have obtained is $\Omega(n^3 \log n)$.

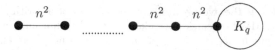

Fig. 2. Example graph with a path containing p edges and a complete graph on q vertices

In the following, we present the example graphs leading to this expected optimization time and discuss the main arguments of the run time analysis. The full proof is much longer than the proof of the upper bounds in Section 2. Since the lower bound does not match the general upper bound, we do not present the complete proof.

The idea of the example is to connect a path with p expensive edges of weight n^2 with a complete graph on $q = n - p$ vertices with cheap edges of weight 1. Figure 1 shows such graphs. Here we focus on the case $q = n/2$ and $p = n/2$. The complete graph ensures a number of $\Theta(n^2)$ edges. The probability of flipping k path edges (or shortly p-edges) is bounded above by $\binom{n/2}{k}(1/m)^k \leq (n/m)^k = O(n^{-k})$.

We call a graph a c-graph if it consists of a connected component on the complete subgraph and $n/2 + 1 - c$ p-edges. Note that c-graphs which do not contain a cycle are Pareto optimal. There are $c - 1$ different 1-bit flips to produce $(c - 1)$-graphs.

After random initialization, it is very likely we obtain a c-graph where $c \geq n/6$. As seen above, an offspring of a c-graph can be a c'-graph, but it is very likely that $c - c' = O(1)$. Hence, at the first point in time where the population contains a c''-graph where $c'' \leq n/10$ it is very likely that the population size is $\Theta(n)$ (note that c-graphs can only be replaced by other c-graphs). Afterwards, we have to produce a 1-graph. Let $c(P)$ be the smallest c such that the population P contains a c-graph. We can hope to reduce $c(P)$ only if we select for mutation a c'-graph where c' is close to $c(P)$. Hence, the waiting time for a good selection is $\Theta(n)$. Then there are at most $c(P) - 1$ good 1-bit flips with an expected waiting time of $\Omega(m/c(P))$. This motivates the $\Omega(mn \log n) = \Omega(n^3 \log n)$ bound. The complete proof is more involved since it may happen that the population contains a graph with c components which is not a c-graph.

Summarizing, SEMO has the advantage of working with 1-bit flips while the (1+1) EA is mainly based on 2-bit flips. However, SEMO has to cope with an increase in the population size leading to a waiting time until an appropriate search point is chosen for mutation.

3.5 Experimental Results

The theoretical results are asymptotic ones. They reveal differences for worst-case instances and large m. We add experimental results that show what happens for typical instances and reasonable m. In order to compare randomized algorithms on perhaps randomly chosen instances, one may compare the average run times, but these values can be highly influenced by outliers. We have no hypothesis about the probability distribution describing the random run time for constant input length. Hence, only parameter-free statistical tests can be applied. We apply the Mann-Whitney test (MWT) that ranks all observed run times. Small ranks correspond to small run times. If the average rank of the results of algorithm A_1 are smaller than those of A_2, MWT decides how likely it can be that such a difference or a larger one can occur under the assumption that A_1 is not more efficient than A_2. If the corresponding p-value is at most 0.05, we call the result significant; if it is at most 0.01, we call it very significant; and if it is at most 0.001, we call it highly significant. The statistical evaluation has been performed with the software SPSS (Version 11.5; see www.spss.com). The tables contain the considered class of graphs, the average rank (AR) of different algorithms and the p-value for the hypothesis that the algorithm with the smaller AR-value is likely to be faster.

The experiments consider the following graph classes.

- $uniform_n$: these are complete graphs with $m = \binom{n}{2}$ edges, and the weights are chosen independently and uniformly at random from $\{1, \ldots, n\}$.
- $uniformbd_n$: each possible edge is chosen with probability $3/n$, leading to a small average degree of 3; unconnected graphs are rejected and the construction is repeated; the weights of existing edges are chosen as for $uniform_n$.
- $plane_n$: the n vertices are placed randomly on the points of the two-dimensional grid $\{1, \ldots, n\} \times \{1, \ldots, n\}$; the weight of an edge is the rounded Euclidean distance between the vertices.
- $planebd_n$: the n vertices are placed as for $plane_n$, but each edge is considered only with probability $3/n$ as for $uniformbd_n$.

These graph classes reflect different choices of weights (one non-metric and one metric) and the possibility of dense and sparse graphs. Our algorithms are RLS, (1+1) EA, SEMO, and GSEMO. The index z denotes the case where the initial search point is the empty edge set (or all-zero string). Without an index the initial search point is chosen uniformly at random. The run times of RLS and the (1+1) EA denote the number of fitness evaluations until a minimum spanning tree is constructed. The run times of SEMO and GSEMO denote the number of rounds until, in one experiment, P contains a minimum spanning tree or until $f(P)$ equals the Pareto front. In each experiment the compared algorithms are considered for 100 runs leading, to an average rank of 100.5.

We have analyzed the influence of the initial search point. First, we have considered the time until the Pareto front is computed. The results are shown in Table 1.

Result 1 *In 23 out of 24 experiments the variant starting with the empty edge set has the smaller AR-value. Only eight results are significant, among them five very significant and two of these highly significant.*

If we are only interested in the computation of a minimum spanning tree, one may expect that one sometimes computes a minimum spanning tree without computing the empty edge set. Indeed, the influence of the choice of the initial search point gets smaller. For the classes $uniform_n$, $n = 4i$ and $3 \leq i \leq 11$, there is no real difference between $SEMO_z$ and SEMO, while the AR values of GSEMO are in eight of the nine experiments smaller than for $GSEMO_z$. For the classes $plane_n$, $n = 4i$ and $3 \leq i \leq 11$, $SEMO_z$ beats SEMO (seven cases) and $GSEMO_z$ beats GSEMO (seven cases). We do not show the results in detail since they are not significant

Table 1. Comparison of SEMO and GSEMO with different initial solutions

Class	AR $SEMO_z$	AR SEMO	p-value	AR $GSEMO_z$	AR GSEMO	p-value
$uniform_{12}$	92.76	108.25	0.058	89.35	111.66	**0.006**
$uniform_{16}$	83.51	117.49	**< 0.001**	91.28	109.72	**0.024**
$uniform_{20}$	99.12	101.89	0.735	94.21	106.80	0.124
$uniform_{24}$	98.01	102.99	0.543	93.65	107.35	0.094
$uniform_{28}$	94.62	106.38	0.151	94.48	106.52	0.141
$uniform_{32}$	91.24	109.76	**0.024**	96.76	104.24	0.361
$plane_{12}$	81.61	119.39	**< 0.001**	88.14	112.86	**0.003**
$plane_{16}$	94.51	106.49	0.143	89.38	111.63	**0.007**
$plane_{20}$	97.17	103.83	0.416	95.15	105.85	0.191
$plane_{24}$	93.33	107.67	0.080	103.11	97.89	0.524
$plane_{28}$	90.58	110.43	**0.015**	93.09	107.91	0.070
$plane_{32}$	94.55	106.45	0.146	97.44	103.56	0.455

(with the exception of three out of 36 cases). The remaining experiments consider the more general case of an initial search point chosen uniformly at random.

We have not considered the worst-case instances for RLS and the (1+1) EA presented by Neumann and Wegener (2004). This would be unfair for these algorithms. Nevertheless, the experiments of Briest et al. (2004) have indicated that, for n and m of reasonable size, dense random graphs are even harder than the asymptotic worst-case examples. This leads to the conjecture that SEMO beats RLS and GSEMO beats its counterpart (1+1) EA. Here, the run time measures the rounds until a minimum spanning tree is constructed. Table 2 proves that our conjecture holds for the considered cases. Note that the average rank of 100 runs of one algorithm is at least 50.5. In several experiments the AR value of SEMO or GSEMO comes close to this value. For $n \geq 20$, all values are at most 51.6 and for small values of n the AR values are smaller than 60.

Result 2 *It is highly significant for all considered graph classes and graph sizes that SEMO outperforms RLS and GSEMO outperforms the (1+1) EA.*

The theoretical analysis of the algorithms gives values of $O(m^2 \log n)$ for RLS and the (1+1) EA, and $O(mn^2)$ for SEMO and GSEMO (if the weights are reasonably bounded). For complete graphs, $m = \Theta(n^2)$, and we get values $n^4 \log n$ vs. n^4. For sparse graphs, $m = \Theta(n)$, and we get values $n^2 \log n$ vs. n^3. Although these are only upper bounds, one may expect different results for the sparse graphs from uniformbd$_n$ and planebd$_n$. Table 3 shows that this is indeed the case.

Result 3 *It is highly significant for* uniformbd$_n$ *with* $n \geq 24$ *and for* planebd$_n$ *with* $n \geq 16$ *(for the considered values of n), that RLS outperforms SEMO. Similar*

Table 2. Comparison on complete uniform and complete geometric instances

Class	AR RLS	AR SEMO	p-value	AR (1+1) EA	AR GSEMO	p-value
$uniform_{12}$	146.36	54.64	< 0.001	147.79	53.32	< 0.001
$uniform_{16}$	148.45	52.55	< 0.001	149.28	51.72	< 0.001
$uniform_{20}$	149.74	51.26	< 0.001	149.40	51.60	< 0.001
$uniform_{24}$	150.00	51.00	< 0.001	150.29	50.71	< 0.001
$uniform_{28}$	150.40	50.60	< 0.001	150.23	50.77	< 0.001
$uniform_{32}$	150.50	50.50	< 0.001	150.50	50.50	< 0.001
$plane_{12}$	141.43	59.58	< 0.001	145.04	55.96	< 0.001
$plane_{16}$	144.25	56.75	< 0.001	148.28	52.72	< 0.001
$plane_{20}$	149.47	51.53	< 0.001	149.54	51.46	< 0.001
$plane_{24}$	149.95	51.05	< 0.001	149.89	51.11	< 0.001
$plane_{28}$	150.40	50.60	< 0.001	150.36	50.64	< 0.001
$plane_{32}$	150.34	50.66	< 0.001	150.28	50.72	< 0.001

results hold for the (1+1) EA and GSEMO, but they are highly significant only for large values of n, namely $n \geq 32$ for both graph classes.

Note that the last group of experiments considers values of n up to 100.

Conclusions

It has been investigated whether the multiobjective variant of a single-objective optimization problem can lead to more efficient optimization processes. We have shown that this is indeed the case for two of the best known combinatorial optimization problems. In the case of the single-source shortest paths problem, we have pointed out that the run time of the optimization process can be reduced drastically by using a multiobjective fitness function. For the minimum spanning tree problem the multiobjective approach is superior on randomly chosen dense graphs. For sparsely connected graphs it is better to use the single-objective variant of the problem. The results are obtained by a rigorous asymptotic analysis of the expected optimization time and by experiments on graphs of reasonable size.

Acknowledgement

The authors thank Dirk Sudholt, who performed the statistical tests with the SPSS software.

Table 3. Comparison on instances with bounded average degree

Class	AR RLS	AR SEMO	p-value	AR (1+1) EA	AR GSEMO	p-value
$uniformbd_{12}$	91.91	109.09	**0.036**	101.44	99.57	0.819
$uniformbd_{16}$	90.62	110.39	**0.016**	103.54	97.46	0.458
$uniformbd_{20}$	89.79	111.22	**0.009**	98.98	102.02	0.710
$uniformbd_{24}$	73.19	127.82	< 0.001	91.53	109.47	**0.028**
$uniformbd_{28}$	78.01	122.99	< 0.001	93.03	107.98	0.068
$uniformbd_{32}$	77.92	123.08	< 0.001	80.85	120.15	**< 0.001**
$uniformbd_{40}$	73.02	127.98	< 0.001	84.37	116.63	**< 0.001**
$uniformbd_{60}$	65.40	135.60	< 0.001	71.22	129.78	**< 0.001**
$uniformbd_{80}$	56.70	144.30	< 0.001	58.72	142.28	**< 0.001**
$uniformbd_{100}$	54.99	146.01	< 0.001	58.47	142.53	**< 0.001**
$planebd_{12}$	97.56	103.45	0.472	105.24	95.77	0.247
$planebd_{16}$	81.88	119.13	< 0.001	96.79	104.22	0.364
$planebd_{20}$	81.06	119.95	< 0.001	101.70	99.30	0.769
$planebd_{24}$	84.45	116.55	< 0.001	86.52	114.48	**0.001**
$planebd_{28}$	81.94	119.06	< 0.001	88.45	112.55	**0.003**
$planebd_{32}$	71.53	129.47	< 0.001	80.86	120.14	**< 0.001**
$planebd_{40}$	67.18	133.82	< 0.001	74.57	126.44	**< 0.001**
$planebd_{60}$	56.59	144.41	< 0.001	60.69	140.31	**< 0.001**
$planebd_{80}$	52.98	148.02	< 0.001	59.60	141.40	**< 0.001**
$planebd_{100}$	52.21	148.79	< 0.001	52.30	148.70	**< 0.001**

References

[1] Briest, P., Brockhoff, D., Degener, B., Englert, M., Gunia, C., Heering, O., Jansen, T., Leifhelm, M., Plociennik, K., Röglin, H., Schweer, A., Sudholt, D., Tannenbaum, S., and Wegener, I. (2004). Experimental supplements to the theoretical analysis of EAs on problems from combinatorial optimization. In Proc. of the 8th Int. Conf. on Parallel Problem Solving from Nature (PPSN VIII). LNCS 3242, 21–30.

[2] Coello Coello, C. A., Van Veldhuizen, D. A., and Lamont, G. B. (2002). *Evolutionary Algorithms for Solving Multi-Objective Problems*. Kluwer Academic Publishers, New York.

[3] Cormen, T., Leiserson, C., Rivest, R., and Stein, C. (2001). *Introduction to Algorithms*. 2nd Edition, McGraw Hill, New York.

[4] Deb, K. (2001). *Multi-Objective Optimization Using Evolutionary Algorithms*. Wiley, Chichester.

[5] Droste, S., Jansen, T., and Wegener, I. (2002). On the analysis of the (1+1) evolutionary algorithm. Theoretical Computer Science 276, 51–81.

[6] Ehrgott, M. (2000). Approximation algorithms for combinatorial multicriteria optimization problems. Int. Transactions in Operational Research 7, 5–31.

[7] Giel, O. (2003). Expected runtimes of a simple multi-objective evolutionary algorithm. In Proc. of the 2003 Congress of Evolutionary Computation (CEC), 1918–1925.

[8] Giel, O. and Wegener, I. (2003). Evolutionary algorithms and the maximum matching problem. In Proc. of the 20th Ann. Symp. on Theoretical Aspects of Computer Science (STACS). LNCS 2607, 415–426.

[9] Hamacher, H. W. and Ruhe, G. (1994). On spanning tree problems with multiple objectives. Annals of Operations Research 52, 209–230.

[10] Jensen, M. T. (2004). Helper-objectives: using multi-objective evolutionary algorithms for single-objective optimisation. Journal of Mathematical Modelling and Algorithms, 3(4), 323–347.

[11] Knowles, J. D., Watson, R. A., and Corne, D. W. (2001). Reducing local optima in single-objective problems by multi-objectivization. In Proc. of Evolutionary Multi-Criterion Optimization (EMO 2001), LNCS 1993, 269–283.

[12] Laumanns, M., Thiele, L., Zitzler, F., Welzl, E., and Deb, K. (2002). Running time analysis of multi-objective evolutionary algorithms on a simple discrete optimization problem. Proc. of the 7th Int. Conf. on Parallel Problems Solving from Nature (PPSN VII). LNCS 2439, 44–53.

[13] Neumann, F. (2004). Expected run times of a simple evolutionary algorithm for the multi-objective minimum spanning tree problem. In Proc. of the 8th. Int. Conf. on Parallel Problem Solving from Nature (PPSN VIII). LNCS 3242, 80–89.

[14] Neumann, F. and Wegener, I. (2004). Randomized local search, evolutionary algorithms, and the minimum spanning tree problem. In Proc. of Genetic and Evolutionary Computation Conference (GECCO 2004). LNCS 3102, 713–724.

[15] Raidl, G. R. and Julstrom, B. A. (2003). Edge sets: an effective evolutionary coding of spanning trees. IEEE Trans. on Evolutionary Computation 7, 225–239.

[16] Scharnow, J., Tinnefeld, K., and Wegener, I. (2002). Fitness landscapes based on sorting and shortest paths problems. Proc. of the 7th Conf. on Parallel Problem Solving from Nature (PPSN VII). LNCS 2439, 54–63.

[17] Zhou, G. and Gen, M. (1999). Genetic algorithm approach on multicriteria minimum spanning tree problem. European Journal of Operational Research 114, 141–152.

[18] Zitzler, E., Laumanns, M., Thiele, L., Fonseca, C. M., and Grunert da Fonseca, V. (2003). Performance assessment of multi-objective optimizers: An analysis and review. IEEE Trans. on Evolutionary Computation 7, 117–132.

Modes of Problem Solving with Multiple Objectives: Implications for Interpreting the Pareto Set and for Decision Making

Julia Handl[1] and Joshua Knowles[2]

[1] Faculty of Life Sciences, University of Manchester, UK
    j.handl@manchester.ac.uk
[2] School of Computer Science, University of Manchester, UK
    j.knowles@manchester.ac.uk

Summary. This chapter identifies five distinct modes in which multiobjective optimization is used to solve practical optimization problems. Implications for the interpretation and analysis of the resulting Pareto front, and for decision making, are discussed, and each mode is illustrated using application examples taken from recent research.

Introduction

Much research in multiobjective optimization (MOO) follows what might be called a standard *mode* of problem solving in which (i) the problem with its several objectives defined is taken as granted,[3] (ii) a Pareto optimal front (or set) approximation is generated, and (iii) decision making is carried out to select a preferred or subjectively best solution. There are variations on this in which decision making is partially carried out before optimization, or interactively during it, but these distinctions do not alter the main assumptions that the objectives are predefined at the beginning of the problem-solving process and that human (expert) decision-maker preferences are the basis for selecting the 'final' solution.

Undoubtedly, this standard mode of problem solving with MOO is highly successful and general, and will continue to be central to numerous applications (see the next section for more details). But, we also observe that this mode of problem solving is not the only one being used in practice. Rather, it is evident that multiobjective optimization is now being used in unconventional ways to produce certain effects or benefits. Indeed, in this book, other chapters describe novel uses of MOO, which do not necessarily begin with a well-defined MOO problem and end with human decision making. Perhaps the most well-known example of this is the idea that constrained optimization problems may be treated as multiobjective problems, as

---

[3] The multiobjective knapsack problem is an example of such a well-defined multiobjective problem that may be taken as a starting point.

in chapter 3, and see also [16]. But the other chapters in this book provide plenty
of evidence that this idea is finding currency more widely too.

Our thesis, then, is that MOO gets put to use in practical problem-solving applications in a variety of subtly different ways, which we call *modes*. The modes
really capture the specific reason why the problem has been formulated with multiple objectives, and what job each of the objectives is doing. In some modes, the
interpretation of what the Pareto front/set means is different from the standard and
this can be understood by appreciating the specific problem formulation. Moreover,
in some of the modes, human decision making for solution selection is not appropriate or is not the only valid approach; sometimes automatic solution selection
(automated decision making) is possible.

In this chapter, we explore this thesis with reference to a variety of concrete applications from our own research. We identify five different modes of using MOO, and
discuss in each case an application and the potential for automatic solution selection
in that mode. Our applications include clustering, unsupervised feature selection,
semi-supervised classification, instrument configuration and the travelling salesman
problem. The chapter is organized so that each of the five numbered sections corresponds to one mode of MOO, and this introduction section and a summary section
are not numbered.

1 Standard Multiobjective Optimization

The different modes of using multiobjective optimization (MOO) that we identify
in this chapter categorize what job the objectives are doing. In some uses of MOO,
as we shall see, the objectives are put to work to achieve specific effects and are
not really an intrinsic part of the basic problem formulation. But in the first mode
we identify—what we call 'standard' multiobjective optimization—the objectives do
indeed serve to define the basic problem to be solved.

Thus, in standard multiobjective optimization all objectives are distinct, measurable criteria, each expressing some aspect requiring optimization. Assuming all
important criteria have been included as objectives, one may be unsure about their
relative importance to a decision maker (and/or about how to normalize them), but
one can be certain that the 'ideal' solution will be Pareto optimal (for the reasons,
see the Introduction chapter of this book). Thus, using an approach that generates
a Pareto front (approximation), a decision maker can learn something about the
conflicts between the objectives and the space of possible solutions, and, through
careful inspection, may subsequently select a single preferred solution. As the name
suggests, this is all very standard, and the reader may refer to any textbook on multiobjective optimization or multicriterion decision making for further information on
the many existing techniques for 'solving' these problems (generating a Pareto front
approximation) and selecting a final preferred solution (e.g., see [12, 35, 44, 49]).

As an example of this commonly used mode of MOO, we consider the problem
of instrument configuration.

1.1 Application Example: Instrument Configuration

Context

The task of instrument configuration is an optimization problem encountered in a wide range of different disciplines where complex scientific apparatus is employed for measurement or analysis purposes. Certain aspects of such instruments' output may depend on a set of configuration parameters, and, with a growing number of such parameters, an enumeration and deterministic probing of all possible settings may quickly become impossible. Some typical aspects affected by parameter settings may be the instrument's accuracy, the "cost" of an experiment (e.g., in financial terms) and its throughput (i.e., the time consumed for obtaining the output). Some of these aspects may be strongly conflicting, and, consequently, certain instrument optimization problems are best modelled as multiobjective problems, and have recently been tackled using explicit multiobjective optimization approaches. A prominent example is the use of multiobjective optimization in radiation therapy treatment planning, where an intrinsic conflict exists between maximizing the tumoricidal dose (that which disrupts the tumor) and minimizing the dose to the surrounding healthy tissue, especially in the critical organs [13, 43]. More recently, multiobjective approaches have also been used for the configuration of instruments employed in biochemical research, and one such example from the second author's own research [40, 39] is discussed in the following.

Multiobjective Instrument Configuration

In [40], a series of experiments directed at improving the bandwidth and fidelity of metabolomics[4] studies is described. A combination of mass spectrometry and chromatography can be used to perform metabolic analyses, but the scientific instruments available have generally been designed for other applications; so an optimization of the instrument configuration was undertaken. About a dozen real-valued parameters comprising the configuration of a gas chromatography time-of-flight mass spectrometer (GC-TOF MS) were varied, while a fixed biological sample was assayed. Three objectives in the optimization were considered: (i) maximizing the signal-to-noise ratio in the chromatogram, that is, the output signal of the instrument; (ii) maximizing the number of 'true' peaks in the chromatogram; and (iii) minimizing the processing time — the time for the instrument to analyse one sample. In the experiments reported, a multiobjective evolutionary algorithm (MOEA), the Pareto-Envelope-based Selection Algorithm (PESA-II [7]), was used to perform the optimization and some 180 GC-TOF MS configurations were assayed in all. At the end, a particular configuration that offered the preferred compromise solution was chosen by an expert DM from the estimated Pareto front obtained (see Figure 1). This solution represented a threefold increase in the number of peaks observed (and thus a similar increase in the number of compounds detected) coupled with acceptable signal-to-noise, and a small increase in the potential throughput (i.e., a reduction in the processing time of a sample), over the hand-tuned configuration usually employed.

---

[4] Metabolomics [27] is the science of monitoring metabolic processes in biological cells or samples, and it relies on instruments capable of detecting the concentration of thousands of different chemical compounds present in a sample.

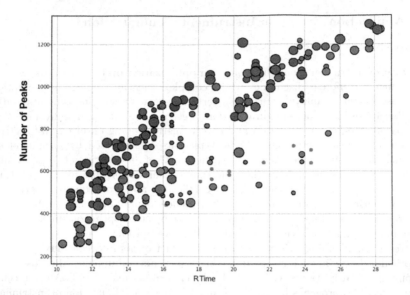

Fig. 1. Evolution of GC-TOF conditions for the optimal separation of typical serum metabolites. The diagram shows peak number and run time for each trial separation. The generation number is encoded in the size of the symbol (larger = later)

1.2 Potential for solution selection

In this mode of standard multiobjective optimization, all objectives express important criteria to be optimized. A priori, it is often not known how these different criteria should be normalized or how they will trade off one against the other. This is the reason behind generating a Pareto front and deferring the selection of a desirable solution until the end. This final selection of a solution will depend very largely on the particular preferences of a human decision maker. Consequently, in this mode, automatic selection of good solutions from the Pareto front is not likely to yield satisfying results and, indeed, most work on methods for selecting solutions has been related to *aiding* a decision maker only (see references above). Nevertheless, some recent work [8, 5] has suggested that the most promising solutions of a multiobjective optimization problem may often be located at a 'knee' in the Pareto front and that automatic identification (and even constriction of the search during the optimization process) may therefore be possible. The rationale of 'knee' points as good solutions in the Pareto front is that these solutions simultaneously score relatively well under all of the objectives considered (compared to neighbouring points in objective space).

2 Counterbalance for Bias

The second mode we identify is where MOO is used as a tool to counterbalance a measurement bias affecting an objective function. Mathematically, this setting can be described as follows, assuming just one (primary) objective to be optimized:

$$f(\mathbf{x}) = f'(\mathbf{x}) + m(g(\mathbf{x})), \tag{1}$$

where f' is an ideal (i.e., unknown), unbiased measure of the primary objective, $m(g(\mathbf{x}))$ is a bias term where m is an unknown but monotone function of a measurable function g, and f is the measurable but biased sum of the two.

Given that f' is thought to provide an objective assessment of the quality of a solution, it would be desirable to minimize $f'(\mathbf{x})$ as follows:

$$\text{minimize } f'(\mathbf{x}) = f(\mathbf{x}) - m(g(\mathbf{x})); \tag{2}$$

but since m is unknown it is not possible to formulate the problem in this way. However, the problem may, instead, be formulated as

$$\text{`minimize'} \ (f(\mathbf{x}), -(g(\mathbf{x}))), \tag{3}$$
$$\text{with } \mathbf{x} = (x_1, x_2, \ldots, x_n) \in X,$$

in terms of two measurable objectives. Hence, the framework of MOO is used as a means of introducing an additional objective, g, to counterbalance the bias of the primary objective.[5]

The set of Pareto optimal solutions will certainly contain the desired solution since each Pareto optimum is the best value of $f(x)$ given a fixed value of $g(x)$. In this scenario, selection of the best solution does not usually depend on preferences, but on the estimation of the biases. In some applications, the biases may be estimated using Monte Carlo methods, and this may help us identify the best solution in the Pareto front.

The task of unsupervised feature selection is a good example of a problem in which this mode of multiobjective optimization is useful, as will be explained in the following.

2.1 Application Example: Unsupervised Feature Selection

Context

Machine learning in high-dimensional feature spaces can be very difficult for a number of reasons: the computational cost of processing many dimensions; the presence of many redundant or noise-dominated features; and the curse of dimensionality [3], which means that the number of data may need to grow quickly as dimension increases for learning or pattern recognition to work. For these reasons, dimensionality reduction techniques are often employed, so that the dimension of the data seen by the learning algorithm is reduced.

---

[5] N.B. the equations above can be generalized to more than one primary objective where necessary.

One way of dimensionality reduction is subset selection, also referred to as feature selection [18]. Here, a lower-dimensional projection of the data is obtained by selecting a subset of the original features and discarding the remaining ones. Hence, in this approach each feature in the reduced feature space corresponds directly to a distinct feature in the original high-dimensional feature space. The main advantage of the approach lies in its ease of interpretation: the method directly returns the variables that are relevant for a given classification task.

We consider here the case of *unsupervised* feature selection, that is, dimensionality reduction in the face of unlabelled input data. This particular area of machine learning has been seriously addressed only relatively recently in the literature [9, 11, 17, 32, 37, 38, 41, 45, 47]. Performance in unsupervised classification is typically measured as the ability of a clustering to reveal 'interesting' groupings (clusters) in a given data set E. Good feature subsets can be found by an iterative search, where many candidate subsets are evaluated. In a 'wrapper' approach to evaluating the feature subsets, a clustering algorithm is applied to a candidate feature subspace, and the quality of the resulting clustering solution is evaluated using internal cluster validation techniques. Here, the term 'internal' signifies the fact that no external information, that is, information about the (true) class memberships of individual data items (if known), is used for the evaluation of individual clustering solutions. Cluster validation techniques have been designed specifically in order to allow for the selection of the best clustering solution out of a set of partitionings obtained on the same set of data but generated by different algorithms or corresponding to different numbers of clusters. Crucially, however, they are not originally aimed at accurately comparing partitionings obtained in different feature subspaces; and therein lies the problem with their use in feature subset selection.

To elaborate, internal cluster validation techniques are generally based on some form of distance computation in feature space, and this is problematic for their use in feature selection as it automatically induces a bias of these measures with respect to the dimensionality of the feature space. The existence of this bias is related to the fact that when moving to high dimensions, the histogram of distances between items in a data space changes: the mean of the histogram tends to increase and the variance of the histogram tends to decrease. In other words, the distances between all pairs of points tend to become highly similar, and (depending on the specific form of the validation technique) this causes a bias to low or high dimensions. For example, many cluster validation techniques consider ratios between intra-cluster compactness and inter-cluster separation, the values of which draw closer for high dimensions. Consequently, these validation techniques are biased towards low dimensions, and thus clustering solutions in a higher-dimensional space that are actually better than solutions in a lower-dimensional space may be overlooked.

Such biases complicate the validation of clustering results across subspaces of different dimensionalities. If the natural dimensionality bias is not accounted for, a wrapper-based feature selection method will always favour extreme feature spaces (i.e., the lowest- or highest-dimensional feature spaces available). Previously suggested remedies for this issue are the ad hoc normalization of the evaluation function and the cross-projection technique proposed by Dy and Brodley [11]. Normalization requires the selection of an appropriate scaling factor, which is usually expected to be a function of the feature cardinality d_F [11, 32]. This may reduce the bias or overcompensate for it, but will not usually remove it cleanly. The cross-projection technique proposed by Dy and Brodley [11] attempts to reduce the cardinality-specific bias by considering pairs of clustering solutions, each derived in a different feature subspace, and comparing each of them in both of these subspaces. This relation can be used for pairwise comparisons between features sets, but it is not transitive, which makes its use in global optimization techniques problematic.

Multiobjective Unsupervised Feature Selection

Recently, in [21, 32, 38], the application of Pareto multiobjective optimization algorithms has been suggested to deal with the bias by considering feature cardinality as a separate objective, as in equation 3. In [21], the multiobjective evolutionary algorithm PESA-II was used to optimize two objectives. The first of these was an internal validation technique, the Silhouette Width [42], which assesses the compactness of the clusters identified, as well as the separation between them, and which was to be maximized. Experimental analysis of the Silhouette Width revealed that it is biased towards feature subspaces of low cardinality, and the feature cardinality was therefore introduced as a second objective to be maximized. Experiments on an extensive test suite demonstrated the suitability of this approach to overcome the problem of bias in unsupervised feature selection. In particular, the method performed very well at identifying the optimal feature subspace as a part of the estimated Pareto front, and outperformed alternative methods in this respect. Also, differently from traditional greedy approaches, the algorithm was able to simultaneously uncover distinct partitionings embedded in subspaces of different cardinality.

2.2 Potential for solution selection

In this particular mode of multiobjective optimization, one of the two objectives is introduced merely as a way of counterbalancing the bias of the primary objective. As discussed above, this approach is useful only due to the difficulty of quantifying this bias exactly, which would otherwise enable one to formulate the problem as a single-objective optimization problem. It is therefore particularly material to explore for this problem whether, once a complete Pareto front has been obtained, it might then be possible to single out the best solution(s) from this set of solutions.

A possible way of doing solution selection for this mode has been suggested in [21]. The approach is based on the assumption that the same bias that affects the objective evaluation of the candidate subspaces and the corresponding partitionings is also present when evaluating completely unstructured data. Furthermore, it takes advantage of the discrete nature of the secondary objective.

The methodology proposed in [21] for solution selection works as follows. First, the feature selection algorithm is applied to the data at hand in order to obtain an estimated Pareto front, which we term the 'solution front'. Second, the minimum and maximum bounds of the original data (in each feature) are determined, and uniformly random data is generated within these bounds of the original data. This 'control data' is then subjected to the same procedure of feature selection as the original data. The resulting estimated Pareto front is referred to as the 'control front'. The solutions obtained in the solution and in the control front for a given feature cardinality are then directly comparable: we can therefore score solutions in the solution front by their distance from the corresponding solution in the control front. This score is then plotted as a function of the cardinality of the feature set and

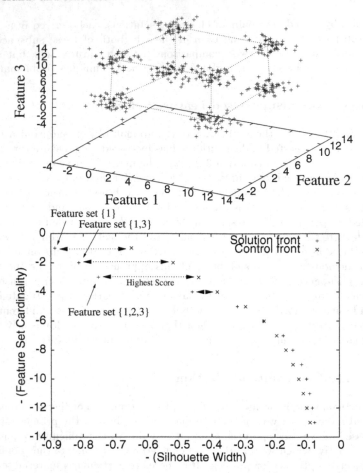

Fig. 2. (Top) Plot of Square3d, a 13-feature data set, containing eight clusters arranged in a cube pattern in the first three dimensions, and Gaussian noise in the remaining ten dimensions. (Bottom) Solution front and control front obtained on this data. The distance between the solution and the control point obtained for a given feature cardinality can serve as an indicator of quality. In our method of solution selection, the solution point with the maximum distance from its control point is selected as the best solution

the maximum value (often corresponding to a 'knee' in the solution front) is selected as the best solution. This methodology is illustrated in Figure 2. The performance of this approach was evaluated in [21] and promising results were obtained.

3 Multiple Source Integration

In the third mode, MOO is used to integrate noisy data from multiple sources. Hence, in this setting, it is used as an alternative to an a priori or an a posteriori integration technique. The problems where this approach is used are originally single-objective. However, multiple noisy views of the data need to be integrated, as their combined use may yield better results than the use of data from a single information source.

Mathematically, this setting can be described by a set of objective functions:

$$f_1(\mathbf{x}) = f_1'(\mathbf{x}) + \bar{n}_1 \tag{4}$$

$$\vdots$$

$$f_m(\mathbf{x}) = f_m'(\mathbf{x}) + \bar{n}_m,$$

where the function value of each objective function f_i is equal to the value of an ideal function f_i' with some unknown random noise $\bar{n}_i$ on it, for $i \in 1, \ldots, m$. In some cases, the f' are all identical, e.g., if the 'views' of the data arise from the same types of measurement but taken at different times. By formulating the problem as

$$\text{'minimize' } \mathbf{z} = \mathbf{f}(\mathbf{x}) = (f_1(\mathbf{x}), f_2(\mathbf{x}), \ldots, f_m(\mathbf{x})) \tag{5}$$

$$\text{with } \mathbf{x} = (x_1, x_2, \ldots, x_n) \in X,$$

and finding the Pareto optima, the impact of the noise may be reduced, if it is reasonably uncorrelated with the solution space X. The Pareto set corresponding to minimizing the f is not necessarily that which minimizes the f'; therefore it is not guaranteed that the desired solution will be among the Pareto optima.

An example of this type of problem is given in the following.

3.1 Application Example: Semi-supervised Classification

Context

In certain classification scenarios it may be desirable to combine the strengths of unsupervised and supervised classification techniques, that is, to exploit both limited prior knowledge of class labels *and* knowledge of the underlying data distribution in feature space; semi-supervised approaches aim to do this. Through the combined use of labelled and unlabelled data it becomes possible to give a degree of external guidance to the classification algorithm, while still permitting intrinsic structure in the data to be taken into account. This is considered particularly useful when dealing with data sets consisting of a large number of unlabelled data items and relatively few labelled ones (a scenario typically encountered in application domains where the categorization of individual data items is accompanied by high computational, analytical or experimental costs) and, more generally, in the case of very limited prior knowledge. For example, in cases where the classes within a particular data set are only partially known, additional ones may be identified by taking the data distribution into account. Also, due to the combination of two fundamentally different sources of information, semi-supervised approaches would be expected to be more robust than both unsupervised and supervised approaches, and may be less sensitive towards both annotation errors and the occlusion of structure in the data due to noise.

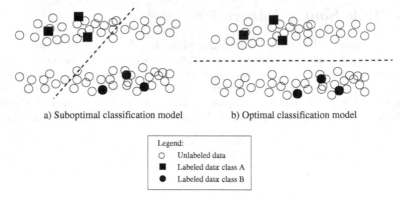

a) Suboptimal classification model   b) Optimal classification model

Legend:
○   Unlabeled data
■   Labeled data class A
●   Labeled data class B

Fig. 3. Illustration of the fundamental idea behind semi-supervised classification. The unlabelled data points can help us avoid suboptimal solutions and identify the classification model that is optimal with respect to the given data. Semi-supervision is closely related to the notion of transductive inference [6], where labelled data is used to classify a set of known but unlabelled data rather than to obtain a general model or to classify future 'unseen' data

In the case of sparse labelled 'training data', the supervised classification problem on this data will, usually, be underdetermined, and the models resulting from an entirely supervised analysis may therefore be meaningless. On the other hand, an entirely unsupervised analysis may produce a partitioning not consistent with the class labels available, and may therefore be of little significance to the user. Semi-supervised classification aims to find a solution to these classification problems that is consistent both with the data distribution (internal knowledge) and prior information about class memberships or related constraints (external knowledge), as illustrated in Figure 3.

Different approaches to the integration of these two sources of information exist, and these fundamentally differ in the underlying algorithms and their bias to one or the other of the two types of information. Many existing methods are based on classification approaches initially designed to take into account only one of the two information sources, which have then be adapted to modify candidate solutions to comply with the additional information available. Examples are methods based on established supervised classifiers, such as support vector machines; these classifiers are trained on the labelled data, but decision boundaries between classes are 'shifted' into areas of low densities, as measured across the unlabelled data [4, 31]. An alternative approach is the use of established clustering methods, such as k-means or agglomerative algorithms [2, 26]; evidently, these are principally guided by the unlabelled data, but the labelled data may be used to bias the search towards clusters consistent with the class labels (e.g., through initialization or the introduction of constraints).

Multiobjective Semi-supervised Clustering and Feature Selection

Fundamentally, semi-supervision calls for the integration of both unsupervised and supervised components into the classification process. The use of Pareto multiobjective optimization provides the means to avoid the need for hard constraints or a fixed weighting between unsupervised and supervised objectives when integrating these components—and because the Pareto approach generates a set of solutions offering different balances between optimizing the supervised and unsupervised objectives, it promises greater consistency across different data sets—and to be less affected by annotation errors (that is, faulty class labels in the supervised part of the data). However, identifying the best solution amongst the set returned may, of course, be a challenge.

In [22, 23], multiobjective approaches to the problems of semi-supervised clustering and semi-supervised feature selection have been proposed. Both methods are quite similar in essence: a multiobjective evolutionary algorithm is used to optimize candidate solutions. In the case of clustering, each candidate solution represents a partitioning of the data set. In the case of feature selection, each candidate solution represents a feature subset associated with a target number of clusters used for k-means partitioning of the data in this feature subspace. In both applications, unsupervised and supervised information are captured by individual objectives that are used to score the quality of these candidate solutions. In semi-supervised clustering, the objectives used are an internal validation technique (the Silhouette Width [42]) and an external validation technique (the Adjusted Rand Index [28]). In semi-supervised feature selection there are two unsupervised objectives, namely the Silhouette Width and subspace cardinality (in order to counterbalance the cardinality bias of the Silhouette Width, as discussed in Section 3). The Adjusted Rand Index is used as a third supervised objective.

In [22, 23], these multiobjective approaches to clustering and feature selection were compared to a range of alternative single-objective methods including 'pure' unsupervised and supervised approaches, as well as linear and nonlinear combinations of the two approaches. In both applications, a clear advantage of the multiobjective approach could be observed.

3.2 Potential for solution selection

In some applications of this type, automatic selection may be possible, namely in those types where priority information regarding the different information sources is available. For example, in the semi-supervision problem considered above, the class labels may sometimes be seen as a more reliable source of information than the unsupervised information provided by internal validation techniques (e.g., if they have been provided by human experts). Thus, those solutions with the highest score under the supervised objective may be selected from the estimated Pareto front.

In [22], this way of solution selection was investigated and it was shown that the solutions returned are different (and superior) to those identified by a purely supervised approach. This may seem surprising, but can be explained through the problem of underdetermination mentioned above: due to the sparseness of the class labels available a variety of models consistent with the class labels exist (see Figure 4). The use of semi-supervision allows one to select the one that corresponds to the highest degree of structure in the feature space, and therefore yields results superior to those obtained by a purely supervised approach.

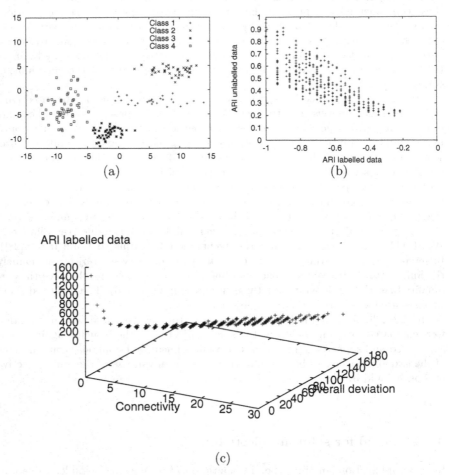

Fig. 4. (a) Simple two-dimensional data set with no conflict between external and internal information (clusters do not overlap). The classes of a small number of the data points (randomly selected) are used for a semi-supervised clustering. (b) Score under the external objective versus objective quality of the partitioning (as established by the Adjusted Rand Index between the clustering solution considered and the correct partitioning of the data). (c) Three-dimensional estimated Pareto front returned by multiobjective semi-supervised clustering

This solution selection approach will not be optimal if there are annotation errors, however. But, in that case, the shape of the estimated Pareto front can provide evidence for or against there being annotation errors, and can help us select the most satisfactory solution. If, by taking a solution which compromises slightly on the supervised objective a large improvement in the unsupervised objective occurs, this can be taken as evidence for some class annotation error(s) in the input data.

In this case, a solution nearly but not quite optimal under the supervised objective may be preferred.

Further examples of this mode of MOO will be needed to see, more generally, whether the shape of the Pareto front tends to give reliable information about good solutions, and whether or not effective solution selection procedures can usually be designed.

4 Performance Approximation by Proxies

The fourth mode of MOO comprises those applications in which the 'real', underlying objective of the problem, $f'(\mathbf{x}, \mathbf{y})$, is a function of both the solution $\mathbf{x}$ and some 'hidden' variables $\mathbf{y}$ that are not available during optimization. For example, in training a supervised classifier, $\mathbf{y}$ refers to the generalization ability of the classifier on future data (which may be estimated using a test set after the optimization, though the classifier must not be trained using these examples).

Since the function f' is not suitable for use in the optimization process (because $\mathbf{y}$ is unavailable), it needs to be replaced by 'proxy' objectives $f_i(\mathbf{x})$, which are functions of $\mathbf{x}$ only. Often, such 'proxy' objectives only capture certain aspects of a good solution, and different proxies are complementary with respect to each other. Thus, it should be expected that the desired solution(s) will score relatively highly under all of the 'proxy' objectives, and an MOO approach may therefore be suitable, although the desired solution cannot be guaranteed to be in the associated set of Pareto optima.

The difference between this context and that of standard MOO (as introduced above) may seem unclear to some readers. However, the distinction is clear: in the case of standard MOO, the objective functions have primacy, i.e. they define the Pareto set; e.g., the concept of a 'best car' does not exist *per se* but, given a search space, a set of 'best cars' is induced by the objectives chosen. In contrast, in the context of proxy objectives, it is the solution that has primacy, and the objectives are only a means of orienting the search in order to discover this solution; e.g., the real structure of a protein exists, and one may try and find it by employing a number of different energy/cost functions.

An example of this type of problem is given in the following.

4.1 Application Example: Clustering

Context

Data clustering is a problem where informal definitions are usually used to capture the essence of the problem [10, 14, 29], and is thus a problem where any single 'objective' measure of quality is only a proxy of some intangible concept of a good clustering. Arabie et al. [1] define clustering as

> "Those methods concerned in some way with the identification of homogeneous groups of objects",

while Everitt [14] documents several different definitions including this one of a cluster:

"A cluster is a set of entities that are alike, and entities from different clusters are not alike".

Because of these rather informal definitions (in the most respected texts in the field), it can be difficult even to assess clustering algorithms fairly, after the fact. But, for this reason, Milligan [36] argues that the assessment of clustering algorithms should be based on simulation and Monte Carlo studies that "allow the researcher to know the exact cluster structure underlying the data". In such a scenario "no doubt exists as to the true clustering, or the extent to which any given clustering algorithm has recovered this structure". By this means, an assessment based on external knowledge is achieved, which means that the suitability of the problem formulation (as reflected by the choice of the clustering criterion), rather than the performance of an algorithm at optimizing one particular problem formulation only, is taken into account.

Different formulations of the clustering problem vary in the optimization criterion used internally by the clustering algorithm (where obviously correct class information cannot be used). Importantly, though, most existing clustering methods attempt, explicitly or otherwise, to optimize just one such internal criterion—and it is this confinement to a particular cluster property that explains the fundamental discrepancies observable between the solutions produced by different algorithms on the same data and that will cause a clustering method to fail (as judged by means of external knowledge) in a context where the criterion employed is inappropriate.

Multiobjective Clustering

In practice, the problem of choosing an appropriate clustering objective (viz. algorithm) can be alleviated through the application and comparison of multiple clustering methods, or through the a posteriori combination of different clustering results by means of ensemble methods [46, 48]. However, a more principled approach may be the consideration of clustering as a multiobjective optimization problem, as suggested in [15]. This approach has been explored using evolutionary algorithms in our recent research [19, 20, 24] and results indicate very competitive performance against a range of algorithms, including modern ensemble approaches. Figure 5 illustrates the basic principle behind the method.

4.2 Potential for Solution Selection

Similarly to the first mode of multiobjective optimization, solution selection in the context of 'proxies' will usually require user inspection and/or prior knowledge. If the 'proxies' have been chosen such that they reflect different, complementary and conflicting properties of the solution required, 'good' solutions can probably be expected to be located towards the centre of the Pareto front (rather than at its extremes). Again, automatic methods may be employed to discover regions of high local curvature, which will indicate solutions that compare particularly favourably to all of their neighbouring solutions. For example, in the context of our work on multiobjective clustering, a fully automatic procedure was developed for identifying the best solution from the estimated Pareto front. This method has been tested extensively and selects consistently good solutions across data of different dimension, cluster number and other variables [24].

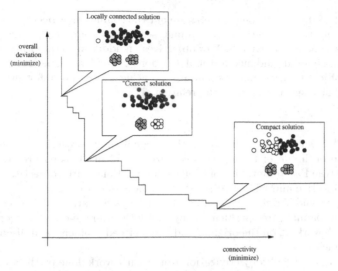

Fig. 5. The correct clustering solution often corresponds to a trade-off between two or more clustering objectives. A Pareto front (depicted as a line) and three different Pareto optimal clustering solutions are shown for a simple three-cluster data set, plotted in two-objective space. The solution at the top left is generated by an algorithm like single-link agglomerative clustering, which minimizes connectivity only. The solution at the bottom right is generated by an algorithm like k-means, which minimizes overall deviation only. (Connectivity and overall deviation are defined in [19].) The correct solution is situated somewhere between these two solutions, and so it would not usually be discovered by either method. A multiobjective approach considering the trade-offs between the two objectives should be able to access this solution much more readily. For sake of clarity, the approximation set in this example contains only solutions for $k = 3$. More generally, the number of clusters can also be kept dynamic — in this case an approximation set is obtained in which the number of clusters varies along the estimated Pareto front

5 Multiobjectivization

The fifth and final category identified is where MOO may be used solely as a way to obtain improved search 'guidance' in what is essentially a single-objective problem.[6] Assuming a single objective that is measurable, a problem may still be difficult because of its search landscape. There are at least two difficulties in search landscapes that can potentially be mitigated by 'multiobjectivization': (i) where a problem involves frustration (or epistasis), which causes excessive local optima in the search landscape; (ii) where the search landscape contains regions offering little or no objective function gradient. In the first case, decomposition of the primary objective into several different functions (each function either defined over all of the variables

---

[6] N.B., there is no reason why multiobjectivization cannot also be generalized to the case where the original problem is itself multiobjective.

or a subset of them) may help to separate out the conflicting aspects of the problem, thus reducing the number of local optima 'seen' by a search algorithm [33]. In the second case, the use of extra 'helper objectives' in addition to the primary objective may provide helpful guidance in the flat regions of the landscape [30, 33].

Multiobjectivization may potentially be achieved by any reformulation of the problem that respects the following relation [33]:

$$\forall \mathbf{x}^{opt} \in X \; \exists \mathbf{x}^* \in X, \; x^* = x^{opt}, \tag{6}$$

where $\mathbf{x}^{opt}$ is an optimal solution to the original single-objective problem and $\mathbf{x}^*$ is a Pareto optimum of the multiobjectivized problem. This ensures that at least one of the true Pareto optimal solutions will be optimal with respect to the original primary objective and will correspond to the best solution.

Neumann and Wegener use a kind of multiobjectivization in their chapter (this book) for a spanning tree problem. They derive the average-case time complexity of a simple EA working on the original and transformed problem, and discuss why the latter is much less.

An example of multiobjectivization from earlier work done by the second author (and collaborators) is given in the following.

5.1 Application Example: Travelling Salesman Problem

We are given a set $C = \{c_1, c_2, \ldots, c_N\}$ of *cities* and for each pair $\{c_i, c_j\}$ of distinct cities there is a *distance* $d(c_i, c_j)$. Our goal is to find an ordering π of the cities that minimizes the quantity

$$\sum_{i=1}^{N-1} d(c_{\pi(i)}, c_{\pi(i+1)}) + d(c_{\pi(N)}, c_{\pi(1)}). \tag{7}$$

In [34], a two-objective version of the problem was formulated as follows:

$$\text{"minimize" } \mathbf{f}(\pi, a, b) = (f_1(\pi, a, b), f_2(\pi, a, b))$$
$$\text{where } f_1(\pi, a, b) = \sum_{i=\pi^{-1}(a)}^{\pi^{-1}(b)-1} d(c_{\pi(i)}, c_{\pi(i+1)})$$
$$\text{and } f_2(\pi, a, b) = \sum_{i=\pi^{-1}(b)}^{N-1} d(c_{\pi(i)}, c_{\pi(i+1)}) + \tag{8}$$
$$\sum_{i=1}^{\pi^{-1}(a)-1} d(c_{\pi(i)}, c_{\pi(i+1)}) + d(c_{\pi(N)}, c_{\pi(1)})$$

where a and b are two cities specified a priori, and if $\pi(a) < \pi(b)$ they are swapped (see Figure 6). It is intended that a and b be chosen arbitrarily. Notice that the sum of the two objectives is the same as the quantity to be minimized in (7). This ensures that the optimum of (7) is coincident with at least one of the Pareto optima of (8), as required by our definition of multiobjectivizing above.

It was shown in [34] that this formulation was preferable to the original one when considering the performance of a simple hill climber versus a multiobjective hill climber, the Pareto archived evolution strategy (PAES), and when genetic algorithm performance was compared with that of a multiobjective evolutionary algorithm.

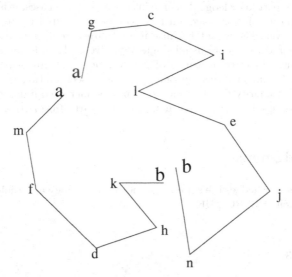

Fig. 6. A TSP tour split into two sub-tours

5.2 Potential for Solution Selection

In the above mode of multiobjective optimization, solution selection is straightforward: we are, originally, given a single-objective formulation of the optimization problem at hand, and additional objectives are introduced only as a means of facilitating the search process. It is therefore justified to assume that the solutions scoring best under the original single objective will provide us with the best solution.

Summary

Multiobjective optimization is sometimes put to use in unconventional ways, which demands further explanation and investigation. In this chapter, five distinct motivations for the use of multiobjective optimization were identified, and an application example for each one of them was provided. It is not intended that this classification is not absolute or final, and it is certainly true that some applications can be understood in terms of more than one of our identified modes of MOO in practice. Nevertheless, we hope that identifying these modes gives a clearer picture of what is happening in some of these problems and hence what the solutions in the Pareto front actually signify. We have already used the same classification to help shed light on some other applications of MOO in the field of computational biology [25] and hope it may prove useful in seeing how MOO is used in future applications as well.

An important element of the work presented in this chapter concerns solution selection: we have found that understanding the modes of MOO also helps us understand different approaches to selecting 'final' solutions. Although a decision maker's preferences or expert knowledge do come into play in many cases, it is also true that effective automatic solution selection has been demonstrated in some applications already. This is very significant because if multiobjective optimization approaches are to be compared with more usual single-objective problem formulations, one has to be able to use objective measures of performance; and this usually means that some objective means of returning one single proposed solution is required. More than that, the popularity of MOO methods may be increased if decision aid (to the point of automatic selection) were a more integral part of using these methods.

Acknowledgment

JK gratefully acknowledges the support of the Biotechnology and Biological Sciences Research Council (BBSRC), UK.

References

[1] P. Arabie, L. J. Hubert, and G. D. Soete. *Clustering and Classification*. World Scientific, New Jersey, NJ, 1996.

[2] S. Basu, A. Banerjee, and R. Mooney. Semi-supervised clustering by seeding. In *Proceedings of the 19th International Conference on Machine Learning*, pages 19–26. ACM Press, New York, NY, 2002.

[3] R. Bellman. *Adaptive Control Processes: A Guided Tour*. Princeton University Press, Princeton, NJ, 1961.

[4] A. Blum and T. Mitchell. Combining labeled and unlabeled data with co-training. In *Proceedings of the Conference on Computational Learning Theory*. ACM Press, New York, NY, 1998.

[5] J. Branke, K. Deb, H. Dierolf, and M. Osswald. Finding knees in multi-objective optimization. In *Proceedings of the Eighth International Conference on Parallel Problem Solving from Nature*, pages 722–731. Springer-Verlag, Berlin, Germany, 2004.

[6] O. Chapelle, V. Vapnik, and J. Weston. Transductive inference for estimating values of functions. In S. A. Solla, T. K. Leen, and K.-R. Müller, editors, *Neural Information Processing Systems (NIPS)*, pages 421–427. The MIT Press, 1999. ISBN 0-262-19450-3. URL http://nips.djvuzone.org/djvu/nips12/0421.djvu.

[7] D. W. Corne, N. R. Jerram, J. D. Knowles, and M. J. Oates. PESA-II: region-based selection in evolutionary multiobjective optimization. In *Proceedings of the Genetic and Evolutionary Computation Conference*, pages 283–290. ACM Press, New York, NJ, 2001.

[8] I. Das. On characterizing the 'knee' of the Pareto curve based on normal-boundary intersection. *Structural Optimization*, 18:107–115, 1999.

[9] M. Dash and H. Liu. Handling large unsupervised data via dimensionality reduction. In *Proceedings of the ACM SIGMOD Workshop on Research Numbers in Data Mining and Knowledge Discovery*, 1999. http://www.almaden.ibm.com/cs/dmkd/.

[10] R. O. Duda, P. E. Hart, and D. G. Stork. *Pattern Classification, second edition.* John Wiley and Son Ltd, London, UK, 2001.

[11] J. G. Dy and C. E. Brodley. Feature selection for unsupervised learning. *Journal of Machine Learning Research*, 5(5):845–889, 2004.

[12] M. Ehrgott. *Multicriteria Optimization.* Springer-Verlag, Berlin, Germany, 2000.

[13] M. Ehrgott and R. Johnston. Optimisation of beam directions in intensity modulated radiation therapy planning. *OR Spectrum*, 25(2):251–264, 2003.

[14] B. S. Everitt. *Cluster Analysis.* Edward Arnold, London, UK, 1993.

[15] A. Ferligoj and V. Batagelj. Direct multicriterion clustering. *Journal of Classification*, 9:43–61, 1992.

[16] R. Fletcher and S. Leyffer. Nonlinear programming without a penalty function. *Mathematical Programming*, 91(2):239–269, 2002.

[17] D. Guo, M. Gahegan, D. Peuquet, and A. MacEachren. Breaking down dimensionality: an effective feature selection method for high-dimensional clustering. In *Proceedings of the Third SIAM International Conference on Data Mining*, pages 29–42. SIAM Press, San Francisco, CA, 2003.

[18] I. Guyon and A. Elisseeff. An introduction to variable and feature selection. *Journal of Machine Learning Research*, 3(3):1157–1182, 2002.

[19] J. Handl and J. Knowles. Exploiting the Trade-Off—The Benefits of Multiple Objectives in Data Clustering. In C. A. Coello Coello, A. Hernández Aguirre, and E. Zitzler, editors, *Evolutionary Multi-Criterion Optimization. Third International Conference, EMO 2005*, pages 547–560, Guanajuato, México, Mar. 2005. Springer. Lecture Notes in Computer Science Vol. 3410.

[20] J. Handl and J. Knowles. Multiobjective clustering around medoids. In D. Corne, Z. Michalewicz, B. McKay, G. Eiben, D. Fogel, C. Fonseca, G. Greenwood, G. Raidl, K. C. Tan, and A. Zalzala, editors, *Proceedings of the 2005 IEEE Congress on Evolutionary Computation*, volume 1, pages 632–639, Edinburgh, Scotland, UK, 2-5 Sept. 2005. IEEE Press. ISBN 0-7803-9363-5. URL `http://ieeexplore.ieee.org/servlet/opac?punumber=10417&isvol=1`.

[21] J. Handl and J. Knowles. Feature subset selection in unsupervised learning via multiobjective optimization. *International Journal on Computational Intelligence Research*, 2(3):217–238, 2006.

[22] J. Handl and J. Knowles. On semi-supervised clustering via multiobjective optimization. Technical Report TR-COMPSYSBIO-2006-02, Manchester Interdisciplinary Biocentre, University of Manchester, UK, 2006. http://dbk.ch.umist.ac.uk/handl/publications.html.

[23] J. Handl and J. Knowles. Semi-supervised feature selection via multiobjective optimization. Technical Report TR-COMPSYSBIO-2006-03, Manchester Interdisciplinary Biocentre, University of Manchester, UK, 2006. http://dbk.ch.umist.ac.uk/handl/publications.html.

[24] J. Handl and J. Knowles. An evolutionary approach to multiobjective clustering. *IEEE Transactions on Evolutionary Computation*, 11(1):56–76, 2007.

[25] J. Handl and J. Knowles. Multiobjective optimization in bioinformatics and computational biology. *IEEE/ACM Transactions on Computational Biology*, 4 (2), 2007.

[26] D. Hanisch, A. Zien, R. Zimmer, and T. Lengauer. Co-clustering of biological networks and gene expression data. *Bioinformatics*, 18:145–154, 2002.

[27] G. G. Harrigan and R. Goodacre. *Metabolic Profiling: Its Role in Biomarker Discovery and Gene Function Analysis*. Kluwer Academic Publishers, London, UK, 2003.

[28] L. Hubert and P. Arabie. Comparing partitions. *Journal of Classification*, 2: 193–218, 1985.

[29] A. K. Jain and R. C. Dubes. *Algorithms for Clustering Data*. Prentice Hall, Englewood Cliffs, NJ, 1988.

[30] M. T. Jensen. Guiding single-objective optimization using multi-objective methods. In *Applications of Evolutionary Computation*, pages 268–279. Springer-Verlag, Berlin, Germany, 2003.

[31] T. Joachims. Transductive inference for text classification using support vector machines. In *Proceedings of 16th International Conference on Machine Learning*, pages 200–209. Morgan Kaufmann Publishers, San Francisco, CA, 1999.

[32] D. Kim. Structural risk minimization on decision trees using an evolutionary multiobjective optimization. In *Proceedings of the Seventh European Conference on Genetic Programming*, pages 338–348. Springer-Verlag, Berlin, Germany, 2004.

[33] J. D. Knowles, R. A. Watson, and D. W. Corne. Reducing local optima in single-objective problems by multi-objectivization. In *Proceedings of the First International Conference on Evolutionary Multi-Criterion Optimization*, pages 269–283. Springer-Verlag, Berlin, Germany, 2001.

[34] J. D. Knowles, R. A. Watson, and D. W. Corne. Reducing Local Optima in Single-Objective Problems by Multi-objectivization. In E. Zitzler, K. Deb, L. Thiele, C. A. C. Coello, and D. Corne, editors, *First International Conference on Evolutionary Multi-Criterion Optimization*, pages 268–282. Springer-Verlag. Lecture Notes in Computer Science No. 1993, 2001.

[35] K. Miettinen. *Nonlinear Multiobjective Optimization*. Kluwer Academic Publishers, 1999.

[36] G. W. Milligan. Clustering validation: results and implications for applied analyses. In *Clustering and Classification*, chapter 10, pages 341–367. World Scientific, New Jersey, NJ, 1996.

[37] S. Mitra. Computational intelligence in bioinformatics. *Transactions on Rough Sets*, pages 134–152, 2005.

[38] M. Morita, R. Sabourin, F. Bortolozzi, and C. Y. Suen. Unsupervised feature selection using multi-objective genetic algorithms for handwritten word recognition. In *Proceedings of the Seventh International Conference on Document Analysis and Recognition*, pages 666–671. IEEE Press, New York, NY, 2003.

[39] S. O'Hagan, W. B. Dunn, D. Broadhurst, R. Williams, J. Ashworth, M. Cameron, J. Knowles, and D. B. Kell. Closed-loop, multi-objective optimisation of two-dimensional gas chromatography (gcxgc-tof-ms) for serum metabolomics. *Analytical Chemistry*, 79:464–476, 2007.

[40] S. O'Hagan, W. B. Dunn, M. Brown, J. D. Knowles, and D. B. Kell. Closed-loop, multiobjective optimization of analytical instrumentation: gas chromatography/time-of-flight mass spectrometry of the metabolomes of human serum and of yeast fermentations. *Analytical Chemistry*, 77(1):290–303, 2005.

[41] J. M. Pena, J. A. Lozano, P. Larranaga, and I. Inza. Dimensionality reduction in unsupervised learning of conditional Gaussian networks. *IEEE Transactions on Pattern Analysis and Machine Intelligence*, 23(6):590–603, 2001.

[42] P. J. Rousseeuw. Silhouettes: a graphical aid to the interpretation and validation of cluster analysis. *Journal of Computational and Applied Mathematics*,
20:53–65, 1987.

[43] E. Schreibmann, M. Lahanas, L. Xing, and D. Baltas. Multiobjective evolutionary optimization of the number of beams, their orientations and weights
for intensity-modulated radiation therapy. *Physics in Medicine and Biology*, 49
(5):747–770, 2004.

[44] Y. Siskos and A. Spyridakos. Intelligent multicriteria decision support:
Overview and perspectives. *European Journal of Operational Research*, 113
(2):236–246, 1999.

[45] N. Sondberg-Madsen, C. Thomsen, and J. M. Pena. Unsupervised feature subset selection. In *Proceedings of the Workshop on
Probabilistic Graphical Models for Classification*, pages 71–82, 2003.
http://www.sc.ehu.es/ccwbayes/ecml-pkdd-03-workshop/call.htm.

[46] A. Strehl and J. Ghosh. Cluster ensembles — a knowledge reuse framework
for combining multiple partitions. *Journal on Machine Learning Research*, 3:
583–617, 2002.

[47] L. Talavera. Feature selection as a preprocessing step for hierarchical clustering.
In *Proceedings of the Sixteenth International Conference on Machine Learning*,
pages 389–39. Morgan Kaufmann, San Francisco, CA, 1999.

[48] A. Topchy, A. K. Jain, and W. Punch. A mixture model for clustering ensembles. In *Proceedings of the SIAM International Conference on Data Mining*,
pages 379–390. SIAM, Lake Buena Vista, FL, 2004.

[49] P. Vincke. *Multicriteria decision-aid*. Wiley New York, 1992.

Part II

Machine Learning with Multiple Objectives

Multiobjective Supervised Learning

Jonathan E. Fieldsend and Richard M. Everson

School of Engineering, Computer Science and Mathematics
University of Exeter, Exeter EX4 4QF, UK
[J.E.Fieldsend,R.M.Everson]@exeter.ac.uk

Summary. This chapter sets out a number of the popular areas in multiobjective supervised learning. It gives empirical examples of model complexity optimization and competing error terms, and presents the recent advances in multi-class receiver operating characteristic analysis enabled by multiobjective optimization. It concludes by highlighting some specific areas of interest/concern when dealing with multiobjective supervised learning problems, and sets out future areas of potential research.

1 Introduction: What Is Supervised Learning?

A common task in machine learning is to learn the functional relationship between inputs and outputs. The inputs $\mathbf{x}$ are generally vectors of features, which may be discrete, continuous or mixed. The output is typically a scalar y, the *target*. If y is a continuous variable then the problem is known as a regression problem; for example, $\mathbf{x}$ might be a vector of rainfall measurements and y might be the height of a river at a particular place. Conversely, if y is a discrete variable then the problem is known as a classification problem: y here indicates into which class the observations $\mathbf{x}$ fall; a common example is medical diagnosis in which $\mathbf{x}$ is a vector of physiological measurements for a particular person and y indicates whether the person has a particular disease ($y = 1$, say) or not ($y = 0$).

During supervised learning the machine is equipped with a set of *training data* comprising pairs $\{\mathbf{x}_n, y_n\}_{n=1}^N$ of features and targets which are assumed to be representative of the process being modelled. If the mapping $\mathbf{x} \mapsto y$ is successfully learned, then the learned function can be used to make predictions of the target for features whose target is unknown.

A number of problems arise in supervised learning. On the data side there are issues of how well the training data actually represents the generating process (if important relationships are not represented, they cannot be learnt), and of whether the generating process is stationary or not (i.e., whether the problem itself changes over time). Perhaps the most important question facing the supervised learner is how to prevent *overfitting*; that is, how to ensure that the learned function models the underlying relationship between features and targets, without modelling any noise present in the training data.

On the function induction side there are the problems of choosing *a priori* which specific model/family of models to use and of determining how complex a representation to allow. There is also the issue of which error term to use during the training/learning process in order to generate the model with the best generalization ability or other related properties. Finally, there is the issue of which subset of inputs/features to induce the model from.

The chapter proceeds as follows. In Section 2 a more formal definition of supervised learning is provided; this is followed by a number of sections giving empirical examples of evolutionary multiobjective optimization (EMOO) in the supervised learning domain. Section 3 presents an example of regularization using EMOO. Section 4 discusses competing error terms, and gives examples from both two-class and multi-class Receiver Operating Characteristic analysis. The chapter concludes with a brief discussion of issues arising in the domain, highlighting potential areas for further work and current unanswered questions.

2 Different Formulations of Supervised Learning

More formally, assume we are given a model f which predicts an output $\hat{y}$ (i.e., class membership probabilities or real-valued regression prediction) based on a feature vector $\mathbf{x}$ and model parameters $\mathbf{w}$, so that

$$\hat{y} = f(\mathbf{x}, \mathbf{w}). \tag{1}$$

The model may be quite simple, such as a linear regressor, but frequently very flexible, nonlinear models such as neural networks or support vector machines (SVMs) are used.

Supervised learning techniques try to find a parameterization $\mathbf{w}$ that minimizes some error E over all feature-target pairs:

$$\hat{\mathbf{w}} = \arg \min_{\mathbf{w} \in W} E(f(\mathbf{x}, \mathbf{w}), g(\mathbf{x})) \qquad \forall \mathbf{x} \in \aleph, \tag{2}$$

where $g(\mathbf{x})$ is an oracle function that tells the true target value for every input, $\aleph$ is the set of all valid feature vectors, and W is the set of all feasible model parameterizations. Typically one does not have access to $\aleph$ or the oracle function; rather one has a subset of (often noisy) observations in the form a data set $D = \{\mathbf{x}_n, y_n\}_{n=1}^N$, so that practical supervised learning involves minimizing

$$\hat{\mathbf{w}} = \arg \min_{\mathbf{w} \in W} E(f(\mathbf{x}_n, \mathbf{w}), y_n) \qquad n = 1, \ldots, N. \tag{3}$$

Usually the training pairs are assumed to be independently and identically distributed (i.i.d.) draws from a generating distribution, so that the error becomes the sum of the error for each training pair:

$$E = \sum_{n=1}^N \tilde{E}(f(\mathbf{x}_n, \mathbf{w}), y_n). \tag{4}$$

In regression problems, the error function is usually taken as the squared error, $\tilde{E} = (f(\mathbf{x}, \mathbf{w}) - y_n)^2$, which is tantamount to assuming that the observations are corrupted by Gaussian-distributed noise, although if the noise is not Gaussian distributed some other error function will be appropriate. For classification problems, the cross-entropy is the error function that corresponds to maximizing the likelihood of the data [2].

As noted above, the full range of training examples is generally unavailable, and one has to be content with a finite (and often small) training set. An inevitable consequence of this is that the models with a high degree of flexibility are able to fit the peculiarities of the particular training data set rather than the general trends in the data. Clearly, optimization to a particular data set inhibits generalization and renders predictions on unseen data poor. A common way to tackle this overfitting is to regularize the error function by augmenting (3) with a regularization term that penalizes overly complex models; thus, the error to be minimized becomes

$$E = E_{data} + \alpha E_{reg}, \tag{5}$$

where E_{data} is the data error function (equations (3) and (4)), E_{reg} is a penalty that increases as the model becomes more complex, and α is a regularization parameter.[1] A widespread choice for E_{reg} is the weight-decay penalty:

$$E_{reg} = \|\mathbf{w}\|^2. \tag{6}$$

This penalty function penalizes models with large parameters—models with large nonlinear terms that are therefore likely to be overfitting. Although the form of the penalty may at first sight appear somewhat ad hoc, it is naturally interpreted from a Bayesian point of view as a prior probability over the parameters [25, 26, 2]. Other complexity-penalizing functions such as Minimum Description Length (MDL), the Bayesian Information Criterion (BIC) and the Akaike Information Criterion (AIC) have been proposed [see, e.g., 5].

Conventionally, the overall error is minimized and the regularization parameter determined by cross-validation. It is shown in the next section how choosing the model parameters may be viewed as a two-objective optimization problem.

3 Regularization by Multiobjective Optimization

Arguably one of the more fruitful avenues investigated so far by the EMOO community in multiobjective supervised learning is complexity model optimization (see, e.g., [14, 23, 28] for recent work and overviews). As noted earlier, there tends to be a tendency, especially when using models with high representation capability, to overfit a model parameterization to the training data, leading to poor generalization ability. A textbook example of this would be when using neural networks (NNs). Given enough activation units, NNs are universal approximators, allowing sufficient complexity within the model to permit them to model any deterministic underlying generation process. However, determining the appropriate complexity a

---

[1] Other approaches to controlling overfitting have been to use pruning algorithms to remove nodes [24], other complexity loss functions [32] and topology selection methods [31].

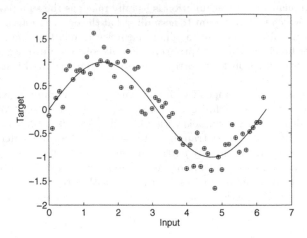

Fig. 1. Noisy sine wave training data (dots), with noiseless generating function shown with the solid line. Noise drawn from a Gaussian with zero mean and 0.3 standard deviation. 63 training data points (input values drawn at intervals of 0.1 from 0 to 2π)

priori for a problem so as not to overfit the data at hand is a persistent problem. As noted above, in statistical machine learning this overfitting is typically tackled by the use of weight decay regularization. This approach requires the determination of the regularization parameter α on the weighting of this penalty. The use of EMOO, in contrast, allows optimization over all complexities. As such the problem can be cast as bi-objective for EMOO, with the first objective being the minimization of the error function (in the regression problems shown here, the mean square error), and the second objective being the minimization of model complexity (here, the sum of the squared weights of a multi-layer perceptron (MLP) neural network; equation 6).

A simple example is now provided. The problem is the regression of a noisy sine wave, using the training data illustrated in Figure 1, with circles denoting the training data and the line representing the continuous (noiseless) generating process. Using a simple greedy (1+1)-evolution strategy (ES), as described by [13, 15], one can discover the networks corresponding to the estimated Pareto front shown in Figure 2 with dots.

A general (1+1)-ES is given in Algorithm 1. In the implementation of this general algorithm in this section an initial nondominated set of points was generated by training an MLP (with one input unit, 50 hidden units and one output unit) using the quasi-Newton method [2, 27] and evaluating its objectives every 50 epochs, up to 5,000 epochs (Algorithm 1 line 1). The ES was run for 50,000 generations (line 2), with a probability of weight mutation of 0.1, and mutation being formed of additive draws from a zero mean Gaussian with standard deviation of 0.2 (line 4). In the implementation here, **update()** (line 6) ensures F always contains the best current

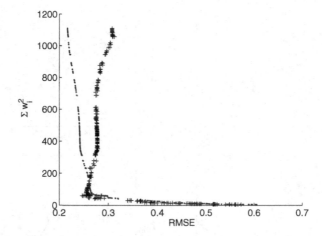

Fig. 2. Estimated Pareto optimal front of NNs (dots) and the same NNs evaluated on a validation set from the same generation and noise process (crosses); note the switch back effect in the lower left corner

---

Algorithm 1 A general 'greedy' (1+1)-ES scheme for multiobjective optimization in supervised learning, where **e** is the set of error evaluations on model parameterization **w** and data **x**. Error terms to be minimized without loss of generality

---

Inputs:

T	*Number of generations*
M	*Number of initial samples of* **w**

1: $F := \texttt{initialize}(\mathbf{x}, M)$ *Initial estimate of front*
2: **for** $t := 1 : T$
3:      $\mathbf{w} := \texttt{select}(F)$ *Select from archive*
4:      $\mathbf{w}' := \texttt{perturb}(\mathbf{w})$ *Perturb/mutate parameters*
5:      $\mathbf{e} := \texttt{evaluate}(\mathbf{x}, \mathbf{w}',)$ *Evaluate error functions*
6:      $F := \texttt{update}(F, \mathbf{w}')$ *Update archive*
7: **end**

---

estimate of the true Pareto front by storing an unconstrained nondominated set of the best parameterizations found so far.

   Figure 3a shows the regression lines of the model with the lowest complexity (summed squared weights) from the estimated Pareto front, with Figures 3b–h showing the regression lines of models with consecutively higher complexity levels (sampled at roughly every fiftieth element of the estimated true Pareto front shown in Figure 2).

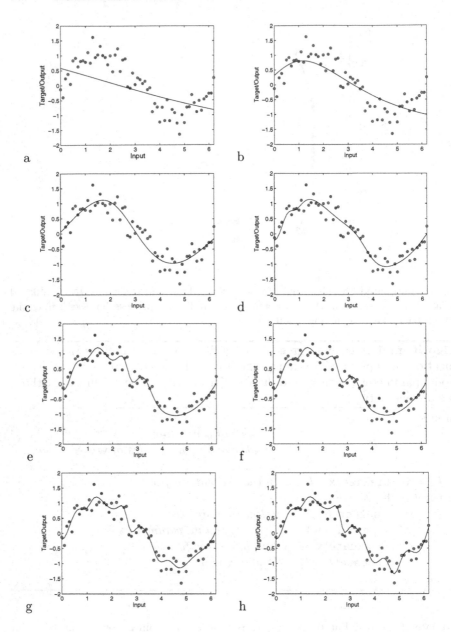

Fig. 3. Regression lines of the estimated Pareto optimal NNs on the training data. Plots (a)–(h) show models sampled regularly from estimated Pareto front, from lowest complexity to highest complexity

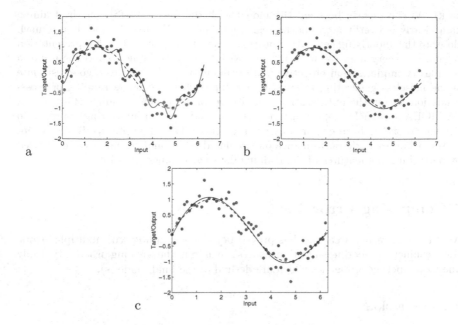

Fig. 4. a) Regression line of model with lowest RMSE on training data from estimated Pareto set. b) Regression line of model with lowest RMSE on validation data from estimated Pareto set (on training data). c) Regression line of ensemble of ten models with lowest RMSE on validation data from estimated Pareto set (on training data)

The models span the spectrum from severe underfitting (such as the almost straight lines in Figure 3a) to severe overfitting (such as the wiggly lines shown in Figure 3h). This range of model types is to be expected from the optimization objectives. The problem still arises as to how to choose an operating model from the set at the end of the optimization run. One approach discussed in [14] is to evaluate the set on a second validation set of data and note at which point the complexity/accuracy curve 'switches back'. This is shown in Figure 2 by crosses, where a validation set of the same size as the training set is used, from the same noisy generating process. A prominent 'switch back' point can be seen in the lower left-hand corner, which would lead one to choose either the model with the lowest root mean squared error (RMSE) in this area or to use an equal-weighted ensemble of models from this region.

Figure 4 shows the regression line of various approaches with a solid line – in all cases the dashed line shows the underlying noiseless generating process. Figure 4a is the model with the lowest RMSE on the training data (i.e., the model corresponding to the leftmost point in Figure 2), Figure 4b is the model with the lowest RMSE on the validation data (i.e., the model corresponding to bottom left of the 'switch back' cross in Figure 2), and Figure 4c is the average regression line of the ten models with the lowest validation error (the models at the knee of the switch back). As can clearly be seen, the model with the lowest RMSE on the training data clearly is overfitting, but the regression lines in Figures 4b and 4c are much closer to the underlying generating process. Even though the network representation capability is very high (50 hidden units, with only 63 training points) the use of a complexity minimization objective and a validation set has led to a good estimate of the noiseless generating process. Other approaches like bootstrapping or cross-validation during the optimization can also be employed for a similar effect within the MOEA approach; see, for instance, [15]. In addition, an interesting approach to regularization has been explored by Gräning et al. [17], who optimize Receiver Operating Characteristic performance on simulated additional training sets generated by perturbing the features of the training data with Gaussian noise.

4 Competing Error Terms

Another application area that has proved popular is training with multiple errors. Here, conflicting 'goodness-of-fit' measures are used in the learning process, typically due to competing properties which are desired of the final model(s).

4.1 Regression

In the area of regression this has been in the formulation of a trade-off between different measures of goodness-of-fit. For instance, using EMOO methods, one may optimize with respect to one measure (e.g., RMSE or absolute error) and also with respect to the distributional properties of this principal error measure [1, 9]. In the regression field, EMOO methods have also been used to optimize multiple 'application specific error terms', for instance, in financial applications, the return on investment of predicting an asset price. It is difficult to train a model using this term by itself, but used in conjunction with a goodness-of-fit error measure, it can ensure that you have models that accurately predict the signal *and* are profitable [13, 29].

4.2 Classification

In classification problems, the task is to allocate new or previously unseen examples **x** to one of two or more classes (categories) C_j. This is generally based on a model, or set of models, induced from some existing corpus of data whose true classes are known already. The misclassification rate (proportion of data which is labelled with an incorrect class by the classifier) is typically taken as a measure of classifier accuracy, and as the objective to be minimized. However, when there is an imbalance in the cardinality of each distinct class in a set of data, for training and/or testing, the total misclassification rate can be misleading. For instance, in a two class problem, it is trivial to get a 10% misclassification rate if the sizes of the two classes are in a 9:1 ratio.

In order to deal with class imbalance, Receiver Operating Characteristic Analysis (ROC) is typically used in the 2-class classifier optimization. This analysis traces out the true positive rate (the proportion of correct assignments to the principal class by the model) against the false positive rate (the proportion of incorrect assignments

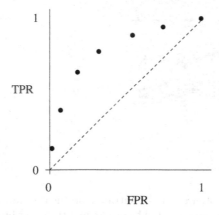

Fig. 5. ROC example. Points denote the different TPR/FPR combinations possible from a classifier by using a different threshold on the model output (or alternatively using different model parameters)

of the second class to the principal class by the model) by varying the classification threshold of the model (if the model outputs a probability of assignment, or a score) or the parameters of the model itself. This visualization shows the trade-off between the accuracy in classifying the two separate classes for a particular model – as illustrated in Figure 5. The best possible classifier would operate in the top left of the plot, with a TPR of 1 and an FPR of 0. The dashed line denotes the random allocation line, the expected performance of a classifier which allocates class labels to data at random (at some fixed ratio). Any classifier operating *below* this line is performing *worse* than randomly; the operating point of such a classifier can be reflected through the random allocation line simply by switching the class labels the classifier has assigned to the data.

The plotting of classifier performance in the TPR/FPR plane also allows a user to evaluate models given different costs of misclassification. For example, in medical diagnosis, the cost of misclassifying a patient by saying he does not have a cancer when he does is far more costly than saying he has a cancer when he does not. The latter error will be detected with a biopsy sample; the former error may not be detected before the cancer progresses to a more dangerous state.

The area under the ROC curve (AUC), which lies between 0 and 1, is often used as a single value to compare classifiers. As explained by Hand and Till [18], this measures a classifier's ability to separate two classes over the range of possible costs. The Gini coefficient is also used, which is twice the area between the curve and the random allocation line.

ROC analysis obviously lends itself to optimization with EMOO methods, with the TPR and FPR being cast as two separate objectives.[2] The example in Figure 6a shows the decision boundaries formed by radial basis function (RBF) neural network

---

[2] Alternatively one could maximize the AUC as a function of a set of solutions, if one were careful as to how the set was updated, as discussed later.

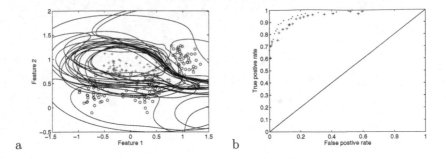

Fig. 6. a) Decision contours of RBF networks on the estimated optimal ROC front; training data with one class denoted by circles and the other by crosses. b) Estimated optimal ROC front on 250 training data pairs (denoted by dots) and their evaluation on 1,000 testing data pairs (denoted by crosses)

classifiers on the test problem from [10], optimized using a simple (1+1)-ES.[3] Figure 6b in turn shows the estimated optimal ROC curve on the 250 training data points (shown with dots on the plot), and their evaluation on 1000 testing data points (shown with crosses on the plot).

Interestingly, although not shown here (but available in [10]), synthetic ROC problems are perhaps the only supervised learning problems for which the true Pareto front can be determined and the performance of the optimized solutions compared to it. This is because with a synthetic classification problem one can determine the exact posterior probability of any feature vector, and therefore trace out the ROC curve of a Bayes rule classifier (the best possible). Without knowing the generating process one cannot know where the true Pareto front lies for any classification problem, and therefore how close any particular model is to it. However, the downside to this is that when optimizing a classifier based on training data, you only actually have access to an estimate of the posterior probability, not the true posterior probability (otherwise you would not need a classifier in the first place). As such the estimated ROC curve may actually seem *above* the known optimal curve. This problem of noise and uncertainty (which is apparent in most if not all supervised learning problems) is one of the principal areas needing additional research in multi-objective supervised learning, and can be the source of over-optimistic assessments of performance.

It is also worth noting that in 2-class ROC optimization the granularity of the front is limited by the cardinality of the dataset —therefore the use of unconstrained archives is often appropriate.

---

[3] The RBFs contained ten units with Gaussian kernels, optimized in the fashion discussed in [11] using a (1+1)-ES for 5,000 generations with a probability of mutation of 0.1 and a variance of additive Gaussian mutation of 0.2.

4.3 Separating Classes

An early formulation of the multi-class ROC problem was proposed by [18], who introduced a generalization of the AUC. In summary, their M measure is the average of the pairwise AUCs between the $Q(Q-1)/2$ pairs of classes. More precisely, Hand and Till show that the AUC is the probability, denoted $\hat{A}(k|j)$, that a randomly drawn member of class $\mathcal{C}_k$ will have a lower estimated probability of belonging to class $\mathcal{C}_j$ than a randomly drawn member of $\mathcal{C}_j$. Clearly, a classifier able to separate $\mathcal{C}_k$ from $\mathcal{C}_j$ has large $\hat{A}(k|j)$, whereas if it makes assignments no better than chance, then $\hat{A}(k|j) = 1/2$. Except in the two-class problem, $\hat{A}(k|j) \neq \hat{A}(j|k)$, and exchanging class labels does not alter their separability, so the classifier's ability to separate $\mathcal{C}_j$ and $\mathcal{C}_k$ is measured by $\hat{A}(j,k) = [\hat{A}(k|j) + \hat{A}(j|k)]/2$. Hand and Till then define overall performance of a classifier as

$$M = \frac{2}{Q(Q-1)} \sum_{j<k} \hat{A}(j,k), \tag{7}$$

where Q is the number of classes. M thus measures the average ability of a classifier to separate classes, although it considers the pairwise performances of the classifier rather than the full Pareto front. Hand and Till also describe the measure for a classifier with fixed parameters, rather than for a parameterized family of classifiers, as done in the next section of this chapter. A natural generalization is to consider the multiobjective maximization (for a parameterized family) of the $Q(Q-1)$ pairwise $\hat{A}(j,k)$. In fact, this leads to a simple algorithm for the maximization of M itself, which is now described.

The key to maximizing M is that it is possible to find a set E of parameters $\mathbf{w}$ that together maximize M. Consequently, if the addition of a proposed parameter vector $\mathbf{w}'$ to E increases any one of the $\hat{A}(j,k)$, it automatically increases M; since an unrestricted set of parameters is kept, no other elements of E need be deleted, so the other $\hat{A}(j,k)$ are, at worst, not decreased. This leads to the straightforward procedure outlined in Algorithm 2. As for the multiobjective evolutionary algorithm, it maintains an archive E of solutions. At each stage, a randomly selected member of E is perturbed and the M measure of the archive plus $\mathbf{w}'$ evaluated; if the addition of $\mathbf{w}'$ increases M then $\mathbf{w}'$ is retained (line 6 of Algorithm 2), and any parameters which now do not contribute to M are removed (lines 7–9).

When maximizing M over a family of classifiers, several ROC curves for individual classifiers generally contribute to the composite ROC curve for the family. Example ROC curves for eight classifiers resulting from the optimization of M for synthetic data using the probabilistic k-nn classifier [20] are shown in Figure 7. For each pair of classes the axes of each panel are C_{kk}, the true positive rate for $\mathcal{C}_k$, and C_{kj}, the rate at which misclassifications of $\mathcal{C}_k$ examples are classified as $\mathcal{C}_j$. Each 'ROC curve' corresponds to a distinct $\mathbf{w} = \{k, \beta\}$ parameter value,[4] and the optimized M is achieved by the envelope of these curves. Evaluation of the $\hat{A}(k|j)$ that contribute to M can be performed by applying the method described by Hanley and McNeil [19] and Hand and Till [18, page 174] for calculating the AUC for a single classifier to the envelope of the ROC curves.

As Figure 7 shows, after optimization only eight distinct (k, β) combinations contribute to the optimized $M \approx 0.991$, although during optimization up to 20 parameter combinations were involved. Selection of the operating parameters on the

---

[4] The probabilistic k-nn classifier is discussed further in Section 4.4.

Algorithm 2 Evolutionary optimization of Hand and Till's M measure

---

Inputs:
    T *Number of generations*

    1: $E := \text{initialize}()$
    2: **for** $t := 1 : T$
    3:       $\mathbf{w} := \text{select}(E)$
    4:       $\mathbf{w}' := \text{perturb}(\mathbf{w})$ *Perturb parameters*
    5:       **if** $M(E \cup \mathbf{w}') > M(E)$
    6:             $E := E \cup \mathbf{w}'$ *Insert* $\mathbf{w}'$
    7:             **for** $\mathbf{u} \in E$
    8:                   **if** $M(E) = M(E \setminus \mathbf{u})$
    9:                         $E := E \setminus \mathbf{u}$ *Remove redundant elements*
   10:                   end
   11:             end
   12:       end
   13: end

---

basis of the $\hat{A}(j,k)$ is possible; it is emphasized that the $\hat{A}(j,k)$ summarize the *overall* pairwise separability rather than permitting specific choices to be made between particular misclassification rates. Additional information is available through examination of the families of pairwise trade-off curves, such as those displayed in Figure 7.

As the optimized M measures the ability of a particular family of classifiers to separate classes, it may be used for comparing classifiers. Table 1 shows the optimized M and number of distinct models (distinct parameter values) contributing to M for a number of standard machine learning data sets taken from the UCI repository [4]. The two-class Ionosphere data is well known to be easily classified, and M (actually the AUC here) is correspondingly high with only three distinct parameter sets for the k-nn classifier and four sets for the MLP. The Image data can be well separated, but only with the use of 13 parameter sets for k-nn; again, better separation is achieved by the more flexible MLP, but at the expense of many more models. The DNA data with only three classes but 180 features requires 181 (k, β) combinations for optimal separation. In contrast, even after optimization the Satimage data cannot be well separated with k-nn classifiers. Results are not presented for the MLP classification of the Abalone, Satimage and DNA datasets because the computation of the $\hat{A}(j|k)$ for envelopes of individual classifiers becomes exorbitantly expensive with many samples and models.

In summary, although M provides a global measure of a classifier's performance on a particular dataset and identifies a relatively small number of optimal parameter sets, the question of how to select an operating point remains. The question arises, for instance, about whether a single operating point selected from a group which together maximizes M would necessarily be as good as a single operating point maximized for M. In the next section a different approach to multi-class ROC optimization is discussed which confronts some of these issues.

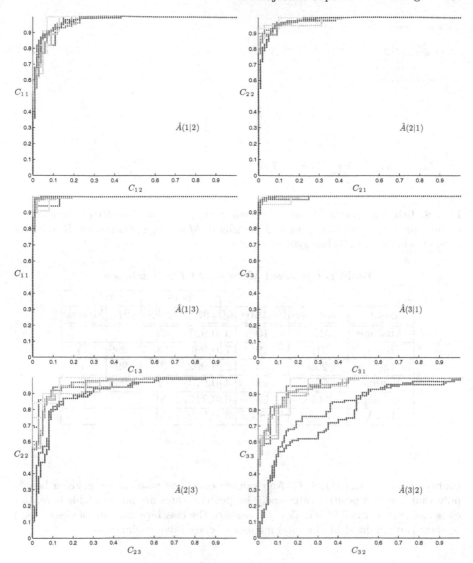

Fig. 7. Pairwise ROC curves for the k-nn classification of the 3-class synthetic data set. Each row corresponds to a pair of classes. Axes correspond to the true positive rate C_{kk} and the rate at which C_k examples are misclassified as C_j. Each curve corresponds to a distinct parameter combination, so that $\hat{A}(k|j)$ is the area under the envelope of the curves

4.4 Multi-class ROC

The authors have shown recently that with Q-classes (where $Q > 2$) ROC analysis can be extended and cast in terms of minimizing the off-diagonal elements of the

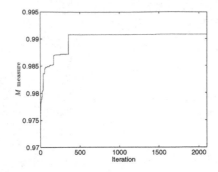 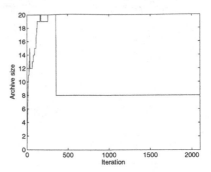

Fig. 8. Left: Growth of M with iteration during optimization. Right: Number of distinct parameter combinations contributing to M during optimization. Results for k-nn classification of 3-class synthetic data

Table 1. Optimized M measure for UCI data sets

Name	Examples	Features	Q	k-nn M	k-nn Models	MLP M	MLP Models
Abalone	3133	10	3	0.927	33		
Image	210	19	7	0.996	13	0.999	25
Ionosphere	200	33	2	0.992	3	0.996	4
Vehicle	564	18	4	0.973	11	0.966	75
Satimage	4435	36	6	0.713	20		
DNA	2000	180	3	0.989	181		

confusion rate matrix [11, 7, 6]. An example confusion rate matrix is given below; note that the true positive rates and false positive rates are not available here, but class assignment rates (where $\mathcal{C}_{i,j}/|\mathcal{C}_i|$ denotes the classification rate of class i data to class j, normalized by the total number of class i data points) are.

$$
\begin{array}{c|cccc}
 & \multicolumn{4}{c}{\text{Predicted Class}} \\
 & 1 & 2 & \ldots & 3 \\
\hline
1 & \frac{\mathcal{C}_{1,1}}{|\mathcal{C}_1|} & \frac{\mathcal{C}_{1,2}}{|\mathcal{C}_1|} & \cdots & \frac{\mathcal{C}_{1,Q}}{|\mathcal{C}_1|} \\
\text{Actual Class} \quad 2 & \frac{\mathcal{C}_{2,1}}{|\mathcal{C}_2|} & \frac{\mathcal{C}_{2,2}}{|\mathcal{C}_2|} & \cdots & \frac{\mathcal{C}_{2,Q}}{|\mathcal{C}_2|} \\
\ldots & \ldots & \ldots & \ldots & \ldots \\
Q & \frac{\mathcal{C}_{Q,1}}{|\mathcal{C}_Q|} & \frac{\mathcal{C}_{Q,2}}{|\mathcal{C}_Q|} & \cdots & \frac{\mathcal{C}_{Q,Q}}{|\mathcal{C}_Q|}
\end{array}
$$

This is therefore a $Q(Q-1)$ objective minimization problem. However, although the dimensionality of the Pareto front/optimal ROC front increases rapidly with the number of classes, as with the 2-class problem, there is a limit (albeit potentially very large) to the number of distinct points on it which is a function of the size of the data set used, and Q.

By extending the Gini coefficient analysis, a random allocation simplex can be used to compare different classifiers in $Q(Q-1)$ dimensional objective space. Classifiers whose off-diagonal confusion rates sum to greater than $Q-1$ are performing worse than average. Additionally, that single model furthest in front of this simplex (closest to the origin) should be the classifier chosen when no misclassification costs are known (for a more extensive discussion see [6]). This also allows the comparison of different classifier *families* for particular supervised learning problems (e.g., k-nearest neighbour classifiers, decision trees, multi-layer perceptrons, radial basis functions, etc.). The comparison can thus concern itself not simply with the cardinality of dominance between the points on the ROC front produced, but also using a measure on the objective space which is meaningful (similar to the volume measure, or S metric, used in general multiobjective optimization, but without scaling, and based on a prespecified region of a hypercube). More formally, [6] have shown that the volume lying between the origin (the perfect classifier) and the random allocation simplex, which also lies in the unit hypercube (where it is feasible for a classifier performing better that average to operate) is

$$\frac{(Q-1)^{Q(Q-1)}}{(Q(Q-1))!} - \frac{Q(Q-1)(Q-2)^{Q(Q-1)}}{(Q(Q-1))!}. \tag{8}$$

This region is denoted here by P. The measure on it, $G()$, is calculated as the proportion of P dominated by elements of the ROC surface (F). Therefore, like the Gini coefficient in two dimensions, $G(F)$ is a measure of how much better than random the elements in a set F are. $\delta(F, F')$ in turn measures how much of P is dominated by the set F but not by the set F'. This can be used to compare two different fronts (generated for instance by two different classifier families) which possibly overlap in parts.

Due to the shape of the region, it is not quite as trivial to calculate its volume as it is to calculate the region used in the S metric [33], as a reference simplex is used, as opposed to a reference point (which is further constrained to lie within a unit hypercube). As such, the region defined by P is a hyper-pyramid, with a $Q(Q-1)$ truncated corners. Monte Carlo sampling of this region can give a good estimate of the volume dominated, and [8] discusses how to do this efficiently.[5]

When using a soft classifier (one that gives a probability of class membership, or a score) it is computationally efficient to assess the effect of a number of different sample cost matrices $\mathbf{c}$ on the misclassification rates for any particular model parameterization.[6] This is because passing the data $\mathbf{x}$ through a classification model can be time consuming, whist transforming this output using different cost matrices allows the evaluation of many different possible misclassification combinations

---

[5] If the reader is considering using this measure to compare multi-class ROC curves, she is advised to consult this technical report, as the probability of randomly generating a sample in the region defined by P, by generating a uniform sample in the unit hypercube, is $\frac{(Q-1)^{Q(Q-1)}}{(Q(Q-1))!} - \frac{Q(Q-1)(Q-2)^{Q(Q-1)}}{(Q(Q-1))!}$, which rapidly becomes become prohibitively small even for small Q. Sampling methods developed in [8] generate random points in P with a probability of $\approx \frac{1}{Q(Q-1)}$.

[6] Assuming linear costs.

Algorithm 3 Converting the general (1+1)-ES scheme for multiobjective optimization in supervised learning (Algorithm 1), into a (1+λ)-ES scheme for ROC optimization (replacing lines 5 and 6 of original algorithm)

Inputs:

    λ *Number of cost samples*

1:       **for** $j := 1 : \lambda$
2:            $\mathbf{c} := \texttt{sample}()$ *Sample costs*
3:            $\mathbf{e} := \texttt{evaluate}(\mathbf{x}, \mathbf{w}', \mathbf{c})$ *Evaluate error functions*
4:            $F := \texttt{update}(F, \mathbf{w}', \mathbf{c})$ *Update archive*
5:       **end**

---

relatively cheaply. The application of Algorithm 1 in this case is better viewed as a $(1 + \lambda)$–ES, with λ cost matrices $\mathbf{c}$ additionally sampled for any particular model parameterization $\mathbf{w}$.[7] As such lines 5 and 6 of Algorithm 1 should be replaced by Algorithm 3.

In the empirical results given below, $\lambda = 50$ different cost matrices are assessed for each model parameterization, drawn from unbiased Dirichlet distributions, with each optimization run lasting $T = 5,000$ generations (therefore 5,000 unique model parameterizations evaluated, each with 50 different cost matrices). The probability of parameter mutation was 0.8, with the mutation being additive draws from a Gaussian distribution with mean 0 and standard deviation 0.2.

Results are given here for the UCI Image, Vehicle and Satimage data sets. Details of data set sizes are given in Table 1, and therefore the objective dimensionalities for these sets in this problem formulation are 42, 12 and 30, respectively. The classification models used are the probabilistic k-nn algorithm, probabilistic k-nn algorithm with tricube kernel [20] and the multinomial logistic regression classifier (MLR) [3]. The probabilistic k-nn classifier is a simple *local* classifier which classifies based on the actual classes of known data in the unlabelled data's immediate locality. It has two parameters, k, the number of neighbours used, and β, which controls the 'strength of association' between neighbours (effectively a way of making closer neighbours more important). The MLR is a simple *global* classifier which separates feature space into different classes with smooth planes, and has $D(Q+1)$ parameters (where D is the number of features, the size of $\mathbf{x}$). The probabilistic k-nn with a tricube kernel has the local classification properties of the probabilistic k-nn, with an additional tendency to push the assignment probability down if the unlabelled sample is 'far' from any labelled data.

Figure 9 shows the signed distance of all points lying on the ROC curve for each classifier from the random allocation simplex; negative numbers mean the operating point is better than random and a value of -1 indicates that the model perfectly classifies the data presented. As can be seen, visually both variants of the probabilistic k-nn classifier seem to do considerably better than the MLR classifier, with all classifiers performing better than random. Table 2 provide the associated G and δ measures, calculated from 10,000 Monte Carlo samples in P. From these it can

---

[7] These sample costs can be straightforwardly and randomly sampled from a Dirichlet distribution; see [6] for a discussion on this.

Image

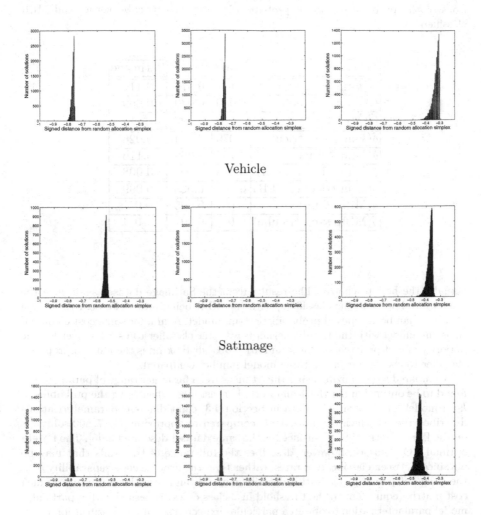

Fig. 9. Distances from the random classifier simplex. Negative distances correspond to models in P. Left: k-nn; Middle: k-nn tricube; Right: MLR. Top: UCI Image data; Middle: UCI Vehicle data; Bottom: UCI Satimage data

be seen that for the Image dataset the probabilistic k-nn model would tend to be the preferred model, with only small portions of its front lying behind that of the probabilistic k-nn tricube model (additionally, the probabilistic k-nn model has the single operating point furthest from the random allocation simplex). A similar result can be seen for the Vehicle data set, although here it is interesting to note that, although the MLR visually seems to underperform compared to the k-nn models, the δ measures show that, for some particular choices of costs, the MLR classifier is

Table 2. Generalized Gini coefficients and exclusively dominated volume comparisons of the probabilistic k-nn, probabilistic k-nn with tricube kernel and MLR classifiers

Measure	Image	Vehicle	Satimage
$G(k\text{-nn})$	0.137	0.073	0.116
$G(k\text{-nn tricube})$	0.080	0.030	0.099
$G(\text{MLR})$	≈ 0	0.009	≈ 0
$\delta(k\text{-nn}, k\text{-nn tricube})$	0.070	0.044	0.026
$\delta(k\text{-nn}, \text{MLR})$	0.137	0.068	0.116
$\delta(k\text{-nn tricube}, k\text{-nn})$	0.013	0.001	0.008
$\delta(k\text{-nn tricube}, \text{MLR})$	0.080	0.028	0.099
$\delta(\text{MLR}, k\text{-nn})$	0	0.005	0
$\delta(\text{MLR}, k\text{-nn tricube})$	0	0.007	0

actually the best to choose. The results from the Satimage dataset again give the same overall order to the classifiers, with the MLR being totally worse, irrespective of costs, than both types of probabilistic k-nn model. Again, for some cost combinations the model with the tricube kernel is a better classifier to use; however for the majority of cost preferences, the standard probabilistic k-nn is the most appropriate classifier to choose out of the three model families compared.

Compared to optimizing using the M measure, a far larger range of parameters is found to be optimal under this framework. For instance, when using the probabilistic k-nn model for the synthetic data in Section 4.3, eight different parameterizations described the set which maximized M, compared to approximately $7,500$ solutions on the Pareto optimal ROC surface for the same dataset described in [6]. The Pareto optimal ROC surface, however, describes the full range of trade-offs that may be obtained between classification rates, rather than the average class separability over the range of pairwise cost ratios described by M, and also shows the user which cost matrix (equivalent to the threshold in 2-class cases) is needed with a particular model parameterization to obtain a particular expected set of misclassification rates.

5 Discussion

There are a number of other avenues in multiobjective supervised learning which have been explored using EMOO, but the examples presented here provide a reasonable overview of the area. A more general overview can also be found in [22].

However, there are still a large number of open questions in the field of multiobjective supervised learning that are worth highlighting.

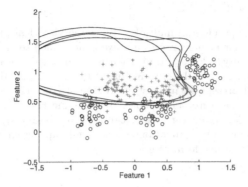

Fig. 10. Example decision boundaries (from RBF classifiers) with identical operating points in ROC space

Hybrid Models

Usually researchers tend to either start a process with a 'traditional' local optimizer (like gradient descent in NNs), or iterate between a local process and an EMOO method. This tends to be the case because the search space is easier traversed (at least to begin with) by local methods, and because, for many of the classifiers/regressors used, the range of parameters to be searched is essentially without limits. As such, EMOO techniques are often used to trace out an estimate of the Pareto front for a problem after a traditional algorithm has supplied a single point on a good estimate of the front. The question of how much search to carry out with local methods and how much time to spend searching with EMOO methods is still an open one.

Overfitting

Unless there is an explicit casting of an objective to minimize complexity, EMOO approaches to optimizing competing errors can be very prone to overfitting. The use of weight decay regularization approaches in hybrid EMOOs may mitigate this somewhat, but to do this they must assume a penalty term independent of the region of objective space, which is a difficult assumption to justify.

Many-to-One Mappings

Perhaps more than other application areas, supervised learning parameter space is full of regions which have identical evaluations in objective space – especially if it is a classification problem. These disjoint plateaus can cause many problems for optimizers, and when using an elite multiobjective optimizer raise the question as to which of the solutions to store if they have the same objective valuations but very *different* input space partitioning. Figure 10 illustrates this with the synthetic classification problem used earlier – the decision contours shown have identical misclassification rates on the data but different decision boundaries.

Noise, Uncertainty, Truth

Arguably, the largest problem in multiobjective supervised learning is the fact that only samples of the generating process are available, which tend to be noisy. Optimizing with uncertainty/robust optimization is an area gaining momentum in the general EMOO community at the current time [21, 30, 12, 16], and supervised learning problems should present an interesting avenue of research. Given the concerns of data mislabelling or feature/sensor noise, and the uncertainty caused when many different models/model parameterizations can lead to the same objective evaluation (on a certain data sample), as mentioned above, all supervised learning problems seem to contain at least one form of uncertainty.

Acknowledgements

The authors would like to thank Michelle Fisher for helpful comments during the development of this chapter, and the anonymous reviewers who helped considerably in its improvement.

References

[1] J. Bi and K. Bennett. Regression Error Characteristic Curves. In *Proceedings of the Twentieth International Conference on Machine Learning (ICML-2003)*, pages 43–50, Washington DC, 2003.

[2] C. Bishop. *Neural Networks for Pattern Recognition*. Oxford University Press, 1995.

[3] C. Bishop. *Neural Networks for Pattern Recognition*. Clarendon Press, Oxford, 1995.

[4] C. Blake and C. Merz. UCI repository of machine learning databases, 1998. URL http://www.ics.uci.edu/~mlearn/MLRepository.html.

[5] R. O. Duda, P. E. Hart, and D. G. Stork. *Pattern Classification*. Wiley-Interscience Publication, 2000.

[6] R. Everson and J. Fieldsend. Multi-class roc analysis from a multi-objective optimisation perspective. *Pattern Recognition Letters*, 2006.

[7] R. Everson and J. Fieldsend. Multi-objective optimization of safety related systems: An application to short term conflict alert. *IEEE Transactions on Evolutionary Computation*, 10(2):187–198, 2006.

[8] J. Fieldsend. A short note on the efficient random sampling of the multi-dimensional pyramid between a simplex and the origin lying in the unit hypercube. Technical Report 419, Department of Computer Science, University of Exeter, August 2005.

[9] J. Fieldsend. Regression error characteristic optimisation of non-linear models. In Y. Jin, editor, *Multi-Objective Machine Learning*, volume 16 of *Studies in Computational Intelligence*, pages 103–123. Springer, 2006.

[10] J. Fieldsend, T. Bailey, R. Everson, W. Krzanowski, D. Partridge, and V. Schetinin. Bayesian inductively learned modules for safety critical systems. In *Proceedings of the 35th Symposium on the Interface: Computing Science and Statistics*, pages 110–125, 2003.

[11] J. Fieldsend and R. Everson. Formulation and comparison of multi-class roc surfaces. In *Proceedings of the 2nd ROCML workshop, part of the 22nd International Conference on Machine Learning (ICML 2005)*, pages 41–48, 2005.

[12] J. Fieldsend and R. Everson. Multi-objective optimisation in the presence of uncertainty. In *Proceedings of the 2005 IEEE Congress on Evolutionary Computation (CEC'05)*, pages 476–483, 2005.

[13] J. Fieldsend and S. Singh. Pareto multi-objective non-linear regression modelling to aid capm analogous forecasting. In *Proceedings of the IEEE International Joint Conference on Neural Networks*, pages 388–393, 2002.

[14] J. Fieldsend and S. Singh. Optimizing forecast model complexity using multi-objective evolutionary algorithms. In C. Coello and G. Lamont, editors, *Applications of Multi-Objective Evolutionary Algorithms*, pages 675–700. World Scientific, 2004.

[15] J. Fieldsend and S. Singh. Pareto evolutionary neural networks. *IEEE Transactions on Neural Networks*, 16(2):338–354, 2005.

[16] C. Goh and K. Tan. Noise handling in evolutionary multi-objective optimization. In *roceedings of the 2006 IEEE Congress on Evolutionary Computation (CEC'06)*, 2006.

[17] L. Gräning, Y. Jin, and B. Sendhoff. Generalization improvement in multi-objective learning. In *2006 International Joint Conference on Neural Networks*, pages 9893–9900, 2006.

[18] D. Hand and R. Till. A simple generalisation of the area under the ROC curve for multiple class classification problems. *Machine Learning*, 45:171–186, 2001.

[19] J. Hanley and B. McNeil. The meaning and use of the area under a receiver operating characteristic (ROC) curve. *Radiology*, 82(143):29–36, 1982.

[20] C. Holmes and N. Adams. A probabilistic nearest neighbour method for statistical pattern recognition. *J. Royal Statistical Society B*, 64:1–12, 2002.

[21] E. Hughes. Evolutionary multi-objective ranking with uncertainty and noise. In *Evolutionary Multi-Criterion Optimization, EMO 2001*, LNCS 1993, pages 329–342, 2001.

[22] Y. Jin, editor. *Multi-objective Machine Learning*. Springer, 2006.

[23] Y. Jin, T. Okabe, and B. Sendhoff. Evolutionary multi-objective optimization approach to constructing neural network ensembles for regression. In C. Coello and G. Lamont, editors, *Applications of Multi-Objective Evolutionary Algorithms*, pages 653–673. World Scientific, 2004.

[24] Y. LeCun, J. Denker, S. Solla, R. Howard, and L. Jackel. Optimal brain damage. In D. Touretzky, editor, *Advances in neural information processing systems II*. Morgan Kaufmann, 1990.

[25] D. J. C. MacKay. Bayesian interpolation. *Neural Computation*, 4(3):415–447, 1992.

[26] D. J. C. MacKay. A practical Bayesian framework for backpropagation networks. *Neural Compuation*, 4(3):448–472, 1992.

[27] I. Nabney. *Netlab: Algorithms for Pattern Recognition*. Springer-Verlag, 2001.

[28] G. Pappa, A. Freitas, and C. Kaestner. Multi-objective algorithms for attribute selection in data mining. In C. Coello and G. Lamont, editors, *Applications of Multi-Objective Evolutionary Algorithms*, pages 603–626. World Scientific, 2004.

[29] F. Schlottmann and D. Seese. Financial applications of multi-objective evolutionary algorithms: recent developments and future research directions. In

C. Coello and G. Lamont, editors, *Applications of Multi-Objective Evolutionary Algorithms*, pages 627–652. World Scientific, 2004.

[30] J. Teich. Pareto-front exploration with uncertain objectives. In *Evolutionary Multi-Criterion Optimization, EMO 2001*, LNCS 1993, pages 314–328, 2001.

[31] J. Utans and J. Moody. Selecting neural network architectures via the prediction risk: application to corporate bond rating prediction. In *Proceedings of the First International Conference on AI applications on Wall Street*, pages 35–41, 1991.

[32] D. Wolpert. On bias plus variance. *Neural Computation*, 9(6):1211–1243, 1997.

[33] E. Zitzler. *Evolutionary Algorithms for Multiobjective Optimization: Methods and Applications*. PhD thesis, Swiss Federal Institute of Technology (ETH), Zurich, Switzerland, 1999.

Reducing Bloat in GP with Multiple Objectives

Stefan Bleuler, Johannes Bader, and Eckart Zitzler

Computer Engineering and Networks Laboratory (TIK), ETH Zurich, Switzerland
[bleuler, bader, zitzler]@tik.ee.ethz.ch

Summary. This chapter investigates the use of multiobjective techniques in genetic programming (GP) in order to evolve compact programs and to reduce the effects caused by bloating. The underlying approach considers the program size as a second, independent objective besides program functionality, and several studies have found this concept to be successful in reducing bloat. Based on one specific algorithm, we demonstrate the principle of multiobjective GP and show how to apply Pareto-based strategies to GP. This approach outperforms four classical strategies to reduce bloat with regard to both convergence speed and size of the produced programs on an even-parity problem. Additionally, we investigate the question of why the Pareto-based strategies can be more effective in reducing bloat than alternative strategies on several test problems. The analysis falsifies the hypothesis that the small but less functional individuals that are kept in the population act as building blocks for larger correct solutions. This leads to the conclusion that the advantages are probably due to the increased diversity in the population.

1 Motivation

The tendency of trees to grow rapidly during a genetic programming (GP) run is well known [16, 26, 2, 6] and may be explained by:

- The bigger trees get, the more code they contain that does not influence the fitness of the individuals. These so-called introns protect the individuals against the destructive effects of the crossover and mutation operators.
- The probability of finding a big tree that achieves a high fitness is greater than of finding a short program with the same behaviour (fitness-causes-bloat theory [19]).
- Removing bigger subtrees is much more likely to destroy the program than removing shorter ones, which leads to a bias for the preservation of long programs (removal-bias theory [25]).

This phenomenon, which is denoted as *bloating*, leads to several problems:

- Trees can grow quadratically [18]; this leads not only to excessive use of CPU time and memory but also makes the evaluation of trees infeasible.
- Smaller solutions usually generalize the training data better than bigger ones [2].
- When trees start to grow rapidly, so does the fraction of the tree constituted of introns. The recombination of individuals therefore usually comprises an exchange of introns, and the fitness of the population does not improve anymore; the GP run stagnates with high probability [2]. Moreover, when the system is bloating, the recombination of individuals may have no effect since introns will be usually exchanged, and this may lead to stagnation.

Therefore, normally at least, an upper limit for the program size is set manually. Several other strategies have been developed to address the problem of bloating, which can roughly be divided into two classes:

- Methods that modify the program structure and/or the genetic operators in order to remove or reduce the factors that cause bloat. Some examples are Automatically Defined Functions (ADFs) [17], Explicitly Defined Introns (EDIs) [2] and Deleting Crossover [5].
- Techniques that incorporate the program size as an additional factor in the selection process, e.g., as a constraint (size limitation) or as a penalty term (Parsimony Pressure [26]).

When bloating occurs, combinations of the different approaches are possible. Nevertheless, both types have certain disadvantages. For methods of the first class, e.g., ADF, EDI or Deleting Crossover, usually knowledge of how the program structure and the genetic operators interact with the effect of bloating is required. A difficulty with some methods of the second class is to optimally set the parameters associated with them, e.g., by choosing an appropriate parsimony factor when applying Constant Parsimony Pressure [26].

Pareto-based methods belong to the second class and have two advantages: They do not rely on problem knowledge and they do not require additional parameters to be set. The idea is to consider the program size as a second, independent objective besides program functionality and apply a Pareto-based method to the resulting bi-objective problem. The algorithm will then always prefer the smaller of two equally performing programs. As an additional side effect, both small but less functional and large but more complete programs will be kept in the population during the evolution. This basic strategy has proved to successfully reduce bloat in several studies [4, 9, 7, 11, 21, 3, 15]. However, it is still an open question why keeping many non-functional small individuals in the population helps in finding small and correct solutions quickly rather than distracting the search. A potential explanation is the increased diversity in the population. Alternatively, small individuals may be partial solutions to the problem, which can be combined by recombination into compact full solutions, thus acting as building blocks.

Using the method proposed in [4] as an example, the present chapter (i) describes how Pareto-based optimization methods can be applied to reduce code growth in GP and (ii) investigates what mechanisms make these methods effective. In the course of this chapter, we briefly discuss various traditional methods against bloat (Section 2.1), give an overview of alternative multiobjective approaches to fight bloat (Section 2.2), describe a particular method in detail (Section 3) and compare it to traditional techniques (Section 3.2), and finally investigate possible reasons for its effectiveness (Section 4).

2 Overview of Existing Approaches

2.1 Traditional Approaches to Reduce Bloat

Towards the end of a GP run introns grow rapidly and comprise almost all of the code while the optimization process stagnates (no fitness improvement anymore) [2]. Thus, the question is why simulated evolution favours programs with large sections of non-functional code over smaller solutions.

One explanation is that GP crossover is inhomogeneous, i.e., it does not exchange code fragments that have the same functionality in both parents. Therefore, crossover most often reduces the fitness of offspring relative to their parents by disrupting valuable code segments or placing them in a different context. Because crossover points are chosen randomly within an individual, the risk of disrupting blocks of functional code can be reduced substantially by adding introns. To keep this process from using too many machine resources, normally a limit on the tree depth or number of nodes is set manually; when an offspring individual exceeds this limit, one of its parents is added to the population instead. However, setting a reasonable limit is difficult. If the limit is too low, GP might not be able to find a solution. If it is too high, the evolution process will slow down because of the immense resource usage, and the chances of finding small solutions are very low. In the following, this setup will be named *Standard GP*. Here, the fitness F_i of individual i is defined as the error E_i of an individual's output compared to the correct solution

$$F_i = E_i,$$

where F_i is to be minimized.

Another obvious mechanism for limiting code size is to penalize larger programs by adding a size-dependent term to their fitness; this is called *Constant Parsimony Pressure* [5, 26]. The fitness of an individual i is calculated by adding the number of edges N_i, weighted with a parsimony factor α, to the regular fitness:

$$F_i = E_i + \alpha \cdot N_i$$

Soule and Foster [26] report that in some runs parsimony pressure drives the entire population to the minimum possible size. With a higher parsimony pressure the probability of a run suffering from this effect increases. This results in a lower probability of finding good solutions.

A third approach to tackling bloat is to optimize the functionality first and the size afterwards [12]. The formula for the fitness of an individual i depends on its own performance. An additional parameter ϵ comes into play; ϵ is the maximum acceptable error and can be set to zero for discrete problems. For fitness assignment the population is divided into two groups:

1. The individuals that have not yet reached an error equal to or smaller than ϵ get a fitness according to their error E_i without any pressure on the size:

$$F_i = E_i + 1 \text{ if } E_i > \epsilon$$

2. The fitness of individuals that have reached an error equal to or smaller than ϵ. The new fitness is calculated using the size N_i of individual i:

$$F_i = 1 - \frac{1}{N_i} \text{ if } E_i \leq \epsilon.$$

An individual with a large tree size will get a fitness near 1 while one with a large tree size will have a fitness closer to 0.

One advantage of this method is that the GP can find good solutions without being hampered since pressure on size is not applied until the individual has already reached the aspired-for performance. In runs where no acceptable solution is found, bloating will continue. Therefore it is useful to additionally set an upper limit on the tree size. In the following we will call this setup *Two Stage* for to the two stages of fitness evaluation.

Similar to this concept is a strategy called *Adaptive Parsimony Pressure*. Zhang and Mühlenbein have proposed an algorithm that varies the parsimony factor α during the optimization process [28]:

$$F_i(g) = E_i(g) + \alpha(g) \cdot C_i(g).$$

$C_i(g)$ stands for the complexity of individual i at generation g. The complexity can be defined in several ways [28], e.g., as the number of nodes in a tree or as normalized size obtained by dividing the individual's size by the maximum size in the population [5]. In contrast to the Two Stage strategy, the fitness function does not depend on the individual's performance but on the best performance in the population at generation g. The parsimony pressure used to calculate the fitness in generation g is increased substantially if the best individual in generation $g-1$ has reached an error below the threshold ϵ:

$$\alpha(g) = \begin{cases} \frac{1}{T^2} \cdot \frac{E_{best}(g-1)}{\hat{C}_{best}(g)} & \text{if } E_{best}(g-1) > \epsilon \\ \frac{1}{T^2} \cdot \frac{1}{E_{best}(g-1) \cdot \hat{C}_{best}(g)} & \text{otherwise.} \end{cases}$$

E_{best} is the error of the best performing individual in the population, T denotes the size of the training set and $\hat{C}_{best}(g)$ is the expected complexity of the best program in the next generation:

$$\hat{C}_{best}(g+1) = C_{best}(g) + \Delta C_{sum}(g),$$

where C_{best} stands for the complexity of the best performing individual in the population and $\Delta C_{sum}(g)$ is recursively defined as

$$\Delta C_{sum}(g) = \frac{1}{2} \left(C_{best}(g) - C_{best}(g-1) + \Delta C_{sum}(g-1) \right)$$

with the following starting value

$$\Delta C_{sum}(0) = 0.$$

The only parameter that has to be set manually is ϵ. Blickle [5] has reported results superior to those of Constant Parsimony Pressure when applying Adaptive Parsimony Pressure to a continuous regression problem, and equal results to those of Constant Parsimony Pressure when using it on a discrete problem.

In summary we can state that traditional methods aggregate the program function and program size in terms of one objective and fix a trade-off between these two criteria by means of a user-defined parameter.

2.2 Multiobjective Approaches

Using Size as Second Objective

Naturally, most optimization problems involve multiple, conflicting objectives which cannot be optimized simultaneously. This type of problem is often tackled by transforming the optimization criteria into a single objective which is then optimized using an appropriate single-objective method. The same is usually done when trying to address the phenomenon of bloat in GP by modifying the fitness evaluation or the selection process. Actually, there are two objectives: i) the functionality of a program and ii) the code size. While the second objective is traditionally converted into a constraint by limiting the size of a program, controlling the code size by adding a penalty term (Parsimony Pressure) corresponds to weighted-sum aggregation. Ranking the objectives, i.e., optimizing the functionality first and the size afterwards (Two Stage strategy), introduces a hierarchy on the objectives which in turn defines a total preorder on the search spaces.

Alternatively, Pareto-based methods can be applied by considering program functionality and program size as independent objectives. In this approach, small but functionally poor programs can coexist with large but good (in terms of functionality) programs, which in turn maintains population diversity during the entire run. It is important to note that only fitness assignment and selection is changed when switching from single-objective to multiobjective GP, while the other operators like mutation and recombination are not influenced. In 2001, three publications independently proposed the idea of using multiobjective methods for reducing bloat in GP, as first mentioned in [23] but not investigated in detail, and showed promising results [4, 9, 7]. In the following years additional studies have successfully used Pareto-based methods for bloat reduction [11, 21, 3, 15]. The remainder of this section summarizes these approaches and the key results of the respective studies. The method proposed in [4] serves as the basis for the analysis presented in this chapter.

Nondomination Tournament [9]

The authors propose a simple selection operator based on nondomination. In this scheme, the selection of one individual works as follows: A comparison set is randomly picked from the population and then candidate solutions are randomly chosen until one is found that is not dominated by any member of the comparison set; this individual is selected. To prevent the method from converging on small but non-functional programs an additional bias towards larger solutions is included in the domination criterion. Two different possibilities are compared: i) Using epsilon dominance on the program size, i.e., depending on the fitness f an individual i may dominate individual j even when it is larger than j ($f_i < f_j$ and $s_i < s_j + \varepsilon$). ii) Redefining the size objective such that it equals a threshold value for all trees that are smaller than this limit.

The approach is compared to standard GP on three symbolic regression problems and on the multiplexer problem. It is demonstrated that much smaller solutions can be found that are of similar quality. The number of fitness evaluations are similar, but due to the smaller average program size the running times are substantially smaller for the multiobjective method. Additionally, the preference for the different variants of the size bias changes with the problem.

FOCUS [7, 11]

Selection in the FOCUS algorithm works by discarding all weakly dominated individuals in the population. In order to promote diversity within the population, a diversity measure is used as a third objective in the optimization. However, unlike in most multiobjective EAs, diversity is measured in the parameter space. A distance measure for GP trees is defined and the average distance to the other members of the population is used as third objective function in the evaluation of a program.

This method is compared to standard GP on three instances of the parity bit problem. The experimental results show that some kind of diversity maintenance is necessary to keep the population from converging to small but non-functional trees. Including the diversity objective, FOCUS was, using fewer function evaluations, able to find correct solutions that are much smaller than those found by standard GP.

POPE-GP [3]

Another study uses NSGA-II [8] as selection operator in another overall method named POPE-GP [3]. Like all state-of-the-art evolutionary multiobjective optimization algorithms NSGA-II employs a diversity mechanism to distribute the individuals in the objective space. This eliminates the problem of convergence to small but non-functional programs. In an assessment on a classification problem, this approach yielded smaller programs with superior generalization compared to standard GP.

Biased Multiobjective Parsimony [21]

The authors argue that in most cases it is much easier for GP to find small non-functional programs than large correct ones, which is the reason why the evolution process can converge to small, non-functional individuals. To avoid this, a bias towards larger solutions is introduced. In contrast to the bias in [9], the user does not specify a target size but rather the relative importance of the two objectives. This is achieved by performing a tournament in which individuals are either compared based on their fitness (program functionality) only or based on their nondominated sorting rank [8]. By setting the probability p for using the fitness as criterion in the tournament, one can adjust the influence of the program size in selection, i.e., higher values of p lead to higher parsimony pressure.

The study uses four classical GP test problems (artificial ant, symbolic regression, multiplexer and even-parity) for comparing the proposed approach to standard GP and to two other single-objective strategies presented in the same publication. The results show that biased multiobjective parsimony is able to reduce the size of the solutions significantly for all problems in comparison to standard GP. But for higher parsimony pressure, which generates a highly significant reduction in program size, the fitness values start to increase compared to standard GP, i.e., the program functionality decreases. Consequently, the parsimony pressure must be carefully chosen to achieve a reduction while maintaining the quality of the solutions. The two single-objective strategies perform similarly to the multiobjective selection.

Other Related Approaches

The method presented in [15] is closely related to [9]. Here the selection consists of picking a random set of individuals and selecting all the nondominated solutions from this set. Instead of replacing the selection operator with a multiobjective version as in most other approaches, Smits et al. [24] propose maintaining an external set of nondominated solutions and adapting the crossover operator to recombine one individual from the normal population with one individual from the nondominated set.

Methods like those presented in this section have successfully been used in several applications [27, 14, 29, 22]. And the same basic idea has found use in areas other than GP. In evolutionary design of classifiers, for example, the same problem with parsimony exists and the multiobjective methods presented for bloat in GP have effectively been applied [20, 10].

Summary of Multiobjective Approaches

Summarizing the studies discussed above, one can state that considering program size as a second objective beside program functionality and applying a Pareto-based optimization method has been highly successful in reducing bloat compared to standard GP. If a pure dominance-based fitness assignment scheme is used, this may lead to convergence to small, non-functional programs, thereby reducing the chance of finding a high-quality solution. The reason is that many more small programs exist than functional ones. Two basic strategies have been proposed to eliminate this problem: i) the introduction of a bias against small programs [9, 21], which gives rise to the difficult problem of correctly setting this bias or ii) the enforcement of diversity in the population with respect to either the parameter space [7, 11] or the objective space [4, 3].

3 A Multiobjective Approach to Reduce Bloat in Detail

This section describes how to apply a multiobjective approach to reducing bloat based on the example of the method presented in [4]. Additionally, it provides an empirical comparison of the multiobjective approach to four alternative strategies for reducing bloat.

3.1 Algorithm

The approach proposed in [4] uses an improved version of the Strength Pareto Evolutionary Algorithm (SPEA) for multiobjective optimization proposed in [32]. Besides the population, SPEA maintains an external set of individuals (archive) which contains the nondominated solutions among all solutions considered so far. The variant implemented here differs from the original SPEA only in the fitness assignment. In SPEA the fitness of an individual in the population depends on the "strengths" of the individual's dominators in the external set, but is independent of the number of solutions this individual dominates or is dominated by within the population. The potential problem arising with this scheme is illustrated in Figure 1. The Pareto-optimal front consists of only four solutions and the second dimension is highly discretized (as is the case for the application considered in Section 3.2, cf. Figure 11). As a consequence, the population is divided into four fitness classes, i.e., clusters which contain solutions having the same fitness. The fitness values only

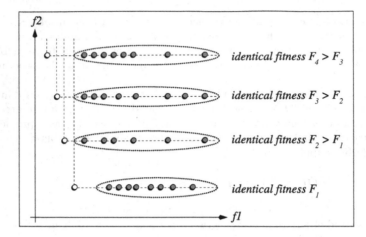

Fig. 1. A problematic situation with the original SPEA fitness assignment scheme in the case of a highly discretized objective space. The white points represent members of the external set while the grey points stand for individuals in the population

among clusters vary, not within clusters. Thereby the selection pressure towards the Pareto-optimal front is reduced substantially and may slow down the evolution process.

To avoid this situation, with the present algorithm both dominating and dominated solutions are taken into account for each individual. In detail, each individual i in the external set $\overline{P}$ *and* the population P is assigned a real value $S(i)$, its strength, representing the number of solutions it dominates:

$$S(i) = |\{j \mid j \in P + \overline{P} \wedge i \succeq j\}|,$$

where $|\cdot|$ denotes the cardinality of a set, $+$ stands for multiset union and the symbol $\succeq$ corresponds to the relation of weak Pareto dominance.[1] The strength of an individual is greater than or equal to 1 as each individual weakly dominates itself. Finally, the fitness $F(i)$ of individual i is calculated on the basis of the following formula:

$$F(i) = \sum_{j \succeq i} S(j).$$

That is, the fitness is determined by the strengths of its dominators. Note again that each individual weakly dominates itself and thus $F(i) \geq S(i)$. In contrast to SPEA, there is no distinction between members of the external set and population members.

It is important to note that fitness is to be minimized here, i.e., low fitness values correspond to high reproduction probabilities. The best fitness value is 1, which means that an individual is neither (weakly) dominated by any other individual nor (weakly) dominates another individual. A low fitness value is assigned to those individuals which

---

[1] A solution weakly dominates another solution if and only if it is not worse in any objective.

i) dominate only few individuals and

ii) are dominated by only few individuals (which in turn dominate only few individuals).

Therefore, not only is the search guided towards the Pareto-optimal front but also a niching mechanism is incorporated based on the concept of Pareto dominance. This enhances population diversity with respect to the objective space and successfully avoids convergence to small but non-functional programs, as will be demonstrated in Section 3.2.

For details of the SPEA implementation we refer the reader to [30]. The clustering procedure is not needed in this study because the size of the external set is unrestricted due to the small number of nondominated solutions emerging with the considered test problem.[2]

3.2 Experiments

In the following, we compare five methods — Standard GP, Constant Parsimony, Adaptive Parsimony, Two Stage and the SPEA variant — by evolving even-parity functions of different arities.

Methodology

The *even-parity* function was chosen because it is commonly used as a GP test problem [16, 26] and the complexity (arity = number of inputs) can be easily adapted to either the available machine resources or the performance of an algorithm. The Boolean *even-k-parity function* of k Boolean arguments returns TRUE if an even number of its Boolean arguments are TRUE, and otherwise returns NIL.

Parity functions are often used to check the accuracy of stored or transmitted binary data in computers because a change in the value of any one of its arguments toggles the value of the function. Because of this sensitivity to its inputs, the parity function is difficult to learn [17]. The training set consist of all 2^k possible input combinations. The error of an individual is measured as the number of input cases for which it did not provide the correct output value. A correct solution to the even-k-parity function is found when the error equals zero. We will call a run successful if it found at least one correct solution. For each setup 100 runs have been performed, and, in the following, usually the average values over 100 runs are reported. If not stated differently, the even-5-parity problem was used. Additionally, in a few runs even-parity functions of higher arities have been evolved.

Parameter Settings

After some test runs with Standard GP we decided to use a population size of 4,000 and a maximum of 200 generations; this setup performed best of all, keeping the product *Generations* $*$ *Popsize* $= 800,000$ constant. All runs were processed up to generation 200, even if they found a correct program before generation 200. We

---

[2] The technique used here is a slight variation of the method later proposed under the name SPEA2 [31] which contains further improvements over SPEA.

set the initial depth for newly created trees to five and, in addition, restricted the maximum allowed depth of trees to 20, which is by far enough to generate correct solutions. It is important to note that only Standard GP and Two Stage runs (if no pressure is applied because no correct solution has been found) are affected by this limit. The other methods manage to keep the tree size so small that no significant part of the population reaches tree depths close to the limit.

The terminal set consists of all inputs $d_0, d_1, ..., d_{k-1}$ to the even-k-parity function. No numerical constants have been used. The function set consists of the following four Boolean functions: $\{AND, OR, IF, NOT\}$. Note that using the same function set without IF makes the task of evolving an even-parity function considerably more difficult. Preliminary tests for Constant Parsimony with different parsimony pressures of 0.001, 0.01, 0.1 and 0.2 showed the best results for $\alpha = 0.01$. This value has been used in all following Constant Parsimony runs.

For Adaptive Parsimony several settings from [5] have been used: The maximum acceptable error ϵ was set to 0.02. $E_i(g)$ was normalized with the maximum possible error. The best error that can be achieved is $E_i(g) = 0$. $C_i(g)$ was defined as the size $N_i(g)$ of an individual i normalized with the maximum size in population $N_{max}(g)$. In order to be able to use the formula given in Section 2 a constant $c = 0.01$ was added to the error measure.

Table 1 summarizes the parameters used for all runs (if not stated differently).

Table 1. Global parameter setting

Population size	4000
Generations	200
Maximum depth	$D_{max} = 20$
Maximum initial depth	$D_{initial} = 5$
Probability of crossover	$p_c = 0.9$
Probability of mutation	$p_m = 0.1$
Tournamentsize	$T = 7$
Reproduction method	Tournament
Function set	$\{AND, OR, IF, NOT\}$
Terminal set	$d_0, d_1, ..., d_{k-1}$
Constant Parsimony Pressure	$\alpha = 0.01$
Threshold (for Adaptive Pars.)	$\epsilon = 0.02$

Results

As expected, all methods have been able to find correct solutions in most of the 100 runs. Table 2 shows the percentage of successful runs, i.e., runs that found at least one correct solution within 200 generations. Two Stage and Standard GP have the same probability of solving the test problem since the fitness function is the same for both unless the concerned individual in Two Stage already represents a correct solution.

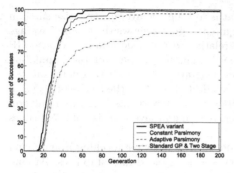

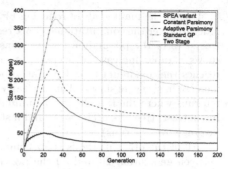

Fig. 2. Comparison of the success rates for the different methods relative to the generations. 100% means that all of the 100 runs found a solution before or in this generation

Fig. 3. Average tree size, mean of 100 runs per method

More information about how fast a method finds correct solutions can be obtained by calculating the probability of a run finding a correct solution within the first k generations. It is computed by summing over the runs that have found a correct solution by generation k. This probability is shown in Figure 2. It is interesting that all methods have found correct solutions before generation 20 in some runs. For all methods the probability of finding the first correct solution in the second half of the run is low. Increasing the arity of the even-parity function from 5 to 7 makes the problem much harder to solve. With an even-7-parity function, Standard GP did not produce one correct solution within 31 runs of 200 generations each. Parsimony was successful in ten and the SPEA variant in 22, out of 31 runs. This shows that keeping smaller trees in the population not only reduces the computational effort but also improves chances of solving the problem. For the even-9-parity function, the SPEA variant was successful within 500 generations in 17 out of 31 runs, and Constant Parsimony in 4 out of 31.

Table 2. Results compared for Standard GP, Two Stage, Constant Parsimony, Adaptive Parsimony and the SPEA variant

Method	Success Rate [%]	Smallest Av. Size	Mean Av. Size	Largest Av. Size
Standard GP	84	324.0	643.2	1701.8
Constant Pars.	100	26.2	52.3	106.9
Adaptive Pars.	99	23.0	87.1	714.9
Two Stage	84	25.7	170.1	867.6
SPEA variant	99	16.8	21.7	37.1

One of the main goals of reducing bloat is to keep the average tree size small in order to lower the computational effort required. Figure 3 shows the mean of average tree sizes in the population for 100 runs relative to the generation. Standard GP shows a rapid increase of average size until a significant part of the population reaches the maximum tree depth at about generation 20. From this point on, the increase in size gets slower. This is clearly an effect of limiting the tree depth. Out of ten runs where the tree depth was unlimited, none showed this saturation pattern. In contrast, tree size grew faster and faster, reaching an average size of 9,764 edges (average over 10 runs).

All of the other methods show common behaviour. After reaching a maximum between generation 20 and 30 the average size is reduced and stabilizes. Around the time when the average size reaches a maximum, the average error reaches a minimum. Maybe it is the general behaviour of algorithms that somehow favour small solutions, at least for discrete problems. An improvement in functionality is first achieved by a large individual and is followed by smaller programs with the same error. At the beginning of a run, when the average error is high, it is easy for evolution to improve functionality and the reduction of the average error is fast. The reduction in size mainly takes place when a lot of individuals have the same fitness. While fitness is changing fast this is not the case. Parsimony pressure with an α of 0.01, for example, mainly distinguishes between programs of equal performance. An individual may be 100 nodes larger than another and compensate for this with classifying only one additional test case correctly. Further investigations would be needed to justify the previously mentioned assumption.

Of more practical relevance is the fact that although the average size development shows a similar pattern for Two Stage, Constant Parsimony, Adaptive Parsimony and the SPEA variant the absolute values differ very much. As can be seen in Figure 3, the proposed SPEA variant has by far the smallest average size throughout the whole run. In generation 200 the average number of edges is down to 21.7; this is less than half of the second smallest average size which was attained by Constant Parsimony. Another important aspect is the range between the highest and the lowest final average size within all runs for one method. Table 2 lists the highest and the lowest final average size that occurred in 100 runs. For the SPEA variant the final average sizes vary only very little. At the other extreme is Two Stage. Some of the Two Stage runs never found a correct solution and therefore never experienced any pressure on tree size. These runs are exactly like Standard GP runs. Adaptive Parsimony performed considerably worse than Constant Parsimony (unlike in [5], where Adaptive Parsimony and Constant Parsimony achieved equal performance), and its final average sizes fell into a large range.

The second main goal when using methods against bloat is to retrieve compact solutions. The question is whether methods that keep the average tree size in the population low also produce small correct solutions. Figures 4 to 8 show a bar for each run. The height of the bar corresponds to the size of the smallest correct solution that was found during the whole run. If no correct solution was found there is no corresponding bar. For calculating the mean and median value only successful runs have been taken into account. It is shown that methods with low average tree sizes like the SPEA variant and Constant Parsimony were not only able to produce correct solutions but also found more compact solutions than methods with a larger average tree size. The average size of the smallest solutions for the SPEA variant is 21.1, which is close to the minimal possible tree size (17) for a solution to the even-5-parity function using the given function set. This ideal solution was found in 22 runs. Every successful run found compact solutions; even the worst run found a solution of size 38. Although Constant Parsimony has a high probability of finding correct solutions within 200 generations, the size of the smallest solutions varies in a wide range. Once again the results of Adaptive Parsimony are worse than those of Constant Parsimony. Especially, the range of the sizes of the smallest solutions is larger with Adaptive Parsimony Pressure.

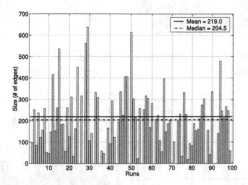

Fig. 4. Standard GP, size of the smallest correct solution

Some insight into why the SPEA variant is more successful than Constant Parsimony can be gained by looking at the distribution of the population in the (size, error)-plane. Figures 9 to 12 show the distribution of the population at generation 30 and 200 both for one representative run of the SPEA variant and one Constant Parsimony run. Each dot in the diagram represents one individual. The two runs for the SPEA variant and Constant Parsimony have been started with the same initial population. While the SPEA variant keeps a set of small individuals with different errors in the population during the whole run, Constant Parsimony moves the entire population towards lower errors and larger sizes. Around generation 30, when the average size reaches a maximum value and the average error a minimum value, parsimony pressure becomes effective and the population is moved back towards smaller sizes. The only small programs that are constantly kept in the population have an error of 16. Into this category also falls the smallest possible program that results from returning one input to the output. It is possible that in the variety of small trees that can be found in populations of the SPEA variant at all stages of the evolution, good building blocks for correct solutions are present.

4 Investigating the Mechanisms of Multiobjective Bloat Reduction

As demonstrated in the previous section and by all the studies described in Section 2.2, Pareto-based multiobjective optimization is successful at reducing bloat.

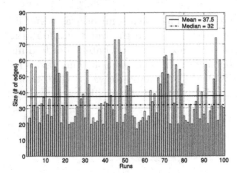

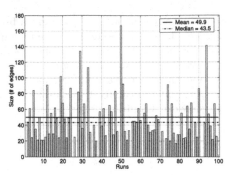

Fig. 5. Constant Parsimony, size of the smallest correct solution

Fig. 6. Two Stage, size of the smallest correct solution

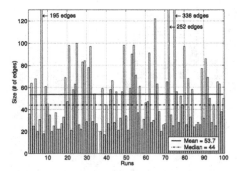

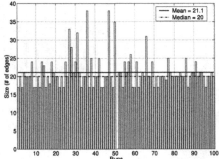

Fig. 7. Adaptive Parsimony, size of the smallest correct solution

Fig. 8. SPEA variant, size of the smallest correct solution

While it is intuitive that maintaining a selection pressure towards smaller programs reduces bloat, it is not obvious why the multiobjective approach is particularly effective compared to alternative strategies. The algorithm used in Section 3 maintains a large portion of small non-functional programs in the population, as no preference for any of the two objectives size, and functionality is applied. This strategy seems to enhance the identification of a compact and correct solution rather than distracting the search algorithm as one would assume. In the following, a hypothesis concerning the cause of the observed behaviour will be presented and analysed.

4.1 Hypotheses

Figure 13 shows the hypothesized minimal-size solution found for the even-5-parity problem with the given operator set. This program is composed of subtrees that are themselves solutions to the parity bit problems for a lower number of inputs and that were often found by the multiobjective approach. This observation, together with the fact that even-parity programs can be obtained by programs for lower a number of bits, leads to the assumption that it was composed of solutions to subproblems by means of recombination. In this scenario, the multiobjective approach may support the existence of such small programs that are not correct solutions but that have relatively good functionality, as they are solutions to subproblems. So, the hypothesis

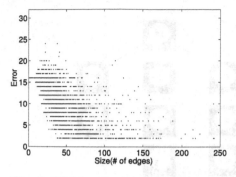

Fig. 9. SPEA variant population at generation 30

Fig. 10. Constant Parsimony population at generation 30

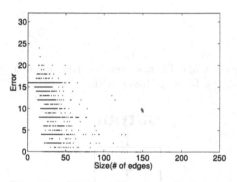
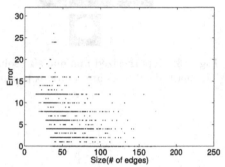

Fig. 11. SPEA variant population at generation 200

Fig. 12. Constant Parsimony population at generation 200

is that the small programs kept in the population act as building blocks for compact and correct solutions.

Alternatively, the advantage of the multiobjective optimization may be based on the increased genetic diversity. The existence of this effect was demonstrated in an empirical study in [1] where single-objective optimization problems were solved by adding objectives and applying a Pareto-based method. A closely related idea is that adding objectives makes a problem easier by removing local optima [13].

If the building block hypothesis describes the dominating factor leading to bloat reduction, the following can be expected to hold:

- The fitness assignment should be able to discriminate between small solutions that are solutions to subproblems or building blocks and random programs of the same size. If this is not the case, the EA will not prefer building blocks over random solutions.
- Through recombinations of solutions to subproblems, the functionality of offspring programs should often be largely better than the parents' functionality.
- Switching off recombination should strongly reduce the effectiveness of the multiobjective method.

In the following, the experimental results on multiple test problems for these effects will be analysed.

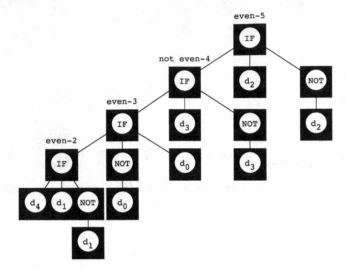

Fig. 13. Hypothesized minimal size solution found for the even-5-parity problem. The marked subtrees are solutions to parity problems of lower arities

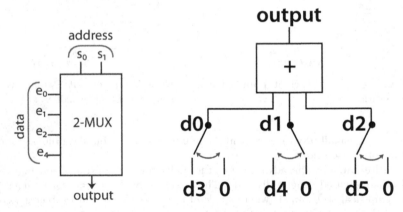

Fig. 14. Multiplexer problem ($k = 6$)   **Fig. 15.** Adder problem ($k = 6$)

4.2 Test Problems

Besides the parity bit problem described in Section 3.2 the following test problems are used for the analysis.

k-Multiplexer

A multiplexer is a device to select one of several data inputs and forward it to the output, cf. Figure 14. The k-multiplexer problem has m binary control inputs and n binary data inputs, where $n = 2^m$ and $k = m + n$. Thus, the feasible values for k are $k = \{3, 6, 11, 20, 37, \ldots\}$

k-Hamming Distance

Here, the task is to calculate the Hamming distance between the first half and the second half of the binary input string of length k. Obviously, k is restricted to being even. This test problem was specifically designed to allow a stepwise buildup of correct programs, since the solutions that calculate the correct Hamming distance for a part of the input string have a relatively high score on the complete problem.

k-Adder

A related test problem is the k-adder where the first $\frac{k}{2}$ bits of the k binary input d_i determine which of the remaining bits are added; cf. Figure 15. Thus, the output is calculated as follows:

$$a = \sum_{i=0}^{k} d_i \cdot d_{i+\frac{k}{2}}.$$ (1)

Operators

The same operators as for the parity bit have been used for the multiplexer problem, namely $NOT, OR, AND,$ and IF. For the Hamming distance and the adder an additional binary plus operator $+$ was introduced.

4.3 Results

This section tests the building block hypothesis described above by analysing experimental results for the different effects that should be observed if the hypothesis holds.

Fitness Discrimination of Small Programs

For the population to contain a significant number of small programs that are solutions to subproblems, they must exhibit better fitness values than random programs of the same size, e.g., when scored on the even-5-parity problem, a solution to the even-3-parity problem should be preferred to a random program of the same size. One characteristic of solutions to subproblems is that they do not use all available inputs. Accordingly, the possible fitness ranges for programs that do not use all of the provided inputs (as do solutions to subproblems) are plotted in Figures 16–19. On the k-parity problem, all solutions that use less than k inputs have equal fitness as they provide the correct result to exactly half of the test cases. Thus, solutions to subproblems or other good building blocks for compact full solutions cannot prevail in the population despite the significant portion of small programs maintained by the Pareto-based selection. Nevertheless, the multiobjective method was highly successful on the parity problems. This gives a first indication against the building block hypothesis. For the other test problems the same problem does not appear, and the Hamming distance problem and the adder problem have been specifically designed such that solutions to subproblems score relatively well on the full problem.

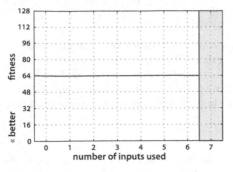

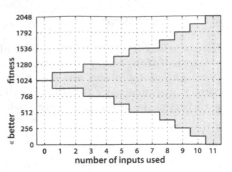

Fig. 16. Fitness ranges for small programs on the even-7-parity problem

Fig. 17. Fitness ranges for small programs on the 11-multiplexer problem

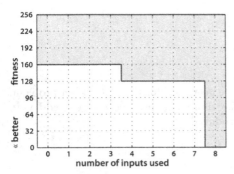

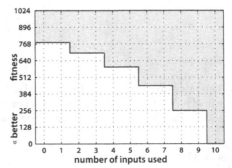

Fig. 18. Fitness ranges for small programs on the 8-Hamming distance problem

Fig. 19. Fitness ranges for small programs on the 10-adder problem

Fitness Changes in Recombination

If the crossover operator successfully combines solutions to subproblems or other building blocks into a full solution, the fitness value of the offspring will be substantially better than the parents' fitness. If such recombinations are a major origin of good solutions, we can expect to often see large fitness increases from parents to offspring. Figures 20–22 show the fitness differences between one parent and one offspring appearing in all recombinations of 25 runs. Large changes in fitness are very rare. This indicates that recombination of building blocks into good programs is extremely rare on our test problems.

Effect of Single-Parent Variation

The hypothesis states that the multiobjective methods maintain more promising building blocks in the population than alternative methods like Constant Parsimony. Thus, its performance should depend more on recombination than that of constant parsimony. Consequently, using single-parent variation, i.e., switching off recombination, should affect the SPEA variant much more than Constant Parsimony. We have tested this on the even-7-parity problem using the parameter settings as described in Table 1, except for the tournament size in Constant Parsimony, which was

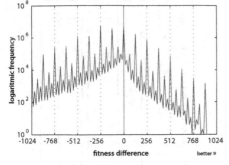

Fig. 20. Fitness differences in recombination on the 11-multiplexer problem

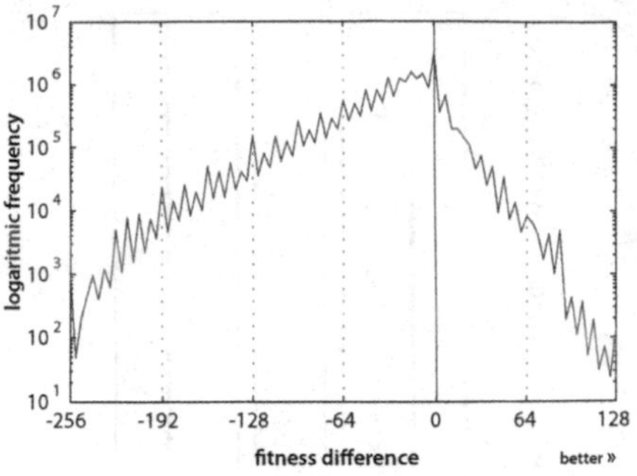

Fig. 21. Fitness differences in recombination on the 8-hamming problem

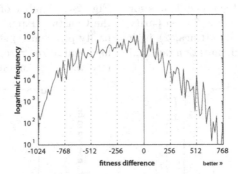

Fig. 22. Fitness differences in recombination on the 10-adder problem

set to 10. In order to compensate for the reduced variation without recombination, the mutation rate was increased to 0.9. Figures 23 and 24 show in how many of the 10 runs a correct solution was found for the three different settings standard ($p_c = 0.9$, $p_m = 0.1$), high mutation ($p_c = 0.9$, $p_m = 0.9$), and no crossover ($p_c = 0$, $p_m = 0.9$). For both methods the number of successful runs is similar and does not change heavily. Another performance indicator is the speed of convergence. Here, both algorithms take much longer to find the first correct solution, as shown in Figures 25 and 26 due to the increased mutation rate, but no significant difference in the influence of crossover exists. Lastly, we compare the sizes of the smallest correct solutions found by the two methods; cf. Figures 27 and 28. Again, the performance is influenced adversely by the increased mutation rate but there is no significant influence of the recombination. In summary, the performance of the SPEA variant is not more dependent on recombination than the performance of Constant Parsimony.

5 Summary

Bloating is a well known problem of variable-length representations, as used in genetic programming, and various strategies have been proposed to address it. A recent

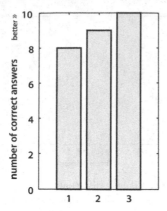

Fig. 23. Number of successful runs for the SPEA variant on the 7-even-parity problem. 1) standard settings, 2) high mutation rate, 3) high mutation rate and no recombination

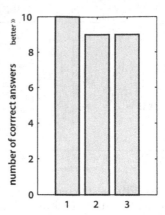

Fig. 24. Number of successful runs for the Constant Parsimony on the 7-even-parity problem. 1) standard settings, 2) high mutation rate, 3) high mutation rate and no recombination

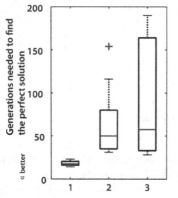

Fig. 25. Generation of the first correct solution for the SPEA variant on the 7-even-parity problem. 1) Standard settings, 2) high mutation rate, 3) high mutation rate and no recombination

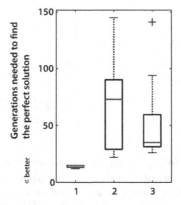

Fig. 26. Generation of the first correct solution for the Constant Parsimony on the 7-even-parity problem. 1) Standard settings, 2) high mutation rate, 3) high mutation rate and no recombination

development is to explicitly use the underlying objectives of program functionality and program size in a multiobjective optimization method. Several variants of this approach have been proposed, all of which successfully reduced code growth compared to standard GP with depth limitation, on a variety of discrete and continuous test problems. Here, we have discussed how to apply Pareto-based multiobjective methods to the problem of bloat on the example of the SPEA variant published in [4]. The experimental validation on the parity-bit test problem showed that this method not only reduces code growth compared to standard GP but also outperforms three alternative methods for bloat control with respect to average size of

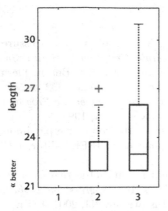
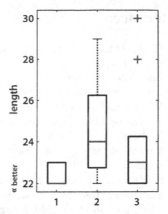
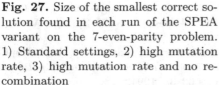

Fig. 27. Size of the smallest correct solution found in each run of the SPEA variant on the 7-even-parity problem. 1) Standard settings, 2) high mutation rate, 3) high mutation rate and no recombination

Fig. 28. Size of the smallest correct solution found in each run of Constant Parsimony on the 7-even-parity problem. 1) Standard settings, 2) high mutation rate, 3) high mutation rate and no recombination

the programs, which is decisive for the overall computational effort, the size of the smallest correct solutions and the best program functionality.

Additionally, we tried to identify the cause for these improvements as it is not obvious why keeping small and non-functional programs in the population can improve the quality of the results rather than distracting the search. We have formulated the hypothesis that small programs may act as building blocks for compact correct solutions. These building blocks could then be combined into compact correct solutions by recombination, whereas alternative methods which do not keep small but non-functional programs in their populations cannot profit from this effect. Several tests revealed evidence against this hypothesis. In particular,

- the multiobjective method is also successful when the fitness of small solutions that may act as building blocks cannot be distinguished from the fitness of random programs of the same size,
- recombination does very rarely leads to the large improvements in fitness that would be expected for successful combinations of building blocks, and
- switching off recombination does not seem to influence the capabilities of the multiobjective approach.

Therefore, we conclude that the positive effects of maintaining small but non-functional programs in the population are mainly due to increased genetic diversity, as described in [1], or the closely related concept of changes in the fitness landscape which induce a lower number of local optima [13]. Additional experiments will be necessary to further verify these conclusions.

One explanation of why maintaining diversity is important with respect to small trees is that the recombination gets more effective the smaller the trees are (as a comparison between Figures 9 and 10 reveals).

References

[1] H. A. Abbass and K. Deb. Searching under Multi-evolutionary Pressures. In C. M. Fonseca et al., editors, *Evolutionary Multi-Criterion Optimization. Second International Conference, EMO 2003*, pages 391–404, Berlin, Germany, 2003. Springer. Lecture Notes in Computer Science. Volume 2632.

[2] W. Banzhaf, F. D. Francone, R. E. Keller, and P. Nordin. *Genetic Programming: An Introduction*. Morgan Kaufmann, San Francisco, CA, 1998.

[3] Y. Bemstein, X. Li, V. Ciesielski, and A. Song. Multiobjective parsimony enforcement for superior generalisation performance. In IEEE, editor, *CEC 04*, pages 83–89, 2004.

[4] S. Bleuler, M. Brack, L. Thiele, and E. Zitzler. Multiobjective Genetic Programming: Reducing Bloat by Using SPEA2. In *Congress on Evolutionary Computation (CEC-2001)*, pages 536–543, Piscataway, NJ, 2001. IEEE.

[5] T. Blickle. Evolving Compact Solutions in Genetic Programming: A Case Study. In H. M. Voigt et al., editors, *PPSN IV*, pages 564–573. Springer-Verlag, 1996.

[6] T. Blickle and L. Thiele. Genetic programming and redundancy. In J. Hopf, editor, *Genetic Algorithms within the Framework of Evolutionary Computation (Workshop at KI-94, Saarbrücken)*, pages 33–38, 1994.

[7] E. D. De Jong, R. A. Watson, and J. B. Pollack. Reducing Bloat and Promoting Diversity using Multi-Objective Methods. In L. Spector et al., editors, *Genetic and Evolutionary Computation Conference (GECCO 2001)*, pages 11–18. Morgan Kaufmann Publishers, 2001.

[8] K. Deb, S. Agrawal, A. Pratap, and T. Meyarivan. A fast elitist non-dominated sorting genetic algorithm for multi-objective optimization: NSGA-II. In M. Schoenauer et al., editors, *Parallel Problem Solving from Nature (PPSN VI)*, Lecture Notes in Computer Science Vol. 1917, pages 849–858. Springer, 2000.

[9] A. Ekárt and S. Z. Németh. Selection Based on the Pareto Nondomination Criterion for Controlling Code Growth in Genetic Programming. *Genetic Programming and Evolvable Machines*, 2:61–73, 2001.

[10] A. Hunter. Expression Inference - Genetic Symbolic Classification Integrated with Non-linear Coefficient Optimisation. In *AISC 02*, LNCS. Springer, 2002.

[11] E. D. D. Jong and J. B. Pollack. Multi-objective methods for tree size control. *Genetic Programming and Evolvable Machines*, 4:211–233, 2003.

[12] T. Kalganova and J. F. Miller. Evolving More Efficient Digital Circuits by Allowing Circuit Layout Evolution and Multi-Objective Fitness. In A. Stoica et al., editors, *Proceedings of the 1st NASA/DoD Workshop on Evolvable Hardware (EH'99)*, pages 54–63, Piscataway, NJ, 1999, 1999. IEEE Computer Society Press.

[13] J. D. Knowles, R. A. Watson, and D. W. Corne. Reducing Local Optima in Single-Objective Problems by Multi-objectivization. In E. Zitzler et al., editors, *Evolutionary Multi-Criterion Optimization (EMO 2001)*, volume 1993 of *Lecture Notes in Computer Science*, pages 269–283, Berlin, 2001. Springer-Verlag.

[14] A. Kordon, E. Jordaan, L. Chew, G. Smits, T. Bruck, K. Haney, and A. Jenings. Biomass Inferential Sensor Based on Ensemble of Models Generated by Genetic Programming. In *GECCO 04*, LNCS, pages 1078–1089. Springer, 2004.

[15] M. Kotanchek, G. Smits, and E. Vladislavleva. Pursuing the Pareto Paradigm Tournaments, Algorithm Variations & Ordinal Optimization. In R. L. Riolo, T. Soule, and B. Worzel, editors, *Genetic Programming Theory and Practice IV*, volume 5 of *Genetic and Evolutionary Computation*, chapter 3. Springer, 2006.

[16] J. R. Koza. *Genetic Programming: On the Programming of Computers by Means of Natural Selection*. MIT Press, Cambridge, MA, USA, 1992.

[17] J. R. Koza. *Genetic Programming II: Automatic Discovery of Reusable Programs*. MIT Press, Cambridge, Massachusetts, 1994.

[18] W. B. Langdon. Quadratic Bloat in Genetic Programming. In D. Whitley et al., editors, *GECCO 2000*, pages 451–458, Las Vegas, Nevada, USA, 10-12 2000. Morgan Kaufmann. ISBN 1-55860-708-0.

[19] W. B. Langdon and R. Poli. Fitness Causes Bloat. In P. K. Chawdhry et al., editors, *Soft Computing in Engineering Design and Manufacturing*, pages 13–22, Godalming, GU7 3DJ, UK, 1997. Springer-Verlag.

[20] X. Llorà, D. E. Goldberg, I. Traus, and E. Bernadó. Accuracy, parsimony, and generality in evolutionary learning systems via multiobjective selection. In *Learning Classifier Systems*, pages 118–142. Springer. Lecture Notes in Artificial Intelligence Vol. 2661, 2002.

[21] L. Panait and S. Luke. Alternative Bloat Control Methods. In *GECCO 04*, LNCS, pages 630–641. Springer, 2004.

[22] D. Parrot, L. Xiandong, and V. Ciesielski. Multi-objective techniques in genetic programming for evolving classifiers. In *CEC 05*, pages 1141–1148. IEEE, 2005.

[23] K. Rodríguez-Vázquez, C. M. Fonseca, and P. J. Fleming. Multiobjective genetic programming: A nonlinear system identification application. In J. R. Koza, editor, *Late Breaking Papers at the 1997 Genetic Programming Conference*, pages 207–212, Stanford University, CA, USA, 13–16 1997. Stanford Bookstore. ISBN 0-18-206995-8.

[24] G. F. Smits and M. Kotanchek. Pareto-Front Exploitation in Symbolic Regression. In *Genetic Programming Theory and Practice II*, volume 8. Springer, 2005.

[25] T. Soule and J. A. Foster. Removal Bias: a New Cause of Code Growth in Tree Based Evolutionary Programming. In *1998 IEEE International Conference on Evolutionary Computation*, pages 781–186, Anchorage, Alaska, USA, 1998. IEEE Press. URL http://citeseer.ist.psu.edu/313655.html.

[26] T. Soule and J. A. Foster. Effects of Code Growth and Parsimony Pressure on Populations in Genetic Programming. *Evoluationary Computation*, 6(4): 293–309, 1999.

[27] M. Streeter and L. A. Becker. Automated Discovery of Numerical Approximation Formulae via Genetic Programming. *Genetic Programming and Evolvable Machines*, 4(3):255–286, 2003.

[28] B.-T. Zhang and H. Mühlenbein. Balancing Accuracy and Parsimony in Genetic Programming. *Evoluationary Computation*, 3(1):17–38, 1995.

[29] Y. Zhang and P. I. Rockett. Evolving optimal feature extraction using multi-objective genetic programming: a methodology and preliminary study on edge detection. In *GECCO 05*, pages 795–802, New York, NY, USA, 2005. ACM Press.

[30] E. Zitzler. *Evolutionary Algorithms for Multiobjective Optimization: Methods and Applications*. PhD thesis, Swiss Federal Institute of Technology (ETH) Zürich, Switzerland, 1999.

[31] E. Zitzler, M. Laumanns, and L. Thiele. SPEA2: Improving the Strength Pareto Evolutionary Algorithm for Multiobjective Optimization. In K. Giannakoglou et al., editors, *Evolutionary Methods for Design, Optimisation and Control with Application to Industrial Problems (EUROGEN 2001)*, pages 95–100. International Center for Numerical Methods in Engineering (CIMNE), 2002.

[32] E. Zitzler and L. Thiele. Multiobjective Evolutionary Algorithms: A Comparative Case Study and the Strength Pareto Approach. *IEEE Transactions on Evolutionary Computation*, 3(4):257–271, 1999.

Multiobjective GP for Human-Understandable Models: A Practical Application

Katya Rodríguez-Vázquez[1] and Peter J. Fleming[2]

[1] Instituto de Investigaciones en Matemáticas Aplicadas y en Sistemas
Universidad Nacional Autónoma de México, México, D.F. 04510
katya@uxdea4.iimas.unam.mx
[2] Automatic Control and Systems Engineering
University of Sheffield, Mappin Street, Sheffield S1 3JD, UK
p.fleming@sheffield.ac.uk

Summary. The work presented in this chapter is concerned with the identification and modelling of nonlinear dynamical systems using multiobjective evolutionary algorithms (MOEAs). This problem involves the processes of structure selection, parameter estimation, model performance and model validation and defines a complex solution space. Evolutionary algorithms (EAs), in particular genetic programming (GP), are found to provide a way of evolving models to solve this identification and modelling problem, and their use is extended to encompass multiobjective functions. Multiobjective genetic programming (MOGP) is then applied to multiple conflicting objectives in order to yield a set of simple and valid human-understandable models which can reproduce the behaviour of a given unknown system.

1 Introduction

In this chapter, we will introduce a multiobjective genetic programming technique applied to a practical problem on identification and modelling nonlinear systems. We do not know the optimal solution or the true Pareto front, but we can produce a set of near optimal solutions according to the user's requirements and present these solutions in a human-understandable way.

2 Genetic Programming

Nature has provided the inspiration for the design of computational algorithms in a variety of ways. These computational processes have taken two main natural systems as their basis: the brain and the the theory of evolution. Evolutionary algorithms (EAs), one of these nature-inspired computational models, optimize by varying a population of data structures and selecting the fittest ones for further variation. The main classification considers three evolutionary algorithms; evolution strategies (ESs) [22], evolutionary programming (EP) [10] and genetic algorithms (GAs) [15]. Genetic programming (GP), popularized by Koza [19], is a branch of GAs we describe in detail, as follows.

GP can be defined as a GA designed to evolve populations of hierarchically structured computer programs according to their performance on a previously specified fitness criterion. The main difference between GP and its predecessor GA is the fact that GP genotypes or individuals are programs which are not fixed in length or size. The maximum depth of the parse tree of the program is specified a priori to constrain the search space, but all solutions up to and including this maximum are considered. When genetic operators operate over the population of tree-structured individuals, the new genotypes differ from their parents in structure (size, shape and contents).

2.1 Genetic Programming Representation

Each hierarchical genotype consists, then, of functions that can be composed recursively from the set of N_F functions from $F = f_1, f_2, ..., f_{N_F}$, and the set of N_T terminals from $T = a_1, a_2, ..., a_{N_T}$. The *function set* can consist of any arithmetic, Boolean, mathematical, or any other more complex functions (routines). The *terminal set* basically contains variables or constant values. To illustrate the hierarchical encoding used for GP, Figure 1 gives a simple example where the operations $+, -, *, \%$ (protected division), IF, $<, >, =$, belong to the function set, and the variables A, B and constants $1, 2, ..., 10$ constitute the terminal set. It is seen in Figure 1, that the number of arguments (arity) taken for each of the functions $+, -, *, \%, \text{IF}, <, >, =$ are $2, 2, 2, 2, 3, 2, 2, 2$, respectively. The function arity is relevant in order for us to create valid hierarchical structures and to select valid substructures for crossing over and mutating.

2.2 Genetic Programming Operators

Selection in GP works in the same way as in a GA; the different representation of individuals used by GP and GAs does not impact upon selection. Thus, any conventional GA selection operator, such as roulette wheel or tournament can be used.

As for the conventional GA, *crossover* is considered the main genetic operator. One of the main differences between GP and the traditional implementation of GA is the fact that GP crossover does not preserve any kind of context in the chromosome. This is due to the fact that the standard crossover defined in [19] exchanges subtrees which are chosen at random in both parents. Koza has pointed out that random subtree crossover maintains diversity in the population because crossing two identical structures will generally create different offspring. This is because the crossover points are, in general, different in the two parents. *Crossover*, a sexual operator, works by first selecting a pair of structures from the current population. Then, a node rooted in each parent is randomly selected. These nodes become the roots for the substructures lying below the crossover point. In the next step, the substructures are exchanged between the parents, producing two new structures which

$$F = \{+, -, *, \%, \mathrm{IF}, >, <, =\}$$

$$T = \{A, B, 1, 2, \ldots, 10\}$$

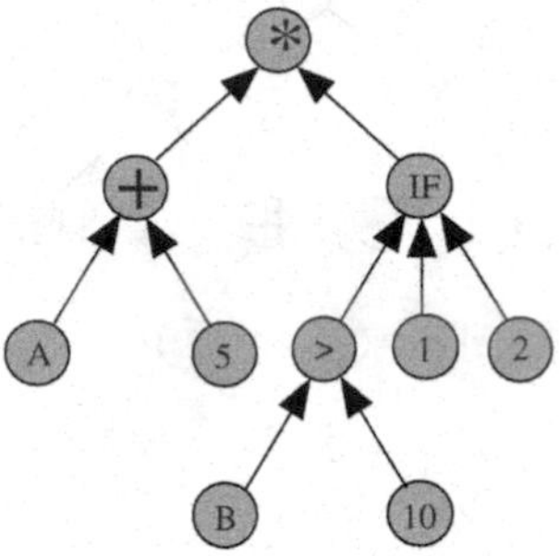

function y = example(A,B)

   if B > 10
        $y = (A + 5) * 1;$
   else
        $y = (A + 5) * 2;$
   end

$$y = \begin{cases} (A + 5) * 1 & \text{if B} > 10 \\ (A + 5) * 2 & \text{otherwise} \end{cases}$$

Fig. 1. An example of a GP individual, tree, function and equivalent program

are usually of different sizes than their parents. Figure 2 illustrates the crossover operation. Note that for GP crossover, the crossover point can be either a terminal or an internal point. In this example, the crossover point in both parents are internal nodes. This means that function nodes are chosen as roots for the substructures to be exchanged. When an internal node is selected, the number of arguments taken by the associated function must be considered in order to exchange a valid substructure.

Mutation is considered a secondary operator. It operates by randomly selecting a node, which can be either a terminal or an internal point, and replacing the associated substructure with a randomly generated subtree up to a maximum size. In a conventional GA, the mutation operator introduces a certain degree of diversity into the population which is beneficial at later generations. In contrast, the GP crossover operation is the mechanism for diversification in the GP population.

This fact is the justification given in [19] for using a 0% mutation probability.[3]

3 Multiobjective Genetic Programming (MOGP)

Conventional genetic programming, and in general evolutionary algorithms, assign a single performance measure to each individual based on the evaluation of a scalar fitness function. However, these methods possess the characteristic of simultaneously

---

[3] However, it is relevant to mention that mutation plays an important role in GP (or any variable encoding) when the neutral efects are considered (more information about this topic can be found in [18, 13]).

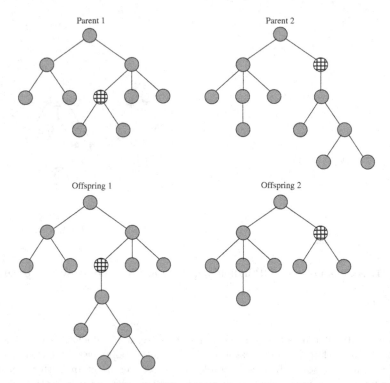

Fig. 2. GP crossover operation

searching for multiple solutions and can furthermore simultaneously evaluate several aspects of the problem to be solved. Therefore, this characteristic makes evolutionary algorithms suitable to solve problems that cannot be seen in a unidimensional space, and cannot be treated by conventional optimization methods.

The field of GP and multicriteria optimization has hardly been explored. A few approaches have been developed to control the tree growth problem of computer programs. Most of these approaches are formulated as aggregating functions with the aim of generating parsimonious computer programs (this fact is related to the well-known *bloating* problem for variable encodings as in genetic programming). Some approaches that use an aggregating function are [16, 2, 25, 26]. More recently, this issue has been handled by means of a multiobjective fitness function (see [9, 5, 24]). However, we will not focus on this topic, which is treated in details in the chapter of Bleuler et al. in this volume. However, it is important to mention that controlling tree size by means of a multiobjective function evaluation can produce simple understandable solutions, easily manageable for the user. In [17], has been proposed a multiobjective fitness function evaluation for structural risk minimization on decision trees, which are evolved by means of genetic programming. Here, a bi-objective function, where one of the two objectives is related to tree size complexity, is evaluated.

But, we can also introduce a multiobjective function into a genetic programming algorithm in order to evolve a solution based on diverse criteria concerning the problem domain. It is well known that in the real world, most problems are described as depending on multiple attributes that must be considered and multiple objectives that have to be satisfied. These multiobjective optimization problems tend to be characterized by a set of alternatives that must be considered equivalent in the absence of information concerning the relevance of each objective relative to the others. Few papers about using a multiobjective fitness function with GP and applying it to practical problems have been published. The first attempt to use multiobjective evolutionary algorithms to evaluate several aspects of a problem on its domain is illustrated in [20]. A multiobjective genetic programming method based on MOGA [11, 12] for a problem of deriving Boolean queries has been introduced in [7]. Here, the authors considered only a bi-objective fitness function.

Zhang and Rockett [27] applied a multiobjective genetic programming approach for the "optimal" feature extraction pre-processing for pattern classification. They used a bi-objective fitness function performance, but they also considered a third objective concerning tree size complexity. Thus, their multiobjective function includes tree complexity measure, Bayes error and misclassification error. In [1], a comparison of fitness performance between a MOGA [11, 12] and NSGA-II [8] Pareto approaches applied to a natural language parsing and tagging problem is presented. Again, as in previous reported papers, the author uses a bi-objective fitness function. The first objective, parsing, is defined as the average probability of the grammar rules used to construct the parsing. The second objective, tagging, is a measure of the total probability of its sequence of tags.

In this chapter, system identification and modelling problems will be introduced. A genetic programming approach combined with a multiobjective fitness function is presented. This multiobjective function considers several aspects of the problem such as model complexity (not tree size complexity but complexity concerning the problem domain), model performance and model validation. The next section introduces details of this problem.

4 System Identification and the Generation of Simple Understandable Models

System identification is defined as the process of constructing a mathematical model from observations and prior knowledge. However, the problem of identifying nonlinear model structures cannot be evaluated using only a single criterion. Identification involves diverse characteristics which have to be considered, such as linearity, degree of nonlinearity, model structure, performance and model validation.

In [23, 24], a tree-structured representation to this problem was introduced. This approach is based on an input-output model that describes the input-output relationship of a system. For representing these systems, Leontaritis and Billings [21] have introduced the well-known NARMAX (Nonlinear AutoRegressive Moving Average with eXogenous inputs) model which is an extended ARMAX description for representing nonlinear systems. This model is given by a nonlinear function F^ℓ

of the output $y(k)$, the input $u(k)$ and the possible noise disturbance $e(k)$. Thus,

$$y(k) = F^\ell(y(k-1), \ldots, y(k-n_y), u(k-d), \ldots, u(k-d-n_u+1),$$
$$e(k-1), \ldots, e(k-n_e)) \tag{1}$$

where n_y, n_u and n_e are the maximum lags considered for the output, input and noise terms, respectively, d is the delay, and ℓ is the degree of nonlinearity of the model structure. Note that if $\ell = 1$, the resulting model is a linear structure. If the disturbance is assumed to be white noise, equation (1) can be simplified to

$$y(k) = F^\ell(y(k-1), \ldots, y(k-n_y), u(k-d), \ldots, u(k-d) - n_u + 1) \tag{2}$$

and the structure becomes a NARX (Nonlinear AutoRegressive with eXogenous inputs) model.

4.1 Polynomial Representation

The NARMAX model is the most general form of input-output model and can be expressed in different ways. [6] has shown that the polynomial NARMAX model is the most common expression which works well in practical applications. Equation (1) can be written in polynomial form as follows,

$$y(k) = \Psi_{yu}^T(k-1)\theta_{yu} + \Psi_{yue}^T(k)\theta_{yue} + \Psi_e^T(k)\theta_e \tag{3}$$

where $\Psi_{yu}^T(k-1)$ includes the constant term and all the output and input terms as well as all possible combinations up to degree ℓ. These terms will be referred to as process terms. The parameters of such terms are in the vector θ_{yu}. The other vectors of monomials are defined likewise. $\Psi_{yue}^T(k)$ and $\Psi_e^T(k)$ will be referred to as noise terms.

However, because the noise $e(k)$ is unknown, equation (3) can be rewritten in the prediction error (PE) form as

$$y(k) = P_i(k)\hat{\theta} + \varepsilon(k) \tag{4}$$

where the residual $\varepsilon(k)$ is defined as

$$\varepsilon(k) = y(k) - \hat{y}(k, \hat{\theta}) \tag{5}$$

and $P_i(k)$ consists of all possible linear output, input and noise terms, and all possible nonlinear terms in the output, input, noise and combined terms. The polynomial model is then nonlinear in the output, input and noise, but linear in the parameters. This set of coefficients is estimated by means of an Extended Least Squares (ELS) algorithm [4] when noise terms are included; otherwise, the traditional LS algorithm can be applied.

4.2 GP Encoding of NARMAX Structures

In this section, the mapping process of NARMAX structures into a GP tree representation is detailed. As mentioned previously, the polynomial form of this structure can be expressed as a tree. Only addition and product functions are required, and associated coefficients are estimated by means of a least squares algorithm. This

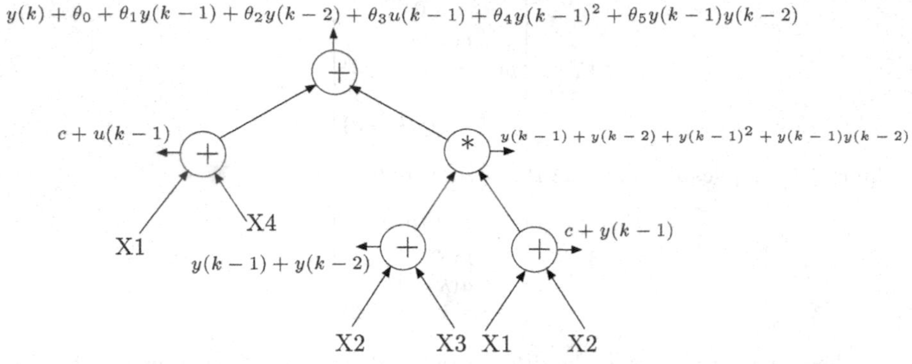

$$y(k) + \theta_0 + \theta_1 y(k-1) + \theta_2 y(k-2) + \theta_3 u(k-1) + \theta_4 y(k-1)^2 + \theta_5 y(k-1)y(k-2)$$

$$(+ \ (+ \ X1 \ X4)(* \ (+ \ X2 \ X3)(+ \ X1 \ X2)))$$

Fig. 3. NARMAX polynomial encoding

process is illustrated in Figure 3. At the root node, the polynomial expression is defined, and an LS is applied based on measured data in order to get the set of coefficients.

Based on equation (4) and Figure 3,

$$
Y = \begin{bmatrix} y(1) \\ y(2) \\ \vdots \\ y(N) \end{bmatrix}
\qquad
P^T = \begin{bmatrix} p_1(k) \\ p_2(k) \\ \vdots \\ p_n(k) \end{bmatrix} = \begin{bmatrix} 1.0 \\ y(k-1) \\ y(k-2) \\ u(k-1) \\ y(k-1)^2 \\ y(k-1)y(k-2) \end{bmatrix}
\qquad
\hat{\theta} = \begin{bmatrix} \theta_1 \\ \theta_2 \\ \vdots \\ \theta_n \end{bmatrix}
$$

and $\hat{\theta}$ is estimated as

$$\hat{\theta} = [P^T P]^{-1} P^T Y, \tag{6}$$

where $\hat{\theta}$ is the model coefficients vector, P is the vector of identified model terms (monomials) and Y is the measured output vector.

The terminal set and the set of functions appropriate for building such models are $\mathbf{T} = \{X_1, X_2, X_3, X_4, X_5\} = \{c, y(k-1), y(k-2), u(k-1), u(k-2)\}$ and $\mathbf{F} = \{\mathtt{ADD}, \mathtt{MULT}, =, +, *\}$; where n_u and n_y (in this case $n_u = n_y = 2$) are maximum lags considered for the output and input terms, respectively. It is important to point out that duplicated rows in matrix $\mathbf{P}$ are deleted before the coefficient estimation and individual evaluation stages (those are eliminated during the decoding process). Thus, redundant terms into the model are removed, but this fact does not mean they are removed from individual GP trees. An example is shown in Figure 4. Decoding the GP expression, the following P matrix is obtained:

$$P_T = \begin{bmatrix} 1.0 \\ y(k-1) \\ y(k-1)y(k-2) \\ u(k-1) \\ y(k-1)y(k-2) \end{bmatrix}.$$

Eliminating duplicated rows, the P matrix is reduced to

$$P_T = \begin{bmatrix} 1.0 \\ y(k-1) \\ u(k-1) \\ y(k-1)y(k-2) \end{bmatrix}.$$

Then, the least squares algorithm is applied using P. As mentioned, if noise terms are involved, the ELS algorithm is applied.

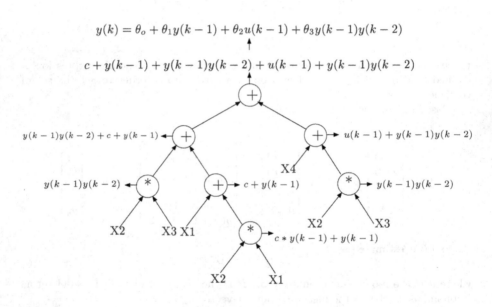

$$(+ \ (+ \ (*X2 \ X1)(+ \ (*X2 \ X1)))(+ \ X4(* \ X2 \ X1)))$$

Fig. 4. NARMAX polynomial encoding (duplicated terms)

5 Multiobjective Fitness Evaluation

In the system identification and modelling procedure, once the input-output data is available, three steps are left: model structure determination, performance and validation. In the conventional identification procedure, these are independently evaluated. For the MOGP approach, each member of the population (a potential candidate model) will be assigned a fitness which is a value that considers the three aspects measured based upon different attributes, as shown in Table 1 and described as follows.

5.1 Model Structure and Parsimony

From the definition of the NARMAX model, it is observed that the number of linear and nonlinear terms, the degree and the maximum lags play an important role in the determination of the model structure. These three model structure attributes are defined as objectives in the multiobjective system identification function. By minimizing these objectives, a parsimonious model structure may be identified which shows good prediction quality and proves to be a valid model by additionally optimizing performance and validation objectives.

Table 1. Objectives used in the MOGP identification procedure

Attribute	Objective	Description
Model complexity	Model size	Number of process and noise terms
	Model degree	Maximum order term
	Model Lag	Maximum lagged input, output and noise terms
Model Performance	Residual variance	Variance of the predictive error between the measured output and the OSAPE
	Long-term prediction error	Variance of the LTPE
Model validation	ACF, CCF and higher-order correlation functions	Correlation based test functions for model with additive noise at the output

5.2 Model Performance

The residual variance is equivalent to the short-term prediction error and it is desirable to estimate also the long-term prediction output of the system. Due to this, these two predictive error measures are defined as the objectives regarding the model performance. The residual variance is calculated by

$$\hat{\sigma}_\varepsilon^2 = \frac{1}{N} \sum_k^N |\hat{y}(k) - y(k)|^2 \qquad (7)$$

where

$$\hat{y} = F^\ell(y(k-1), \ldots, y(k-n_y), u(k-1), \ldots, u(k-n_u), e(k-1), \ldots, e(k-n_e)). \quad (8)$$

The long-term prediction error (LTPE) is computed by using equation (7) where the estimated output is defined as

$$\hat{y} = F^\ell(\hat{y}(k-1), \ldots, \hat{y}(k-n_y), u(k-1), \ldots, u(k-n_u), e(k-1), \ldots, e(k-n_e)). \quad (9)$$

In these equations, $\hat{y}(k)$ denotes the predicted system output.

5.3 Dynamical Model Validation

The use of statistics is a systematic approach to validating the quality of an identified model. In the literature, several approaches to model validation, formulated as statistical hypothesis testing, have been proposed [3]. One of the most common methods are the correlation tests which are used to verify if the residuals are white. In this case, a model is considered valid (the null hypothesis is accepted) if the correlations fail to detect significant dynamics in the residuals.

When the process is nonlinear, additional tests must be performed in order to detect the nonlinearities in the input, output and residuals from a fitted model. These linear and nonlinear correlation-based functions are reviewed, as follows.

Linear Model Validation

Based on the ARMAX model, ideally, the residuals $\varepsilon(t)$ should be reduced to an uncorrelated sequence denoted by $e(t)$ with zero mean and finite variance. Correlation-based model validity tests are used to check if

$$e(t) \approx \varepsilon(t). \qquad (10)$$

This can be done by testing whether all the correlation functions are within the present confidence intervals. When equation (10) is true, the following tests shows that

$$\left. \begin{array}{l} \Phi_{\varepsilon\varepsilon}(\tau) = E[\varepsilon(t-\tau)\varepsilon(t)] = \delta(t) \\ \Phi_{u\varepsilon}(\tau) = E[u(t-\tau)\varepsilon(t)] = 0 \end{array} \quad \forall\tau \right\} \qquad (11)$$

where $\Phi_{\varepsilon\varepsilon}$ and $\Phi_{u\varepsilon}$ are the estimated residual autocorrelation function and the cross-correlation function between the input and the residual, respectively. $\delta(\tau)$ is the Kronecker delta.

The expression $\Phi_{\varepsilon\varepsilon} \neq \delta(\tau)$ is an indication that the process model is correct but the noise model is incorrect, and therefore the residuals are autocorrelated; but they are uncorrelated with the input, such that $\Phi_{u\varepsilon}(\tau) = 0, \forall\tau$. Alternatively, if the noise model is correct but the process model is biased, then the residuals are both autocorrelated such that $\Phi_{\varepsilon\varepsilon}(\tau) \neq \delta(\tau)$, and correlated with the input ($\Phi_{u\varepsilon}(\tau) \neq 0$).

However, the validation of the nonlinear process is not straightforward. These two correlation tests are necessary but not sufficient conditions in order to be confident that an identified nonlinear model is valid. For the purpose of validating nonlinear models, high-order correlation tests are introduced.

Nonlinear Model Validation

For nonlinear systems, it is seen that the nonlinear function $F^\ell(\bullet)$, given by equation (1), will not, in general, satisfy the superposition and homogeneity principles. The model validation tests described by equation (11) can then only detect a subset of possible nonlinear terms which could be presented in the residuals sequence. For the purpose of nonlinear model validation, higher-order correlation tests were proposed by [3]. Based upon the NARMAX model, three correlation tests have been developed in order to detect every possible nonlinear term [3]. Thus, the residuals $\varepsilon(t)$ will be unpredictable from all linear and nonlinear combinations of past inputs and outputs if and only if

$$\left.\begin{array}{ll} \Phi_{\varepsilon\varepsilon}(\tau) = \delta(\tau) & \\ \Phi_{u\varepsilon}(\tau) = 0 & \forall\tau \\ \Phi_{\varepsilon(u\varepsilon)} = E[\varepsilon(t)\varepsilon(t-1-\tau)u(t-1-\tau)] = 0 & \tau \geq 0 \end{array}\right\} \quad (12)$$

However, in the situation where only the system model is involved in the identification procedure and the noise sequence is additive at the output, the NARMAX model can be expressed as equation (2). Here, all the cross-product noise terms are eliminated; this equation is then known as the NARX model. It is seen that parameter estimation of the polynomial representation of equation (2) would therefore require less computational effort compared with the case of the NARMAX model. In the case of NARX model validation, the noise model is not specifically estimated and consequently, the residuals may be coloured. Specific tests are required and the estimated nonlinear model will be unbiased if and only if

$$\left.\begin{array}{ll} \Phi_{u^2\varepsilon^2}(\tau) = 0 & \forall\tau \\ \Phi_{u^2\varepsilon}(\tau) = 0 & \forall\tau \\ \Phi_{u\varepsilon}(\tau) = 0 & \forall\tau \end{array}\right\}. \quad (13)$$

From this equation, the cross-correlation $\Phi_{u^2\varepsilon^2}(\tau)$ detects all process terms in the input and the cross-terms between the input and output. Therefore, the expression $\Phi_{u^2\varepsilon^2}(\tau) = 0$ indicates that the process model is correct and there is no cross-term involved in the model. Otherwise, it would indicate that internal noise cross-terms of the form $u^k(t)e^l(t)$ are missed, where $k =$ even or odd, and $l =$ odd. The correlation functions $\Phi_{u^2\varepsilon}(\tau)$ and $\Phi_{u\varepsilon}(\tau)$ detect odd and even terms in the input, respectively. For all the correlation-based validity tests mentioned above, confidence intervals indicate whether the correlation between variables is significant or not. If N (the number of data points) is large, the standard deviation of the correlation estimate is $1/\sqrt{N}$, where the 95% confidence limits are therefore approximately $\pm 1.96/\sqrt{N}$, assuming a Gaussian distribution. Based on these correlation validation functions, the complete identification procedure is described in the next section.

6 Multiobjective Genetic Programming in Practice

The example presented in this section considers the simple Wiener process. The experiment uses a population of 100 individuals, $P_c = 0.9$ and $P_m = 0.01$; the algorithm was run for 200 generations. In the context of model representation, $n_y = n_u = 10$ and $n_e = 0$ were considered. The differential equation of the linear dynamic part of the simple Wiener process is given as

$$10\dot{v}(t) + v(t) = u(t),\tag{14}$$

and the static nonlinear part is expressed by

$$y(k) = 2 + v(k) + v^2(k).\tag{15}$$

The process described above was executed by a pseudo-random ternary test signal (PRTS) with maximum length 26, amplitude 2 and mean value 1 [14]. The sampling time was $\Delta T = 2$ s and the clock time interval was 10 seconds. Then, $\mathtt{N} = 26 * 5 = 130$ data pairs were used for the identification. The input/output data are shown in Figure 5.

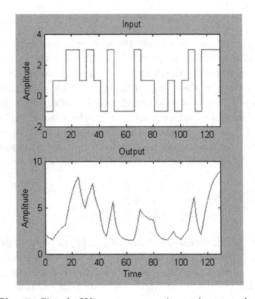

Fig. 5. Simple Wiener process input/output data

Because a NARX representation was used in this example, the objectives were defined and classified as shown in Table 1, where the correlation-based validation tests are defined by equation (13).

Objectives related to model validation were initially ignored. Therefore, only good models in terms of complexity and performance were identified (see Figure 6a). Subsequently, the correlation objective function priorities were modified to be considered as constraints. The target value to be attained was set up to be the 95% confidence limit. The correlation-based validation objectives are contained in a $(2 * \tau + 1)$ element vector. In order to define these functions as scalar, the following operation is introduced:

$$CCF = \max |\mathbf{abs}(\Phi_{xy}(\tau))| \tag{16}$$

where CCF represents a correlation test function and x and y can be input, noise or the vector product of the input and noise. The identification process then evolved through regions of the search space where valid models were located (i.e., regions of the search space where models that satisfy correlation restrictions were located). Ideally, one would expect to obtain a solution (or set of alternatives) with the optimum (minimum or maximum) value in all objectives. However, it is not likely to occur due to the fact that, in general, objectives are in conflict with one another. This fact is shown in Figure 6b, where the validation criteria were treated as constraints. It is seen that in order to obtain valid models, the performance objectives tend to exhibit higher values.

However, this set of models is expected to be robust and applicable to different sets of data (operating conditions).

7 Discussion

As can be seen from Figure 6, a multiobjective genetic programming tool provides a set of potential models which can answer the questions about the complexity, flexibility and validity of a model. It also provides the opportunity of manipulating the family of solutions by changing priorities and goals values of the objective functions depending on the purpose of the identification.

GP is restricted by the number of nodes permissible in a tree, but the search space is still considered extremely large, and its variable size and dynamic representation gives diversity in the population. Thus, it is important to note that MOGP was able to evolve tree structures of variable size which represent models that posses good predictive qualities. An improvement in the performance on both the training and the testing sets of data has been reached by producing models with a slightly higher degree and number of terms.

In Figure 6a., validation objectives are ignored. Then, it is observed that the goals are, in some cases, not reached for these objectives, and models from six to ten terms[4] are plotted. Comparing Figures 6a. and 6b, we see that models possessing better long-term prediction qualities are shown in Figure 6a. However, these models cannot be valid. In Figure 6b, validation criteria have been modified and considered as constraints. This means that these criteria must be satisfied before the optimization of the remaining objectives (the priority is higher for validation). There are some models presented in Figure 6a, which are no longer shown in Figure 6b (models that do not satisfy validation criteria). We are now able to select a simple model (shorter model), or a model presenting the best prediction capabilities but a bit more complex. This final set of models contains simple human-understandable

---

[4] 10 was set as the upper bound for the graph, but this does not mean that there is any other model possessing more than ten terms.

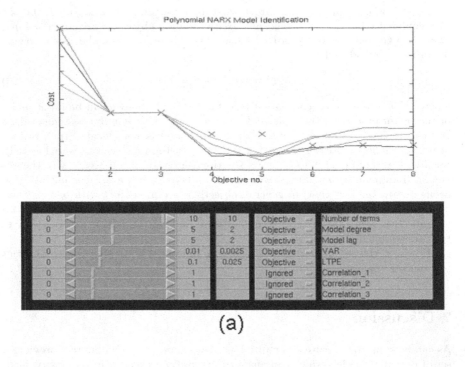

(a)

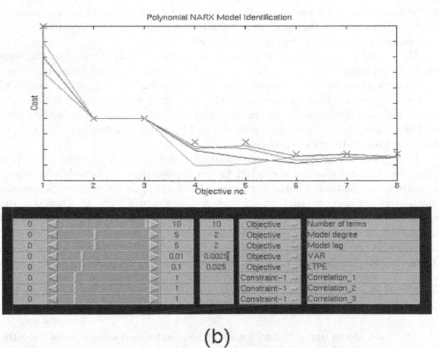

(b)

Fig. 6. MOGP identification procedure. Model validation objectives are considered as hard objectives (constraints): a) without validation measures and b) with validation measures

valid models, which are shown in Tables 2 and 3. Table 2 presents values of the multiobjective function and Table 3 shows the structure of these models. It is important to point out that, although maximum output and input lags were set to 10, only models of lag = 2 were evolved (this was considered as an objective to be minimized). Models produced by MOGP are also compared to models obtained by means of traditional techniques such as stepwise regression[5] and orthogonal least squares (OLS[6]) (see [4, 14] for more details). It can be observed that although both models (produced by stepwise regression and OLS) show similar model complexity, they have a worse performance than models generated by MOGP; also, these models do not satisfy all validation (correlation tests) criteria. It is relevant to point out that stepwise regression and OLS consider only model performance attributes in order to get the model structure by means of the minimization of the VAR objective (residual variance), and the LTPE and validation process are computed after by the modelling process. Thus, we cannot simultaneously optimize all the stages of system identification and modelling by means of traditional methods and only one model can be obtained, not a set of alternatives, as is the case with MOGP.

Table 2. Set of simple human-understandable valid models

Model	Terms	Degree	Lag	VAR	LTPE	Corr. 1	Corr. 2	Corr. 3
1	7	2	2	0.002192	0.02063	✓	✓	✓
2	7	2	2	0.002083	0.02259	✓	✓	✓
3	8	2	2	0.001947	0.01451	✓	✓	✓
4	9	2	2	0.000979	0.01039	✓	✓	✓
Stepwise	7	2	2	0.001608	0.07852	X	✓	✓
OLS	7	2	2	0.005224	0.26808	X	✓	X

8 Concluding Remarks

This chapter has presented an exploration in the area of multiobjective genetic programming (MOGP), providing an alternative to control the search process by restricting the search space and searching the feasible zone of simple human-understandable solutions.

Comparing conventional identification methods with multiobjective evolutionary algorithms, the former do not guarantee that the model can be an acceptable representation of the system. They require a verification process to determine whether an obtained model (not a set) is adequate or not (validation). In contrast, MOGP integrates these three steps in the modelling process, providing a useful tool to generate, in a single step, simple feasible and understandable models.

---

[5] stepwise regression method; see[14] for more details.
[6] orthogonal least squares; see [14] for more details.

Table 3. Structure of simple human-understandable valid models

Term/Model	1	2	3	4	Stepwise	OLS
c	✓	✓	✓	✓	✓	✓
$y(k-1)$	✓	✓	✓	✓	✓	✓
$y(k-2)$	✓	✓	✓	✓	✓	✓
$u(k-1)$		✓		✓		✓
$u(k-2)$	✓	✓	✓			✓
$y(k-1)^2$					✓	
$y(k-2)^2$					✓	
$u(k-1)^2$		✓	✓		✓	✓
$u(k-2)^2$	✓		✓	✓		
$u(k-1)u(k-2)$			✓			
$y(k-1)u(k-1)$		✓			✓	✓
$y(k-2)u(k-1)$	✓		✓	✓		
$y(k-2)u(k-2)$	✓	✓	✓	✓		

Acknowledgements

The authors wish to acknowledge the support of Consejo Nacional de Ciencia y Tecnología (CONACyT-México) through the grant 40602-A and PAPIIT, UNAM under the project IN115806-3. The authors gratefully acknowledge Eng. Fco. J. Cárdenas Flores for his technical support and Dr. Carlos Fonseca for the use of the MOGA framework.

References

[1] Araujo, L. (2006) Multiobjective Genetic Programming for Natural Language Parsing and Tagging, Parallel Problem Solving From Nature PPSN IX, LNCS 4193, Springer-Verlag, pp. 433–442.

[2] Berstein, Y., X. Li, V. Ciesielski and A. Song (2004) Multiobjective Parsimony Enforcement for Superior Generalisation Performance, Congress on Evolutionary Computation CEC 2004, IEEE Press, pp. 83–89.

[3] Billings, S. A. and W. S. F. Voon (1983) Structure Detection and Model Validity Tests in the Identification of Nonlinear Systems. IEE Proceedings Pt. D, 130(4), pp. 193–199.

[4] Billings, S. A. and W. S. F. Voon (1984) Least Square Parameter Estimation Algorithm for Nonlinear Systems. Int. J. Systems Sci., 15(6), pp. 601–615.

[5] Bleuler, S., M. Brack, L. Thiele and E. Zitzler (2001) Multiobjective Genetic Programming: Reducing Bloat Using SPEA2, Congress on Evolutionary Computation CEC 2001, IEEE Press, pp. 536–543.

[6] Chen, S. and S. A. Billings (1989) Representation of Nonlinear Systems: the NARMAX Model. Int. J. Control, 49(3), pp. 1013–1032.

[7] Cordon, O., E. Herrera-Viedma and M. Luque (2006) Evolutionary Learning of Boolean Queries by Multiobjective Genetic Programming, Parallel Problem Solving From Nature PPSN VII, LNCS 2439, Springer-Verlag, pp. 710–719.

[8] Deb, K., A. Pratap, S. Agarwal, and T. Meyarivan (2002) A Fast and Elitist Multiobjective Genetic Algorithm: NSGA–II, IEEE Transactions on Evolutionary Computation, 6(2), pp. 182–197.

[9] De Jong, E. D. and J. B. Pollack (2003) Multiobjective Methods for Tree Size Control, Genetic Programming and Evolvable Machines (4), Kluwer Academic Publisher, pp. 211–233.

[10] Fogel, L. J., A. J. Owens and M. J. Walsh. (1966) Artificial Intelligence Through Simulated Evolution. Wiley Publishing.

[11] Fonseca, C. M. and P. J. Fleming. (1993) An Overview of Evolutionary Algorithms in Multiobjective Optimization, Evolutionary Computation, 3(1), pp. 1–16.

[12] Fonseca, C. M. and P. J. Fleming (1995) Multiobjective Optimization and Multiple Constraint Handling with Evolutionary Algorithms I: A Unified Formulation. Research Report 564, Dept. of Automatic Control and Systems Engineering. University of Sheffield, U.K.

[13] Galvan-Lopez, E. and K. Rodriguez Vazquez (2006) The Importance of Neutral Mutations in GP, Parallel Problem Solving from Nature IX, volume 4193 of LNCS (Runnarson et al., eds.), Pringer-Verlag, pp. 870–879.

[14] Haber, R. and H. Unbenhauen (1990) Structure Identification of Nonlinear Dynamic Systems: A Survey of Input-Output Approaches. Automatica, 26 (4), pp. 651–677.

[15] Holland, J. H. (1975) Adaptation in Natural and Artificial Systems. The University of Michigan Press.

[16] Iba, H., H. de Garis and T. Sato. (1994) Genetic Programming Using a Minimum Description Length. Advances in Genetic Programming (Kinnear, Ed.), MIT Press, pp. 265–284.

[17] Kim, D. (2004) Structural Risk Minimization on Decision Trees Using an Evolutionary Multiobjective Optimization, European Conference on Genetic Programming EuroGP 2004, LNCS 3003, Springer-Verlag, pp. 338–348.

[18] Kimura, M. (1968) Evolutionary Rate at the Molecular Level, Nature (217), pp. 624–626.

[19] Koza, J. R. (1992) Genetic Programming: On the Programming of Computers by Means of Natural Selection. MIT Press.

[20] Langdon, W. B. (1996) Data Structure and Genetic Programming, Advances in Genetic Programming Vol. 2, (Angeline and Kinnear, eds,), Chapter 20, MIT Press, pp. 395–414.

[21] Leontaris, I. J. and S. A. Billings (1985) Input-Output Parametric Models for Nonlinear Systems. Int. J. Control 41(2), 311–341.

[22] Rechenberg, I. (1965) Cybernetic Solution Path of an Experimental Problem. Ministry of Aviation, Royal Aircraft Establishment.

[23] Rodríguez-Vázquez, K., C. M. Fonseca and P. J. Fleming (1997) An Evolutionary Approach to Nonlinear System Identification, to appear in 11th IFAC Symposium on System Identification.

[24] Rodríguez-Vázquez, K., C. M. Fonseca and P. J. Fleming (1997) Multiobjective Genetic Programming: A Nonlinear System Identification Application.

Late Breaking Paper at the 2nd Int. Genetic Programming 97 Conference, Stanford University, pp. 207–212.

[25] Siegel, E. V. and A. D. Chaffe (1996) Genetically Optimizing the Speed of Programs Evolved to Play Tetris, Advances in Genetic Programming Vol. 2 (Angeline and Kinnear, eds.), Chapter 14, MIT Press, pp. 279–298.

[26] Zhang, B. T. and H. Mühlenbein (1996) Adaptive Fitness Functions for Dynamic Growing/Pruning. Advances in Genetic Programming Vol. 2 (Angeline and Kinnear, eds., Chapter 12,, MIT Press, pp. 241–256.

[27] Zhang, Y. and P. I. Rockett (2005) Evolving Optimal Feature Extraction Using Multiobjective Genetic Programming: A Methodology and Preliminary Study on Edge Detection, Genetic and Evolutionary Computation Conference GECCO 2005, ACM Press, pp. 795–802.

Multiobjective Classification Rule Mining

Hisao Ishibuchi, Isao Kuwajima, and Yusuke Nojima

Department of Computer Science and Intelligent Systems
Graduate School of Engineering, Osaka Prefecture University
1-1 Gakuen-cho, Naka-ku, Sakai, Osaka 599-8531, Japan
hisaoi@cs.osakafu-u.ac.jp, kuwajima@ci.cs.osakafu-u.ac.jp,
nojima@cs.osakafu-u.ac.jp

Summary. In this chapter, we discuss the application of evolutionary multiobjective optimization (EMO) to association rule mining. Especially, we focus our attention on classification rule mining in a continuous feature space where the antecedent and consequent parts of each rule are an interval vector and a class label, respectively. First we explain evolutionary multiobjective classification rule mining techniques. Those techniques are roughly categorized into two approaches. In one approach, each classification rule is handled as an individual. An EMO algorithm is used to search for Pareto-optimal rules with respect to some rule evaluation criteria such as support and confidence. In the other approach, each rule set is handled as an individual. An EMO algorithm is used to search for Pareto-optimal rule sets with respect to some rule set evaluation criteria such as accuracy and complexity. Next we explain evolutionary multiobjective rule selection as a post-processing procedure in classification rule mining. Pareto-optimal rule sets are found from a large number of candidate classification rules, which are extracted from a database using an association rule mining technique. Then we examine the effectiveness of evolutionary multiobjective rule selection through computational experiments on some benchmark classification problems. Finally we examine the use of Pareto-optimal and near Pareto-optimal rules as candidate rules in evolutionary multiobjective rule selection.

1 Introduction

Data mining is a very active and rapidly growing research area in the field of computer science. The task of data mining is to extract useful knowledge for human users from a database. Whereas the application of evolutionary computation to data mining is not always easy due to its heavy computation load, especially in the case of a large database ([3], [4], [9]), many evolutionary approaches have been proposed in the literature ([6], [16], [32], [34], [37]). Evolutionary multiobjective optimization (EMO) has also been applied to data mining in some studies ([10]–[12], [17], [20]–[23], [36]). In the field of fuzzy logic, multiobjective formulations have frequently been used for knowledge extraction ([5], [18]–[21], [23], [25], [27], [39], [40]). This is because the interpretability-accuracy trade-off analysis is a very important research issue in the design of fuzzy rule-based systems [5]. Multiobjective formulations have also been used in non-fuzzy genetics-based machine learning ([26], [28], [31]).

Association rule mining [1] is one of the most well-known data mining techniques. In its basic form [1], all association rules satisfying the minimum support and confidence are efficiently extracted from a database. The application of association rule mining to classification problems is often referred to as classification rule mining or associative classification ([29], [30], [33], [38]). Classification rule mining usually consists of two phases: rule discovery and rule selection. In the rule discovery phase, a large number of classification rules are extracted from a database using an association rule mining technique. All classification rules satisfying the minimum support and confidence are usually extracted from a database. Some of the extracted classification rules are selected to design a classifier in the rule selection phase using a heuristic rule sorting criterion. The accuracy of the designed classifier usually depends on the specification of the minimum support and confidence. Their tuning was discussed for classification data mining in [7] and [8].

Whereas the basic form of association rule mining is to extract all association rules that satisfy the minimum support and confidence [1], other rule evaluation measures have also been proposed to quantify the *utility* or *quality* of an association rule. Among them are gain, variance, chi-squared value, entropy gain, Gini, Laplace, lift, and conviction [2]. It is shown in [2] that the best rule according to any of the above-mentioned measures is a Pareto-optimal rule with respect to support and confidence. Motivated by this study, the use of an EMO algorithm was proposed to search for Pareto-optimal classification rules with respect to support and confidence for partial classification ([10]–[12], [36]). Similar formulations were used to search for Pareto-optimal association rules [17] and Pareto-optimal fuzzy association rules [27]. EMO algorithms were also used to search for Pareto-optimal rule sets in classification rule mining ([21], [22]) where the accuracy of rule sets was maximized and their complexity was minimized. The same idea was also used in the multiobjective design of fuzzy rule-based classifiers ([18]–[20], [23], [25]).

In this chapter, we empirically examine the usefulness of evolutionary multi-objective rule selection in classification rule mining in continuous feature spaces through computational experiments on some well-known benchmark data sets from the UCI machine learning repository. We also examine the relation between Pareto-optimal rules and Pareto-optimal rule sets in the classifier design. This examination is performed by depicting selected rules in Pareto-optimal rule sets together with candidate classification rules in the confidence-support space. Our interest is to check whether selected rules in Pareto-optimal rule sets are close to the Pareto front with respect to support and confidence. Then we examine the use of Pareto-optimal and near Pareto-optimal rules as candidate rules in evolutionary multiobjective rule selection.

This chapter is organized as follows. First we explain some basic concepts in classification rule mining in Section 2. Next we explain two approaches in evolutionary multiobjective classification rule mining in Section 3. One approach handles each classification rule as an individual to search for Pareto-optimal rules. In the other approach, each rule set is handled as an individual. An EMO algorithm is used to search for Pareto-optimal rule sets. In Section 4, one method in the latter approach is explained in detail. More specifically, we explain evolutionary multiobjective rule selection as a post-processing procedure in the rule selection phase of classification rule mining in Section 4. Pareto-optimal rule sets are found from a large number of candidate classification rules, which are extracted from a database using an association rule mining technique in the rule discovery phase. In Section 5, we report experimental results on some well-known benchmark data sets. Experimental results demonstrate the usefulness of evolutionary multiobjective rule selection in classification rule mining. The relation between Pareto-optimal rules and Pareto-optimal rule sets is also examined. Then we examine the use of Pareto-optimal and near Pareto-optimal rules as candidate rules in evolutionary multiobjective rule selection in Section 6. Finally we conclude this chapter in Section 7.

2 Classification Rule Mining

Let us assume that we have m training patterns $\mathbf{x}_p = (x_{p1}, x_{p2}, \ldots, x_{pn})$, $p = 1, 2, \ldots, m$ from M classes in the n-dimensional continuous pattern space where x_{pi} is the attribute value of the pth training pattern for the ith attribute. We denote the set of these m training patterns by D. For our pattern classification problem, we use classification rules of the following type:

$$\text{Rule } R_q: \text{If } x_1 \text{ is } A_{q1} \text{ and } \ldots \text{ and } x_n \text{ is } A_{qn} \text{ then class } C_q \text{ with } CF_q, \quad (1)$$

where R_q is the label of the qth rule, $\mathbf{x} = (x_1, x_2, \ldots, x_n)$ is an n-dimensional pattern vector, A_{qi} is an antecedent interval for the ith attribute, C_q is a class label, and CF_q is a rule weight (i.e., certainty grade). We denote the classification rule R_q in (1) as "$\mathbf{A}_q \Rightarrow C_q$" where $\mathbf{A}_q = (A_{q1}, A_{q2}, \ldots, A_{qn})$. Each antecedent condition "x_i is A_{qi}" in (1) means the inclusion relation "$x_i \in A_{qi}$." It should be noted that classification rules of the form in (1) do not always have n antecedent conditions. Some rules may have only a few conditions while others may have many conditions.

In the field of association rule mining, two rule evaluation measures called *support* and *confidence* have often been used ([1], [2]). Let us denote the support count of the classification rule $\mathbf{A}_q \Rightarrow C_q$ by $SUP(\mathbf{A}_q \Rightarrow C_q)$, which is the number of patterns compatible with both the antecedent part $\mathbf{A}_q$ and the consequent class C_q. $SUP(\mathbf{A}_q)$ and $SUP(C_q)$ are also defined in the same manner, and are the number of patterns compatible with $\mathbf{A}_q$ and C_q, respectively. The support of the classification rule $\mathbf{A}_q \Rightarrow C_q$ is defined as

$$Support(\mathbf{A}_q \Rightarrow C_q) = \frac{SUP(\mathbf{A}_q \Rightarrow C_q)}{|D|}, \quad (2)$$

where $|D|$ is the cardinality of the data set D (i.e., $|D| = m$). On the other hand, the confidence of $\mathbf{A}_q \Rightarrow C_q$ is defined as

$$Confidence(\mathbf{A}_q \Rightarrow C_q) = \frac{SUP(\mathbf{A}_q \Rightarrow C_q)}{SUP(\mathbf{A}_q)}. \quad (3)$$

In partial classification ([10]–[12], [36]), the coverage is often used instead of the support:

$$Coverage(\mathbf{A}_q \Rightarrow C_q) = \frac{SUP(\mathbf{A}_q \Rightarrow C_q)}{SUP(C_q)}. \quad (4)$$

Since the consequent class is fixed in partial classification (i.e., since the denominator of (4) is constant), the maximization of the coverage is the same as that of the support.

In classification rule mining ([29], [30], [33], [38]), an association rule mining technique such as Apriori [1] is used in the rule discovery phase to efficiently extract all classification rules that satisfy the minimum support and confidence. These two parameters are prespecified by users. Then a part of extracted classification rules are selected to design a classifier in the rule selection phase.

Let S be a set of selected classification rules. That is, S is a classifier. When a new pattern $\mathbf{x}_p$ is to be classified by S, we choose a single winner rule with the maximum rule weight from among rules compatible with $\mathbf{x}_p$ in S. The confidence of each rule is used as its rule weight in this chapter. The consequent class of the winner rule is assigned to $\mathbf{x}_p$. When multiple compatible rules with different consequent classes have the same maximum rule weight, the classification of $\mathbf{x}_p$ is rejected in evolutionary multiobjective rule selection in this chapter. Only when the accuracy of the rule set finally obtained (i.e., classifier) is to be evaluated, do we use a random tiebreak scheme among those classes with the same maximum rule weight in computational experiments.

3 Evolutionary Multiobjective Rule Mining

Evolutionary multiobjective techniques in classification rule mining can be roughly categorized into two approaches. In one approach, each rule is evaluated according to multiple rule evaluation criteria such as support and confidence. An EMO algorithm is used to search for Pareto-optimal classification rules. In the other approach, each rule set is evaluated according to multiple rule set evaluation criteria such as accuracy and complexity. An EMO algorithm is used to search for Pareto-optimal rule sets. In this section, we explain these two approaches.

3.1 Techniques to Search for Pareto-Optimal Classification Rules

It is shown in [2] that the set of Pareto-optimal rules with respect to support and confidence includes the best rule according to any of the following rule evaluation criteria: gain, variance, chi-squared value, entropy gain, Gini, Laplace, lift, and conviction. Thus it is an important research issue to search for Pareto-optimal rules with respect to support and confidence in association rule mining. The use of NSGA-II ([13], [14]) for this task was proposed by de la Iglesia et al. ([10], [12]) where the following two-objective optimization problem was used for partial classification:

$$\text{Maximize } \{Coverage(R), \ Confidence(R)\}, \tag{5}$$

where R denotes a classification rule. It should be noted that the maximization of the coverage means the maximization of the support since the consequent class is fixed in partial classification. The use of a dissimilarity measure between classification rules instead of the crowding distance in NSGA-II was examined in [11] in order to search for a set of Pareto-optimal classification rules with a large diversity. The Pareto-dominance relation in NSGA-II was modified in [36] in order to search for not only Pareto-optimal classification rules but also dominated (but near Pareto-optimal) classification rules.

Ghosh and Nath [17] used an EMO algorithm to search for Pareto-optimal association rules with respect to *confidence, comprehensibility* and *interestingness*. That is, association rule mining was formulated as a three-objective optimization problem in [17]. A similar three-objective optimization problem was formulated in Kaya [27] where an EMO algorithm was used to search for Pareto-optimal fuzzy association rules with respect to *support, confidence* and *comprehensibility*.

Instead of the two-objective problem in (5) for each class, it is also possible to use the following formulation for all classes:

$$\text{Maximize } \{Support(R), \ Confidence(R)\}. \tag{6}$$

Pareto-optimal rules of this two-objective problem can be searched for by an EMO algorithm. This formulation, however, does not work well when there exists a majority class in the given database (i.e., when the number of patterns from one class is much larger than those from the other classes). For example, the Wisconsin breast cancer data set in the UCI machine learning repository has 458 patterns of benign (Class 1) and 241 patterns of malignant (Class 2). Classification rules of length 3 or less for this data set are shown for each class in the confidence-support space in Fig. 1 (see Section 5 for details of rule extraction in our computational experiments). Pareto-optimal rules for each class with respect to coverage and confidence are depicted by open circles in each plot of Fig. 1. That is, open circles in Fig. 1 show Pareto-optimal rules of the two-objective problem in (5) for each class. When we use the two-objective problem in (6) for all classes (i.e., both classes in Fig. 1) with support and confidence, open circles in Fig. 1 (b) are not Pareto-optimal because they are dominated by some rules in Fig. 1 (a). Of course, open circles in Fig. 1 (b) are Pareto-optimal if we discuss the Pareto optimality for each class with respect to support and confidence. In this case, the maximization of support has the same meaning as the maximization of coverage.

3.2 Techniques to Search for Pareto-Optimal Rule Sets

In classification rule mining ([29], [30], [33], [38]), first an association mining technique such as Apriori [1] is used in the rule discovery phase to efficiently extract all classification rules that satisfy the minimum support and confidence. Then a part of the extracted classification rules are selected using a heuristic rule sorting criterion in the rule selection phase to design a classifier. Evolutionary multiobjective rule selection was proposed in [21] and [22] to search for Pareto-optimal rule sets with respect to accuracy and complexity in the rule selection phase of classification rule mining.

Genetic algorithm-based rule selection was first proposed for the design of accurate and comprehensible fuzzy rule-based classifiers in [24] where a weighted sum fitness function was used to maximize the classification accuracy and minimize the number of fuzzy rules. An EMO algorithm was used to search for Pareto-optimal fuzzy rule-based classifiers with respect to these two objectives in [18]. The total number of antecedent conditions was introduced as the third objective in [19] to minimize not only the number of fuzzy rules but also their length while maximizing the classification accuracy of fuzzy rule-based classifiers. The use of a memetic

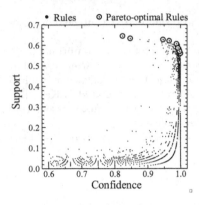   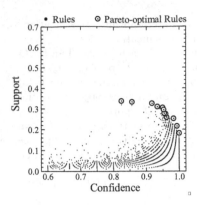

(a) Rules with a Class 1 consequent   (b) Rules with a Class 2 consequent

Fig. 1. Location of rules in the confidence-support space for the Wisconsin breast cancer data set. Pareto-optimal rules with respect to confidence and support for each class are depicted by open circles

EMO algorithm was examined to search for Pareto-optimal fuzzy rule-based classifiers with respect to these three objectives in [25]. Fuzzy rule selection techniques in these studies were used for non-fuzzy classification rule mining in [21] and [22]. The same three-objective problem as in [19] and [25] was handled by multiobjective fuzzy genetics-based machine learning in [23].

4 Evolutionary Multiobjective Rule Selection

Let us assume that we have already extracted N classification rules in the rule discovery phase of classification rule mining. These N classification rules are used as candidate rules in rule selection. Let S be a subset of the N candidate rules (i.e., S is a classifier). We use a binary string of length N to represent S where "1" and "0" mean the inclusion in S and the exclusion from S of the corresponding candidate rule, respectively.

As in our former studies ([21], [22]), we use the following three objectives:

$f_1(S)$: The number of correctly classified training patterns by S,
$f_2(S)$: The number of selected rules in S,
$f_3(S)$: The total number of antecedent conditions over selected rules in S.

The first objective is maximized while the second and third objectives are minimized. The third objective can be viewed as the minimization of the total rule length since the number of antecedent conditions in each rule is often referred to as its rule length. We use NSGA-II ([13], [14]) to search for Pareto-optimal rule sets (i.e., Pareto-optimal subsets of the N candidate rules) with respect to these three objectives. We also use a single-objective genetic algorithm (SOGA) with the $(\mu+\lambda)$-ES generation update mechanism to optimize the weighted sum fitness function of the three objectives for comparison in our computational experiments.

In NSGA-II and SOGA, we use a problem-specific heuristic procedure for decreasing the number of rules in each string. Since the classification of each training pattern is based on a single-winner scheme (i.e., winner-take-all scheme), some rules in a string are used for the classification of many patterns while others are used for the classification of no patterns. When a rule in a string is not used for the classification of any patterns, we can remove this rule from the string without degrading the classification accuracy (i.e., the first objective). At the same time, the removal of such an unnecessary rule improves the second and third objectives. We remove all the unnecessary rules from each string after the first objective is evaluated and before the second and third objectives are evaluated in each generation of NSGA-II and SOGA.

5 Computational Experiments

In this section, first we demonstrate how SOGA can decrease the number of extracted rules and their rule length without severely degrading their classification accuracy through computational experiments on some well-known benchmark data sets in the UCI machine learning repository. Next we show that a large number of nondominated rule sets are obtained by a single run of NSGA-II. We can visualize the accuracy-complexity trade-off using the obtained rule sets. Then we examine the relation between Pareto-optimal rules and Pareto-optimal rule sets by depicting selected rules in the confidence-support space together with candidate rules.

5.1 Conditions of Computational Experiments

We used 12 data sets in Table 1 (though we do not report experimental results on all of these data sets in detail). We did not use incomplete patterns with missing values. All attribute values including discrete attributes were handled as real numbers (e.g., we used 0, 1 and 2 for ternary attributes). The domain of each attribute was divided into multiple intervals using an optimal splitting method [15] based on the class entropy measure [35]. Since the choice of an appropriate number of intervals for each attribute is not easy, we simultaneously used four different partitions with two, three, four, and five intervals (i.e., 14 antecedent intervals in total for each attribute). As a result, various candidate classification rules were examined in the rule discovery phase using overlapping antecedent intervals of various widths for each attribute.

We extracted candidate classification rules with three or fewer antecedent conditions using prespecified values of the minimum support and confidence (in the case of the sonar data set with 60 attributes, we examined candidate rules with just one or two conditions). This restriction on the number of antecedent conditions is to find rule sets with high understandability (i.e., because it is very difficult for human users to intuitively understand long classification rules with many antecedent conditions). This restriction is also to decrease the number of extracted rules. We used a data mining technique, which can be viewed as a variant of Apriori [1], for classification rule mining with the upper bound on the number of antecedent conditions in each rule.

We examined 4×4 combinations of the following four specifications of each threshold for the 12 data sets in Table 1 (in the case of the wine data set, the four values of the minimum support were 4%, 6%, 8% and 10%):

Table 1. Data sets used in computational experiments

Data set	Attributes	Patterns	Classes
Breast W	9	683*	2
Car	6	1728	4
Glass	9	214	6
Heart C	13	297*	5
Iris	4	150	3
Letter	16	20000	26
Nursery	8	12960	5
Sonar	60	208	2
Soybean L	35	266*	19
TicTacToe	9	958	2
Vote	16	232*	2
Wine	13	178	3

∗ Incomplete patterns with missing values are not included

Minimum support: 1%, 2%, 5%, 10%,
Minimum confidence: 60%, 70%, 80%, 90%.

All the extracted classification rules for each combination of the two threshold values were used in evolutionary rule selection as candidate rules. NSGA-II was executed with the following parameter values:

Population size: 200 strings,
Crossover probability: 0.9 (uniform crossover),
Mutation probability: 0.05 (for $0 \rightarrow 1$) and $1/N$ (for $1 \rightarrow 0$) where N is the string length,
Termination conditions: 1,000 generations.

We also used SOGA with the same parameter values to maximize the following weighted sum fitness function:

$$\text{Maximize} \quad f(S) = w_1 \cdot f_1(S) - w_2 \cdot f_2(S) - w_3 \cdot f_3(S), \tag{7}$$

where $\mathbf{w} = (w_1, w_2, w_3)$ is a nonnegative weight vector, which was specified as $\mathbf{w} = (2, 1, 1)$ in our computational experiments.

The classification accuracy on test patterns of candidate rules and selected rules was examined by iterating the twofold cross-validation procedure with 50% training patterns and 50% test patterns five times for each data set (i.e., 5×2 CV). We report average results over its five iterations in the next subsection. In some computational experiments, we show experimental results of only a single run of NSGA-II. In other computational experiments, all the given patterns in each data set were used as training patterns for examining the relation between Pareto-optimal rules and Pareto-optimal rule sets.

5.2 Experimental Results

First we show some experimental results by SOGA to clearly demonstrate the effect of evolutionary rule selection. Experimental results on the Wisconsin breast cancer data set were summarized in Figs. 2 and 3. Each plot on the right-hand side was obtained by applying SOGA to candidate classification rules in the corresponding plot on the left-hand side. For example, about six rules were selected by SOGA in Fig. 2 (b) from thousands of candidate rules in Fig. 2 (a). The deterioration in the classification rates on training patterns by evolutionary rule selection from Fig. 3 (a) to Fig. 3 (b) was less than 1%. When the minimum support was 0.10 (i.e., the rightmost row), the classification rates on training patterns were improved by evolutionary rule selection from Fig. 3 (a) to Fig. 3 (b). The deterioration in the classification rates on test patterns by evolutionary rule selection from Fig. 3 (c) to Fig. 3 (d) was about 1%–2%. The average rule length was decreased by evolutionary rule selection from about 3 in Fig. 2 (c) to less than 2 in Fig. 2 (d). These observations show that only a small number of simple classification rules were selected by SOGA without severely deteriorating the classification accuracy.

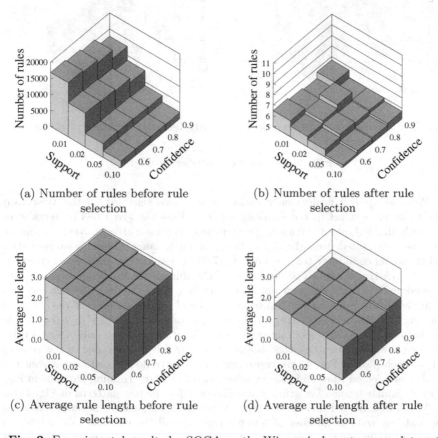

(a) Number of rules before rule selection

(b) Number of rules after selection

(c) Average rule length before rule selection

(d) Average rule length after rule selection

Fig. 2. Experimental results by SOGA on the Wisconsin breast cancer data set

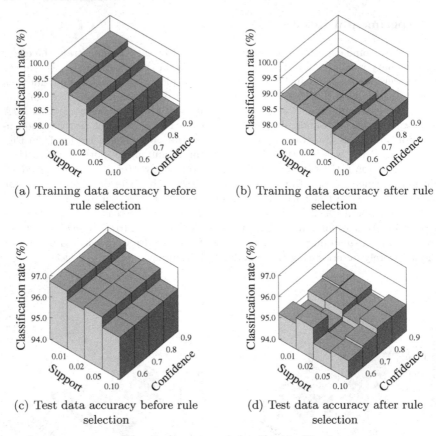

(a) Training data accuracy before rule selection

(b) Training data accuracy after rule selection

(c) Test data accuracy before rule selection

(d) Test data accuracy after rule selection

Fig. 3. Accuracy of classifiers in Fig. 2

We also applied evolutionary multiobjective rule selection to the Wisconsin breast cancer data set in the following manner. First the given 683 patterns were randomly divided into 342 training patterns and 341 test patterns. Next, candidate rules were extracted from the 342 training patterns using minimum support 0.01 and minimum confidence 0.6. As a result, 17,070 classification rules were extracted. Then NSGA-II was applied to the extracted classification rules. From its single run, 20 nondominated rule sets were obtained. Finally each of the obtained rule sets was evaluated for the training and test patterns. The classification rates of the obtained rule sets are shown in Fig. 4 (a) for the training patterns and in Fig. 4 (b) for the test patterns. Some of the obtained rule sets (i.e., rule sets with only a single rule) are not shown because their classification rates are out of the range of the vertical axis of each plot in Fig. 4. We can observe a clear trade-off relation between the number of selected rules and the classification rates on the training patterns in Fig. 4 (a). A similar trade-off relation is also observed for the test patterns in Fig. 4 (b).

While we observed very similar trade-off relations between the accuracy on training patterns and the number of selected rules for all the examined data sets, we obtained totally different results on test patterns. For example, experimental results on the Cleveland heart disease data set are shown in Fig. 5 where we observe a clear deterioration in the accuracy on test patterns due to the increase in the number of selected rules (i.e., the overfitting of selected rules to training patterns).

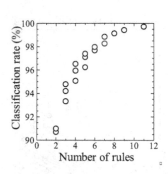

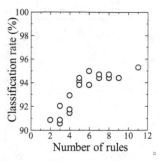

(a) Classification rates on training data   (b) Classification rates on test data

Fig. 4. Experimental results by NSGA-II on the Wisconsin breast cancer data set

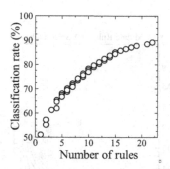

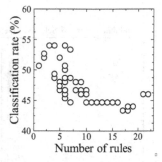

(a) Classification rates on training data   (b) Classification rates on test data

Fig. 5. Experimental results by NSGA-II on the Cleveland heart disease data set

Finally we examined the relation between Pareto-optimal rules and Pareto-optimal rule sets for the Wisconsin breast cancer data set. First we extracted candidate classification rules from all the 683 patterns using minimum support 0.01 and minimum confidence 0.6. Next we applied NSGA-II to the extracted classification rules. Then we chose two rule sets from the obtained nondominated rule sets. One is the most complicated rule set with the highest accuracy on the training patterns. The other is the simplest rule set among rule sets with only two rules. This computational experiment was iterated ten times. Candidate classification rules and selected rules are shown in Fig. 6 for the most accurate rule set and Fig. 7 for the simplest rule set with two rules. It should be noted that these figures include experimental results of ten runs. We can see that Pareto-optimal rule sets do not necessarily consist of only Pareto-optimal rules with respect to confidence and support. We can also see, however, that most of the selected rules are Pareto-optimal or near Pareto-optimal rules in the confidence-support space (especially in Fig. 7).

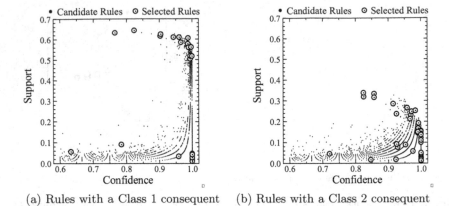

(a) Rules with a Class 1 consequent   (b) Rules with a Class 2 consequent

Fig. 6. Candidate rules and selected rules in the most accurate rule set for the Wisconsin breast cancer data set (experimental results of ten runs are shown)

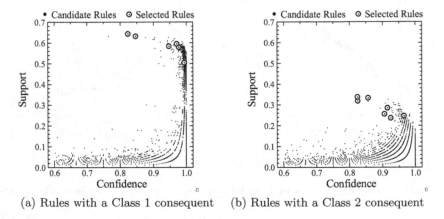

(a) Rules with a Class 1 consequent   (b) Rules with a Class 2 consequent

Fig. 7. Candidate rules and selected rules in the simplest rule set with two rules for the Wisconsin breast cancer data set (experimental results of ten runs are shown)

6 Pareto-Optimal and Near Pareto-Optimal Rules

As shown in Figs. 2 and 3, the number of candidate classification rules and their classification accuracy strongly depend on the choice of the two threshold values: minimum support and confidence. When their values were very small, too many candidate rules were extracted. On the other hand, when their values were too large, many candidate rules were not obtained. Thus the choice of appropriate threshold values is important but difficult. Since Pareto-optimal and near Pareto-optimal rules are likely to be selected as shown in Figs. 6 and 7, it seems a good idea to use only Pareto-optimal and near Pareto-optimal rules as candidate rules. In this section, we examine the effectiveness of this idea.

6.1 Use of Pareto-Optimal Rules as Candidate Rules

In the same manner as in Fig. 4, we applied evolutionary multiobjective rule selection to the Wisconsin breast cancer data set using only Pareto-optimal rules as candidate rules. That is, we chose only Pareto-optimal rules with respect to confidence and support for each class from the generated classification rules in Fig. 4. The chosen Pareto-optimal rules are shown by open circles in Fig. 8. Evolutionary multiobjective rule selection was applied to the Pareto-optimal rules in Fig. 8. Experimental results are shown in Fig. 9. From the comparison between Figs. 4 and 9, we can see that fewer nondominated rule sets were obtained in Fig. 9 than in Fig. 4 (high-accuracy rule sets with many rules were not obtained in Fig. 9 (a)). Whereas the training data accuracy in Fig. 9 (a) was clearly degraded from Fig. 4 (a) by the use of only the Pareto-optimal rules as candidate rules, the test data accuracy was not severely degraded from Fig. 4 (b) to Fig. 9 (b).

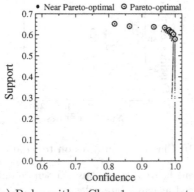 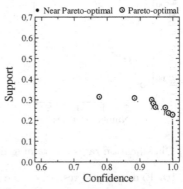

(a) Rules with a Class 1 consequent   (b) Rules with a Class 2 consequent

Fig. 8. Pareto-optimal rules used in Fig. 9 and near Pareto-optimal rules used in Fig. 11. Near Pareto-optimal rules are defined by the ε-dominance with $\varepsilon = 0.01$

In Fig. 10, we show experimental results on the Cleveland heart disease data set where only the Pareto-optimal rules for each class were used as candidate rules. The number of candidate rules was decreased from 15,745 in Fig. 5 to 36 in Fig. 10. The number of obtained nondominated rule sets was also decreased from 42 in Fig. 5 to 13 in Fig. 10. From the comparison between Figs. 5 and 10, we can see that the training data accuracy in Fig. 10 (a) was severely degraded from Fig. 5 (a) by the use of only the Pareto-optimal rules as candidate rules. Rule sets with many rules, which had high training data accuracy in Fig. 5 (a), were not obtained in Fig. 10

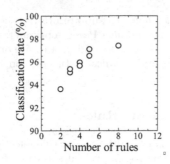
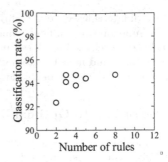

(a) Classification rates on training data   (b) Classification rates on test data

Fig. 9. Experimental results on the Wisconsin breast cancer data set where only Pareto-optimal rules for each class were used as candidate rules

(a). The test data accuracy in Fig. 10 (b), however, was not severely degraded from Fig. 5 (b).

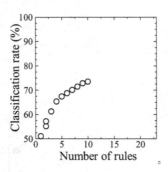
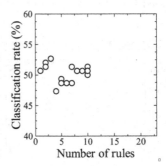

(a) Classification rates on training data   (b) Classification rates on test data

Fig. 10. Experimental results on the Cleveland heart disease data set where only the Pareto-optimal rules for each class were used as candidate rules

6.2 Use of Pareto-Optimal and Near Pareto-Optimal Rules

As shown in the previous subsection, the use of only Pareto-optimal rules as candidate rules had a large effect on the number of candidate rules. In some cases, the number of Pareto-optimal rules was too small. In this subsection, we examine the use of not only Pareto-optimal rules but also near Pareto-optimal rules as candidate rules.

Using a dominance margin ε, we define a modified Pareto dominance in the same manner as [36]. A rule R_i is said to be ε-dominated by another rule R_j when both the inequalities

$$confidence(R_i) + \varepsilon \leq confidence(R_j), \ \ support(R_i) + \varepsilon \leq support(R_j) \qquad (8)$$

hold and at least one of the following two inequalities holds:

$$confidence(R_i) + \varepsilon < confidence(R_j), \ \ support(R_i) + \varepsilon < support(R_j). \qquad (9)$$

When a rule R_i is not dominated by any other rules in the sense of the ε-dominance in (8) and (9), we call R_i an ε-nondominated rule. It should be noted that the ε-dominance with $\varepsilon = 0$ is exactly the same as Pareto dominance. In the other extreme case with $\varepsilon = \infty$, all rules are ε-nondominated rules.

In Fig. 8 of the previous subsection, we showed Pareto-optimal rules (open circles) and near Pareto-optimal rules (small closed circles), which were defined by the ε-dominance with $\varepsilon = 0.01$. Experimental results using Pareto-optimal and near Pareto-optimal rules as candidate rules are shown in Fig. 11. Computational experiments in Fig. 11 ($\varepsilon = 0.01$) were performed in the same manner as in Figs. 4 ($\varepsilon = \infty$) and 9 ($\varepsilon = 0$). From these figures, we can see that an increase in the value of ε leads to an increase in the number of obtained nondominated rule sets. It also improves the training data accuracy (compare Fig. 9 (a) with Fig. 11 (a)).

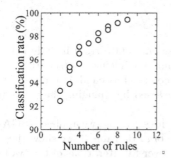 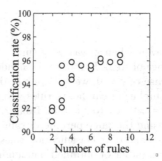

(a) Classification rates on training data   (b) Classification rates on test data

Fig. 11. Experimental results on the Wisconsin breast cancer data set where Pareto-optimal and near Pareto-optimal rules with $\varepsilon = 0.01$ for each class were used as candidate rules

In Table 2, we show the relation between the value of ε and the number of candidate rules (i.e., Pareto-optimal and near Pareto-optimal rules). Near Pareto-optimal rules are defined by the ε-dominance for each value of ε in Table 2. Table 2 shows average results over five iterations of the twofold cross-validation procedure (i.e., ten runs of evolutionary multiobjective rule selection) for each specification of ε for each data set. We used the same parameter specifications in the rule discovery process as in some computational experiments in the previous section:

Minimum support: 1% (4% only for the wine data set),
Minimum confidence: 60%,
Maximum rule length: 3 (2 only for the sonar data set).

We can see from Table 2 that the value of ε has a large effect on the number of candidate rules. In Table 2, many candidate rules were obtained for some data

Table 2. Average number of candidate rules

Data set	Value of ε				
	0	0.001	0.01	0.1	∞
Breast W	18	10083	10312	12573	**16828***
Car	10	667	676	757	**1654***
Glass	240	6755	7098	10597	**13930***
Heart C	65	4318	4543	7789	**14577***
Iris	21	890	893	929	**1140***
Letter	150	646	2479	**5102***	**5102***
Nursery	8	1365	1367	1494	**2572***
Sonar	33	42602	42734	49327	**85780***
Soybean L	4578	32481	36884	57060	**57892***
TicTacToe	14	167	178	1678	**4153***
Vote	8	1344	1363	2146	**4102***
Wine	27	20465	20489	23984	**35530***

* Largest value in each row.

sets even when ε was very small. This is because candidate rules include a large number of weak Pareto-optimal rules with maximum confidence 1.0 (see Fig. 8). In the case of $\varepsilon = 0$, weak Pareto-optimal rules are not used as candidate rules. Thus the number of candidate rules was drastically decreased by changing the value of ε from $\varepsilon = 0.001$ to $\varepsilon = 0$ in Table 2.

Using the same parameter values as in the previous section, we applied NSGA-II to the candidate classification rules for each value of ε. In Table 3, we show the average number of obtained nondominated rule sets over five iterations of the twofold cross-validation procedure. Only when rule sets had different objective vectors, did we count them as different rule sets. Some of different rule sets had the same objective vector. In that case, they were viewed as the same rule set in Table 3. From Table 3, we can see that much fewer nondominated rule sets were obtained in the case of $\varepsilon = 0$ than in the other cases with $\varepsilon > 0$. The average CPU time used for NSGA-II (i.e., used for evolutionary multiobjective rule selection) is shown in Table 4. Our computational experiments were performed using a PC with a Dual Core Pentium D 3.6 GHz processor. From this table, we can see that the CPU time was drastically decreased by using a small value of ε (especially by specifying $\varepsilon = 0$).

As we have already explained, a number of nondominated rule sets were obtained by a single run of NSGA-II for evolutionary multiobjective rule selection in our computational experiments. In each of the ten runs of NSGA-II in five iterations of the twofold cross-validation procedure, we chose the rule set with the highest training data accuracy. That is, we chose the best rule set with respect to the classification rate on training data. Then the classification rate on test data of the chosen rule set was calculated. Average results over ten runs of NSGA-II were summarized in Table 5 for training data and Table 6 for test data. It should be noted that the experimental results in Table 6 are not the best results with respect to the generalization ability since the training data accuracy was used for the choice of a single classifier among obtained nondominated rule sets in each run.

Table 3. Average number of obtained nondominated rule sets

Data set	Value of ε				
	0	0.001	0.01	0.1	∞
Breast W	10.4	16.4	15.8	**16.7***	16.2
Car	8.6	18.1	19.0	**27.7***	23.9
Glass	21.6	**35.7***	35.1	32.8	34.9
Heart C	20.2	52.8	**56.5***	52.4	53.5
Iris	5.0	7.0	7.0	**7.1***	7.0
Letter	61.5	106.3	130.7	**134.3***	**134.3***
Nursery	10.0	10.0	10.2	19.6	**23.2***
Sonar	8.4	17.0	18.1	**19.2***	17.0
Soybean L	48.3	58.7	58.1	57.3	**63.1***
TicTacToe	10.5	25.7	**28.2***	22.6	21.3
Vote	4.0	5.9	6.0	**6.1***	6.0
Wine	8.1	8.6	9.0	9.8	**10.3***

* Largest value in each row

Table 4. Average CPU time for evolutionary multiobjective rule selection (minutes)

Data set	Value of ε				
	0	0.001	0.01	0.1	∞
Breast W	4.0	51.5	135.4	147.5	**170.6***
Car	0.8	8.4	21.4	21.9	**32.1***
Glass	5.7	18.4	40.4	49.9	**55.3***
Heart C	3.1	15.2	20.4	31.3	**51.7***
Iris	1.5	3.5	4.3	4.3	**4.5***
Letter	118.9	191.4	511.9	942.6	**943.7***
Nursery	5.4	90.7	129.4	151.3	**245.7***
Sonar	8.3	92.5	98.0	109.8	**191.0***
Soybean L	33.3	61.7	217.4	276.1	**278.8***
TicTacToe	2.4	4.2	5.3	15.9	**32.9***
Vote	2.7	6.6	8.5	10.5	**13.4***
Wine	5.4	31.6	75.9	90.3	**100.6***

* Largest value in each row

Table 5. Average classification rates on training data of nondominated rule sets with the best training data accuracy

Data set	Value of ε				
	0	0.001	0.01	0.1	∞
Breast W	97.6	**99.7***	**99.7***	**99.7***	**99.7***
Car	69.5	81.7	85.5	**87.7***	84.8
Glass	81.8	97.1	**97.3***	96.9	97.0
Heart C	77.6	96.5	**97.0***	96.0	95.6
Iris	95.6	**98.5***	**98.5***	**98.5***	**98.5***
Letter	53.7	54.6	56.0	**56.3***	**56.3***
Nursery	89.9	89.9	90.0	93.0	**94.0***
Sonar	95.0	**100.0***	**100.0***	**100.0***	**100.0***
Soybean L	90.5	98.3	98.5	**98.9***	**98.9***
TicTacToe	76.0	99.0	99.1	99.7	**99.8***
Vote	97.8	99.7	**99.8***	**99.8***	99.7
Wine	99.5	**100.0***	**100.0***	**100.0***	**100.0***

* Best value in each row.

From Table 5, we can see that the use of only Pareto-optimal rules (i.e., $\varepsilon = 0$) leads to severe deterioration in the training data accuracy in almost all the examined data sets. This is because the number of candidate rules was very small when $\varepsilon = 0$ (see Table 2). We can also see that the highest classification rates on training data were not always obtained from the case where $\varepsilon = \infty$. If we examine all combinations of candidate rules, the highest training data accuracy is always obtained from the case where $\varepsilon = \infty$. This is because candidate rules in this case include all candidate rules in the other cases. The highest training data accuracy, however, was not always obtained from the case where $\varepsilon = \infty$ in Table 5. This means that the best combination of candidate rules was not always found by NSGA-II since the size of the search space was too large (i.e., 2^N where N is the number of candidate rules).

Whereas the training data accuracy was not good in the case of $\varepsilon = 0$ in Table 5, good results were obtained from $\varepsilon = 0$ with respect to the test data accuracy in Table 6. This is because the use of only Pareto-optimal rules as candidate rules prevents the overfitting of selected rule sets to training data (compare Fig. 10 with Fig. 5).

7 Conclusions

In this chapter, first we explained two approaches to evolutionary multiobjective classification rule mining. One is to search for Pareto-optimal rules and the other is to search for Pareto-optimal rule sets. Next we demonstrated the usefulness of evolutionary rule selection as a post-processing procedure in the second phase of

Table 6. Average classification rates on test data of nondominated rule sets with the best training data accuracy in Table 5

Data set	Value of ε				
	0	0.001	0.01	0.1	∞
Breast W	**95.6***	94.3	95.4	95.1	95.0
Car	67.2	80.5	84.3	**86.6***	83.4
Glass	61.9	60.6	**63.2***	60.4	62.7
Heart C	**52.1***	41.5	42.1	40.6	40.5
Iris	**95.4***	**95.4***	95.2	**95.4***	**95.4***
Letter	52.7	53.6	**54.9***	54.8	54.8
Nursery	90.0	90.0	90.1	92.8	**93.8***
Sonar	**74.0***	58.6	62.2	63.7	64.1
Soybean L	**78.4***	65.2	71.7	63.2	69.3
TicTacToe	73.3	98.5	98.4	98.6	**98.9***
Vote	**96.1***	94.7	94.5	94.6	94.2
Wine	**92.8***	90.4	91.3	90.2	90.5

* Best value in each row.

classification rule mining. Then we demonstrated the accuracy-complexity trade-off relation in nondominated rule sets using evolutionary multiobjective rule selection. Finally we examined the use of Pareto-optimal and near Pareto-optimal rules as candidate rules in evolutionary multiobjective rule selection after examining the relation between Pareto-optimal rules and Pareto-optimal rule sets.

The following observations were obtained from computational experiments on some benchmark problems in the UCI machine learning repository.

(1) Clear accuracy-complexity trade-off relations of rule sets were observed on training data for all the examined benchmark problems.

(2) A totally different relation between accuracy and complexity was observed on test data for each benchmark problem. Whereas clear trade-off relations were observed on test data as well as training data for some problems, trade-off relations on training data were reversed on test data for other problems due to the overfitting of rule sets to training data.

(3) Almost all rules comprising small Pareto-optimal rule sets with respect to accuracy and complexity were Pareto-optimal or near Pareto-optimal with respect to confidence and support.

(4) Some rules comprising large Pareto-optimal rule sets were far from the Pareto front with respect to confidence and support.

(5) Restricting candidates to Pareto-optimal and near Pareto-optimal rules using ε-dominance increased the efficiency of evolutionary multiobjective rule selection by decreasing the size of the search space. Whereas this restriction degraded the accuracy of rule sets on training data, the accuracy on test data was not always degraded.

Acknowledgement

This work was partially supported by Grant-in-Aid for Scientific Research on Priority Areas, KAKENHI (18049065), and Grant-in-Aid for Scientific Research (B), KAKENHI (17300075).

References

[1] Agrawal, R., Mannila, H., Srikant, R., Toivonen, H., Verkamo, A. I.: Fast Discovery of Association Rules. In Fayyad, U. M., Piatetsky-Shapiro, G., Smyth, P., Uthurusamy, R. (eds.) Advances in Knowledge Discovery and Data Mining. AAAI Press, Menlo Park (1996) 307–328

[2] Bayardo Jr., R. J., Agrawal, R.: Mining the Most Interesting Rules. Proc. of 5th ACM SIGKDD International Conference on Knowledge Discovery and Data Mining (1999) 145–153

[3] Cano, J. R., Herrera, F., Lozano, M.: Stratification for Scaling up Evolutionary Prototype Selection. Pattern Recognition Letters 26 (2005) 953–963

[4] Cano, J. R., Herrera, F., Lozano, M.: On the Combination of Evolutionary Algorithms and Stratified Strategies for Training Set Selection in Data Mining. Applied Soft Computing 6 (2006) 323–332

[5] Casillas, J., Cordon, O., Herrera, F., Magdalena, L. (eds.): Interpretability Issues in Fuzzy Modeling, Springer, Berlin (2003)

[6] Chiu, C. C., Hsu, P. L.: A Constraint-Based Genetic Algorithm Approach for Mining Classification Rules. IEEE Trans. on Systems, Man, and Cybernetics: Part C — Applications and Reviews 35 (2005) 205–220

[7] Coenen F., Leng, P.: Obtaining Best Parameter Values for Accurate Classification. Proc. of 5th IEEE International Conference on Data Mining (2005) 549–552

[8] Coenen, F., Leng, P., Zhang, L.: Threshold Tuning for Improved Classification Association Rule Mining. Lecture Notes in Artificial Intelligence, Vol. 3518: Advances in Knowledge Discovery and Data Mining — PAKDD 2005. Springer, Berlin (2005) 216–225

[9] Curry, R., Heywood, M. I.: Towards Efficient Training on Large Datasets for Genetic Programming. Lecture Notes in Artificial Intelligence, Vol. 3060: Advances in Artificial Intelligence — Canadian AI 2004. Springer, Berlin (2004) 161–174

[10] de la Iglesia, B., Philpott, M. S., Bagnall, A. J., Rayward-Smith, V. J.: Data Mining Rules using Multi-Objective Evolutionary Algorithms. Proc. of 2003 Congress on Evolutionary Computation (2003) 1552–1559

[11] de la Iglesia, B., Reynolds, A., Rayward-Smith, V. J.: Developments on a Multi-Objective Metaheuristic (MOMH) Algorithm for Finding Interesting Sets of Classification Rules. Lecture Notes in Computer Science, Vol. 3410: Evolutionary Multi-Criterion Optimization — EMO 2005. Springer, Berlin (2005) 826–840

[12] de la Iglesia, B., Richards, G., Philpott, M. S., Rayward-Smith, V. J.: The Application and Effectiveness of a Multi-Objective Metaheuristic Algorithm for Partial Classification. European Journal of Operational Research 169 (2006) 898–917

[13] Deb, K.: Multi-Objective Optimization Using Evolutionary Algorithms. John Wiley & Sons, Chichester (2001)

[14] Deb, K., Pratap, A., Agarwal, S., Meyarivan, T.: A Fast and Elitist Multiobjective Genetic Algorithm: NSGA-II. IEEE Trans. on Evolutionary Computation 6 (2002) 182–197

[15] Elomaa, T., Rousu, J.: General and Efficient Multisplitting of Numerical Attributes. Machine Learning 36 (1999) 201-244

[16] Freitas, A. A.: Data Mining and Knowledge Discovery with Evolutionary Algorithms. Springer, Berlin (2002)

[17] Ghosh, A., Nath, B. T.: Multi-Objective Rule Mining using Genetic Algorithms, Information Sciences 163 (2004) 123–133

[18] Ishibuchi, H., Murata, T., Turksen, I. B.: Single-Objective and Two-Objective Genetic Algorithms for Selecting Linguistic Rules for Pattern Classification Problems. Fuzzy Sets and Systems 89 (1997) 135–150

[19] Ishibuchi, H., Nakashima, T., Murata, T.: Three-Objective Genetics-Based Machine Learning for Linguistic Rule Extraction. Information Sciences 136 (2001) 109–133

[20] Ishibuchi, H., Nakashima, T., Nii, M.: Classification and Modeling with Linguistic Information Granules: Advanced Approaches to Linguistic Data Mining. Springer, Berlin (2004)

[21] Ishibuchi, H., Namba, S.: Evolutionary Multiobjective Knowledge Extraction for High-Dimensional Pattern Classification Problems. Lecture Notes in Computer Science, Vol. 3242: Parallel Problem Solving from Nature — PPSN VIII. Springer, Berlin (2004) 1123–1132

[22] Ishibuchi, H., Nojima, Y.: Accuracy-Complexity Tradeoff Analysis by Multiobjective Rule Selection. Proc. of ICDM 2005 Workshop on Computational Intelligence in Data Mining (2005) 39–48

[23] Ishibuchi, H., Nojima, Y.: Analysis of Interpretability-Accuracy Tradeoff of Fuzzy Systems by Multiobjective Fuzzy Genetics-based Machine Learning. International Journal of Approximate Reasoning 44 (2007) 4–31

[24] Ishibuchi, H., Nozaki, K., Yamamoto, N., Tanaka, H.: Selecting Fuzzy If-Then Rules for Classification Problems using Genetic Algorithms. IEEE Trans. on Fuzzy Systems 3 (1995) 260–270

[25] Ishibuchi, H., Yamamoto, T.: Fuzzy Rule Selection by Multi-Objective Genetic Local Search Algorithms and Rule Evaluation Measures in Data Mining. Fuzzy Sets and Systems 141 (2004) 59–88

[26] Jin, Y. (ed.) Multi-Objective Machine Learning, Springer, Berlin (2006)

[27] Kaya, M.: Multi-Objective Genetic Algorithm based Approaches for Mining Optimized Fuzzy Association Rules. Soft Computing 10 (2006) 578–586

[28] Kupinski, M. A., Anastasio, M. A.: Multiobjective Genetic Optimization of Diagnostic Classifiers with Implications for Generating Receiver Operating Characteristic Curve. IEEE Trans. on Medical Imaging 18 (1999) 675–685

[29] Li, W., Han, J., Pei, J.: CMAR: Accurate and Efficient Classification based on Multiple Class-association Rules. Proc. of 1st IEEE International Conference on Data Mining (2001) 369–376

[30] Liu, B., Hsu, W., Ma, Y.: Integrating Classification and Association Rule Mining. Proc. of 4th International Conference on Knowledge Discovery and Data Mining (1998) 80–86

[31] Llora, X., Goldberg, D. E.: Bounding the Effect of Noise in Multiobjective Learning Classifier Systems. Evolutionary Computation 11 (2003) 278–297

[32] Mitra, S., Pal, S. K., Mitra, P.: Data Mining in Soft Computing Framework: A Survey. IEEE Trans. on Neural Networks 13 (2002) 3–14

[33] Mutter, S., Hall, M., Frank, E.: Using Classification to Evaluate the Output of Confidence-based Association Rule Mining. Lecture Notes in Artificial Intelligence, Vol. 3339: Advances in Artificial Intelligence — AI 2004. Springer, Berlin (2004) 538–549

[34] Pal, S. K., Talwar, V., Mitra, P.: Web Mining in Soft Computing Framework: Relevance, State of the Art and Future Directions. IEEE Trans. on Neural Networks 13 (2002) 1163–1177

[35] Quinlan, J. R.: C4.5: Programs for Machine Learning. Morgan Kaufmann, San Mateo (1993)

[36] Reynolds, A., de la Iglesia, B.: Rule Induction using Multi-Objective Metaheuristics: Encouraging Rule Diversity. Proc. of 2006 International Joint Conference on Neural Networks (2006) 6375–6382

[37] Tan, K. C., Yu, Q., Lee, T. H.: A Distributed Evolutionary Classifier for Knowledge Discovery in Data Mining. IEEE Trans. on Systems, Man, and Cybernetics: Part C — Applications and Reviews 35 (2005) 131–142

[38] Thabtah, F., Cowling, P., Hammoud, S.: Improving Rule Sorting, Predictive Accuracy and Training Time in Associative Classification. Expert Systems with Applications 31 (2006) 414–426

[39] Wang, H., Kwong, S., Jin, Y., Wei, W., Man, K. F.: Multi-Objective Hierarchical Genetic Algorithm for Interpretable Fuzzy Rule-Based Knowledge Extraction. Fuzzy Sets and Systems 149 (2005) 149–186

[40] Wang, H., Kwong, S., Jin, Y., Wei, W., Man, K. F.: Agent-Based Evolutionary Approach for Interpretable Rule-Based Knowledge Extraction. IEEE Trans. on Systems, Man, and Cybernetics: Part C — Applications and Reviews 35 (2005) 143–155

Multiple Objectives in Design and Engineering

Multiple Objective Decision Making

INNOVIZATION: Discovery of Innovative Design Principles Through Multiobjective Evolutionary Optimization

Kalyanmoy Deb* and Aravind Srinivasan

Indian Institute of Technology Kanpur
Kanpur, PIN 208016, India
Email: {deb,aravinds}@iitk.ac.in
http://www.iitk.ac.in/kangal/deb.htm

Summary. In optimization studies, often researchers are interested in finding one or more optimal or near-optimal solutions. In this chapter, we describe a systematic optimization-cum-analysis procedure which performs a task beyond simply finding optimal solutions, but first finds a set of near-Pareto-optimal solutions and then analyses them to unveil salient knowledge about properties which make a solution optimal. The proposed 'innovization' task is explained and its working procedure is illustrated on a number of engineering design tasks. The variety of problems chosen in the chapter and the resulting innovations obtained for each problem amply demonstrate the usefulness of the proposed innovization task. The procedure is a by-product of performing a routine multiobjective optimization for a design task and in our opinion portrays an important process of knowledge discovery which may not be possible to achieve by other means.

Keywords: Innovative design, optimization, engineering design, evolutionary optimization, multiobjective optimization, commonality principles, Pareto-optimal solutions.

1 Introduction

In the context of engineering the design of a system, a product or a process, researchers and practitioners constantly look for innovative solutions. Unfortunately, there exist very few scientific and systematic procedures for achieving such innovations. Goldberg [11] narrates that a competent genetic algorithm – a search and optimization procedure based on natural evolution and natural genetics – can be an effective way to arrive at an innovative design for a single-objective scenario.

---

* Currently occupying the Finnish Distinguished Professor position at the Helsinki School of Economics, Finland (email: Kalyanmoy.Deb@hse.fi)

In this chapter, we extend Goldberg's argument and describe a systematic procedure involving a multiobjective optimization task and perform a subsequent analysis of optimal solutions to arrive at a deeper understanding of the problem, and not simply to find a single optimal (or innovative) solution. In the process of our gaining insights into the problem, the systematic procedure suggested here may often uncover new and innovative design principles which are common to optimal trade-off solutions. Such commonality principles among multiple solutions should provide a reliable procedure for arriving at a 'blueprint' or a 'recipe' for solving the problem in an optimal manner. Through a number of engineering design problems, we describe the proposed 'innovization' procedure and present resulting *innovized* design principles which are useful, not obvious from the appearance of the problem, and not possible to achieve by a single-objective optimization.

In the remainder of the chapter, we describe the importance of considering multiple conflicting objectives in an innovative design task in Section 2. Thereafter, we present the proposed innovization procedure in Section 3. The innovization task is illustrated by applying the procedure to a number of engineering design problems in Sections 4, 5 and 6. Finally, conclusions are made in Section 7.

2 Multiple Conflicting Objectives of Design

The crux of the proposed innovization procedure involves optimization of at least two *conflicting* objectives of a design. When a design is to be achieved for the single goal of minimizing the *size* of a product or of maximizing *output* from it, usually one optimal solution is the target. When optimized, the optimal solution portrays the design, fixes the dimensions, and conveys little else. Although a sensitivity analysis can provide some information about the relative importance of constraints, it only provides local information close to the single optimum solution. Truly speaking, such an optimization task of finding a single optimum design does not often give a designer any understanding deeper than what the optimum solution should look like. After all, how much can a single (albeit optimal) solution in the entire search space of solutions offer?

Let us now think of an optimum design procedure in the context of two or more conflicting goals. Say, we talk about the design of an electric induction motor involving armature radius, wire diameter and number of wiring turns as design variables, and the design goal is to minimize the size of the motor; possibly we shall arrive at a motor which will look small and deliver only a few horsepower (shown as solution A in Figure 1), just enough to run a pump for lifting water to a two-storey building. On the other hand, if we design the motor for the maximum delivered power using the same technology as before, we will arrive at a motor which can deliver, say, a few hundred horsepower, needed to run a compressor in an industrial air-conditioning unit (solution B in Figure 1). However, the size and weight of such a motor will be substantially larger. If we let use a bi-objective optimization method of minimizing size and maximizing delivered power simultaneously, we shall arrive at these two extreme solutions and a number of other intermediate solutions (as shown in the figure) with different trade-offs in size and power, including motors which can be used in an overhead crane to hoist and manoeuvre a load, motors delivering 50 to 70 horsepower which can be used to run a machining centre in a

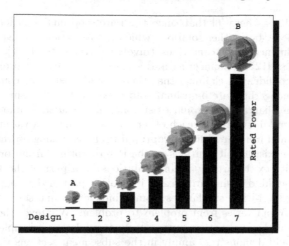

Fig. 1. Trade-off designs show a clear conflict between motor size and power delivered in a range of TEFC three-phase squirrel cage induction motors (data taken from Siemens Ltd. [14]). Despite the differences, are there any similarities in their designs?

factory, and motors delivering about two hundred horsepower which can be used for an industrial exhauster fan.

If we now line up all such motors in worsening order of one of the objectives, say their increased size, in the presence of two conflicting objectives, they would also get sorted in the other objective in an opposite sense (say their increased output). Obtaining such a wide variety of solutions in a single computational effort is itself a significant matter, discussed and demonstrated in various evolutionary multiobjective optimization (EMO) studies in the recent past [5, 3] and in some chapters of this book. Here, we suggest a post-optimality analysis, which should result in a set of innovized principles about the design problem, which we describe next.

After the multiobjective optimization task, we have a set of optimal solutions specifying the design variables and their objective trade-offs. We can now analyse these solutions to investigate if there exist some *common* principles among all or many of these optimal solutions. In the context of the motor design task, it would be interesting to see if all the optimal solutions have an identical wire diameter or an armature diameter proportional or in some relation to the delivered power. If such a relationship between design variables and objective values exists, it is needless to say that it would be of great importance to a designer. Such information will provide a plethora of knowledge on (or a recipe for) how to design the motor in an optimal manner. With such a recipe, the designer can later design a new motor for a new application without resorting to solving a completely new optimization problem again. Moreover, the crucial relationship between design variables and objectives will also provide vital information about the theory of design of a motor, which can bring out limitations and scopes of the existing procedure and spur new and innovative ideas for designing an electric motor.

It is argued elsewhere [6] that since the Pareto-optimal solutions are not any arbitrary solutions, but rather solutions which mathematically must satisfy the so-called Fritz-John necessary conditions (involving gradients of objective and constraint functions) [12] in engineering and scientific systems and problems, we may be reasonably confident in claiming that there would exist some commonalities (or *similarities*) among the Pareto-optimal solutions which will ensure their optimality. On the other hand, there would exist some *dissimilarities* among them which would make them different from each other and place them on various locations on the Pareto-optimal front providing an optimal trade-off among objectives. Whether such similarities exist for all solutions on the Pareto-optimal front, or whether some kind of similarity exist partially among solutions on a part of the Pareto-optimal front and another kind of similarity exists in another part of the front, or there exist hierarchical (or level-wise) similarities (some kinds for all and some sub-kinds for a portion of the front) varies from problem to problem. Whatever the extent of the commonalities, if they exist, they must convey some design principles worth knowing. We argue and demonstrate amply in the subsequent sections that such design principles deciphered from the obtained Pareto-optimal solutions may often bring out new and innovative principles which were unknown earlier. They are also useful in design activities and provide a better understanding of parameter interactions. Since these innovative principles are the outcome of a carefully performed optimization task, we call the procedure an act of 'innovization' – a process of obtaining *innov*ative solutions and design principles through the act of optim*ization*.

3 Innovization Procedure

As described above, the analysis of the optimized solutions will result in worthwhile design principles if the trade-off solutions are really close to the optimal solutions or if they are exactly on the Pareto-optimal front. Since for engineering and complex scientific problem solving we need to use a numerical optimization procedure, and since in such problems the exact optimum is not known a priori, adequate experimentation and verification must first be done to gain confidence about the proximity of the obtained solutions to the actual Pareto-optimal front. In all case studies performed here, we have used the well-known elitist nondominated sorting genetic algorithm, or NSGA-II [7], as the multiobjective optimization tool. NSGA-II begins its search with a random population of solutions and iteratively progresses towards the Pareto-optimal front so that at the end of a simulation run multiple trade-off optimal solutions are obtained simultaneously. For a detail description of NSGA-II, readers are referred to the original study [7]. The NSGA-II solutions are then clustered to identify a few well-distributed solutions. The clustered NSGA-II solutions are then modified by using a local search procedure (we have used Benson's method [1, 5] here). The obtained NSGA-II-cum-local-search solutions are then verified by two independent procedures:

1. The extreme Pareto-optimal solutions are verified by running a single-objective optimization procedure (a genetic algorithm is used here) independently on each objective function, subject to satisfying given constraints.
2. Some intermediate Pareto-optimal solutions are verified by using the normal constraint method (NCM) [15], starting at different locations on the hyperplane constructed using the individual best solutions obtained from the previous step.

When the attainment of optimized solutions and their verifications are made, ideally a data mining strategy must be used to automatically evolve design principles from the combined data of the optimized design variables and their corresponding objective values. By no means is this an easy task, and it is far from being simply a matter of regression over a set of multidimensional data: (i) there may exist multiple relationships, thereby requiring us to find multiple solutions to the problem simultaneously, (ii) a relationship may exist partially in the data set, thereby requiring a clustering procedure to identify which design principles are valid in which clusters, and (iii) since optimized data may not exactly be the optimum data, exact relationships may not be possible to achieve, thereby requiring us to use fuzzy-rule- or rough-set-based approaches. While we are currently pursuing various data mining and machine learning techniques for an automated learning and deciphering of important design principles from optimized data, in this chapter we mainly use visual and statistical comparisons and graph-plotting software for the task. We present the proposed innovization procedure here:

Step 1: Find an individual optimum solution for each of the objectives by using a single-objective GA (or sometimes using NSGA-II by specifying only one objective) or by a classical method. Thereafter, derive and note the *ideal* point.

Step 2: Find the optimized multiobjective front by NSGA-II. Also, obtain and note the *nadir* point[2] from the front.

Step 3: Normalize all objectives using ideal and nadir points and cluster a few solutions $Z^{(k)}$ ($k = 1, 2, \ldots, 10$), preferably in the area of interest to the designer or uniformly along the obtained front.

Step 4: Apply a local search (Benson's method [1] is used here) and obtain the modified optimized front.

Step 5: Perform the normal constraint method (NCM) [15], starting at a few locations, to verify the obtained optimized front. These solutions constitute an optimized front that we can be reasonably confident in.

Step 6: Analyse the solutions for any commonality principles, to be interpreted as plausible innovized relationships.

Since the above innovization procedure is expected to be applied to a problem once and for all, designers may not be especially concerned by the computation time needed to complete the task. However, if needed, the above procedure can be made faster by parallelizing Steps 1, 2, 4 and 5 on a distributed computing machine.

We now illustrate the working of the above innovization procedure on a number of engineering applications.

4 Overhead Crane Manoeuvring

In an optimal operation of an overhead crane, often the performance of a crane operator depends on how quickly (with minimum time of operation) and efficiently

---

[2] It is interesting to note that finding a set of trade-off Pareto-optimal solutions using an evolutionary multiobjective optimization (EMO) procedure is one way of arriving at the nadir point. Finding the nadir point is an important task in the classical multicriterion decision-making approaches, and is reported to be a difficult task [13].

(with minimum power consumption) a task is performed. The task is often to lift a load from an elevation and place it on a truck or a railway wagon by lowering it. Here, we formulate a dynamic model of the trolley and the load (shown in Figure 2), so as to allow the length of the load to vary in the following manner, as a function of position (x) from the starting point of the trolley:

$$l(x) = l_0 + \left(\frac{x}{x_f}\right)^{2\gamma} (l_f - l_0). \tag{1}$$

The length $l(x)$ is computed downwards, and l_0 is the length of the cord connecting the hanging load at $x = 0$ and l_f is the length at $x = x_f$ (the destination of trolley). A condition of $l_f > l_0$ is assumed here. The parameter γ is kept as a variable in the NSGA-II and is allowed to vary within $[-3, 3]$ so that different strategies of lowering the load from the initial point to the destination point are possible, as shown in the right-hand side plot in Figure 2.

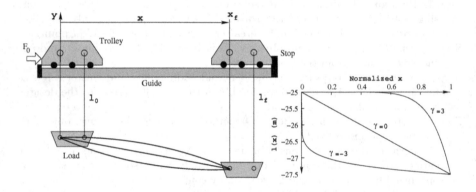

Fig. 2. A schematic of the crane-manoeuvring problem. The right plot shows a true variation with $\gamma = -3$, 0 and 3 with the normalized x (or x/x_f)

The dynamics (differential equations) of the trolley mass and the sway of the load as a function of time are formulated by force balance equations [9]. The terms will involve time-dependent parameters: Applied force F_0, length of cord, angular location of cord from the vertical, position of trolley mass along the guide etc. These equations can then be solved numerically starting with an initial condition of rest, and the simulation continued until the trolley mass reaches the stop at the rightmost position on the guide and the maximum angular displacement of the cord goes below a threshold value α_c. The solution of differential equations will provide the time of operation (the first objective) and the application of step-wise force on the trolley mass will provide the consumed energy (the second objective).

For this problem, NSGA-II considers three sets of variables: (i) magnitude of applied force F_0 (varying in $[100, 1675]$ N, coded in a 10-bit string), (ii) pattern of applying force F_0 (coded in a 500-bit string and applying the force at an interval of $\Delta t = 4$ sec, thereby keeping a maximum of 500×4 or 2,000 sec time for the trolley to reach the destination point), and (iii) γ (treated as a real-valued variable lying within $[-3, 3]$). NSGA-II allows mixed variables (discrete and real) to be handled together. The variable γ is operated by using the simulated binary crossover (SBX) and the polynomial mutation operators [5]. Each of these two operators requires a parameter to be set specifying the extent of the search, and here we use standard values of $\eta_c = 5$ and $\eta_m = 20$ for crossover and mutation, respectively. The time for completing the task is computed by summing the time required for the trolley to reach the destination and the time required for the hanging load to reach a small angle of sway of $\alpha_c = 0.0002$ rad. NSGA-II is run with a population of size 100 and the nondominated solutions found after 1,000 generations are shown in Figure 3. Single-objective EAs are also applied to individually minimize operation time and supplied energy. NSGA-II solutions outperform the single-objective EA solutions in this problem.

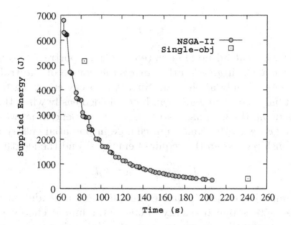

Fig. 3. Nondominated solutions for the crane manoeuvring problem are shown

4.1 Innovized Principles

Table 1 shows the variables (force, pattern of applying force and γ) of a few well-distributed sets of nondominated solutions. It is observed that in most solutions a near-optimal strategy is to apply the load (and hence consumed energy) at the beginning of the task. After sufficient inertia is generated, no further application of load is needed, and the trolley can reach the stop without much velocity, and hence producing little sway of the load. This is an energy-efficient operation of the crane that is somewhat obvious; but the optimized solutions bring out this property as a useful operating condition for solving the crane manoeuvring problem.

Another extremely interesting observation can be made from the variation in γ values of the trade-off near-optimal solutions. Although this variable can take any value within $[-3, 3]$, all solutions seem to have been fixed close to its upper bound (≈ 3.00). A plot of this variation ($\gamma = 3$, marked on the right-hand side plot

Table 1. Near-optimal trade-off solutions arranged in an increasing order of applied force

γ	Force (N)	Pattern
2.98	149.26	111111110010001000000000000000000000000000000000000000
3.00	187.76	11111111001000100000000000000000000000000000
3.00	238.56	111111110010001000000000000000000
3.00	327.86	111110110010011000000000000
2.98	427.93	111110110010011000000
3.00	544.94	11111111001000100
2.87	629.62	111111110010001
2.88	818.99	110111010010001

in Figure 2) reveals that all nondominated solutions portrays a strategy in which the load should not be dropped early during the course of the trolley movement, and should be lowered only at the end. Such a strategy emerging from all optimal solutions is certainly interesting and can be explained easily when this phenomenon is revealed. Assuming the hanging load to be an ideal pendulum, we can equate the supplied energy (E) with the maximum change in potential energy and obtain the following relationship between the required energy E and the length of pendulum l:

$$E = mg(1 - \cos\theta)l. \qquad (2)$$

For a fixed termination condition of a critical $\theta = \alpha_c$, a pendulum having a larger length requires more supplied energy. Thus, for minimal energy consideration a smaller length is better. This is the reason why the optimal strategy for minimum energy-time operation is to keep the length fixed to its lower value (l_0) initially and then lower the load as late as possible.

This problem clearly demonstrates the importance of the proposed innovization task in practice. In the following sections, we show a few more engineering design problems where more concrete and innovative, though yet unknown design principles are revealed.

5 Two-Member Truss Design

We consider a three-variable, two-objective truss design problem, which was originally studied using the ϵ-constraint method [2, 16] and later by an evolutionary approach [5]. The truss (Figure 4) has to carry a certain load without elastic failure. We consider two conflicting objectives of design: (i) minimize total volume of truss members and (ii) minimize the maximum stress developed in both members (AC and BC) due to the application of the 100 kN load. There are three decision variables: cross-sectional area AC (x_1) and BC (x_2) measured in m^2 and the vertical distance between A (or B) and C (y) measured in m. The nonlinear optimization problem is given as follows:

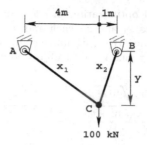

Fig. 4. A two-membered truss structure

$$\text{Minimize } f_1(\mathbf{x}, y) = x_1\sqrt{16 + y^2} + x_2\sqrt{1 + y^2},$$
$$\text{Minimize } f_2(\mathbf{x}, y) = \max(\sigma_{AC}, \sigma_{BC}),$$
$$\text{Subject to } \max(\sigma_{AC}, \sigma_{BC}) \le S_{\max}, \tag{3}$$
$$0 \le x_1, x_2 \le A_{\max},$$
$$1 \le y \le 3.$$

Using the dimensions and loading specified in Figure 4, it can be observed that member AC is subjected to a $20\sqrt{16 + y^2}/y$ kN load and member BC is subjected to an $80\sqrt{1 + y^2}/y$ kN load. The stresses are calculated as follows:

$$\sigma_{AC} = \frac{20\sqrt{16 + y^2}}{yx_1}, \tag{4}$$

$$\sigma_{BC} = \frac{80\sqrt{1 + y^2}}{yx_2}. \tag{5}$$

Here, we limit the stresses to $S_{\max} = 1(10^5)$ kPa and cross-sectional areas to $A_{\max} = 0.01$ m^2. All three variables are treated as real-valued. The simulated binary crossover (SBX) with $\eta_c = 10$ and the polynomial mutation operator with $\eta_m = 50$ are used [5]. All constraints are handled using the constraint-tournament approach developed elsewhere [5]. Figure 5 shows all nondominated solutions obtained by NSGA-II. Although the trade-off between the two objectives is clear from the figure, we perform two other studies to gain confidence about optimality of these solutions. First, we employ a single-objective genetic algorithm to find the optimum of individual objective functions subjected to the constraint and variable bounds. Figure 5 marks these two solutions as '1-obj' solutions. It is evident that the NSGA-II front extends to these two extreme solutions. Next, we use the NCM method [15] with different starting points from a line joining the two extreme solutions. The solution found at the end of each optimization is shown in the figure as well. Since these solutions fall on the NSGA-II front, it gives us confidence that the obtained NSGA-II nondominated solutions are close to the Pareto-optimal front.

5.1 Innovized Principles

Before we discuss the NSGA-II solutions, we perform an exact analysis to find the true Pareto-optimal solutions for this problem. The problem, although simple mathematically, is a typical optimization problem having two resource terms in objectives involving variables x_1 and x_2 each and interlinking them with another variable y. For such problems, the optimum occurs when the identical resource allocation between the two terms in both objective and constraint functions are made:

$$x_1\sqrt{16+y^2} = x_2\sqrt{1+y^2}, \tag{6}$$

$$\frac{20\sqrt{16+y^2}}{yx_1} = \frac{80\sqrt{1+y^2}}{yx_2}. \tag{7}$$

Thus, every optimum solution is expected to satisfy both the above equations, yielding $y = 2$ and $x_1/x_2 = 0.5$. Using $y = 2$ m in the expression for the first (volume) objective, we can also obtain $x_2 = V/2\sqrt{5}$ m^2, where V is the volume (in m^3) of the structure. Substituting these values in the objective functions $V = f_1$ and $S = f_2$, we obtain $SV = 400$ kN, an inverse relationship between the objectives. Thus, the solutions in the Pareto-optimal front are given in terms of volume V, as follows:

$$x_1^* = \frac{V^*}{4\sqrt{5}}\ \text{m}^2, \quad x_2^* = \frac{V^*}{2\sqrt{5}}\ \text{m}^2, \quad y^* = 2\ \text{m}, \quad S^* = 400/V^*\ \text{kPa}.$$

When the variable x_2 reaches its maximum limit, that is, at the transition point T shown in Figure 6, $V_T = 0.04472$ m^3 and $S_T = 8,944.26$ kPa, and x_2 cannot be increased any further.

The inset plot (drawn with a logarithmic scale on both axes) in Figure 5 shows this interesting aspect of the obtained front. There are two distinct behaviours of the optimal front around the transition point T marked in the figure, one spanning from the smallest-volume solution to a volume of about 0.04478 m^3 (point T), and another spanning from this transition point to the smallest stress solution. The extreme

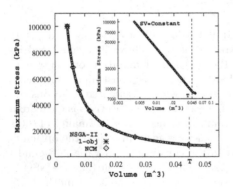

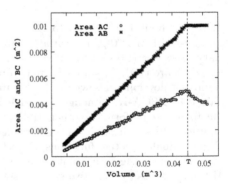

Fig. 5. NSGA-II solutions obtained for the two-member truss structure problem

Fig. 6. Variation of x_1 and x_2 for the truss structure problem

solutions and this intermediate solution, obtained by NSGA-II, are tabulated in Table 2. An investigation of the values of the decision variables reveals the following innovizations:

Table 2. Two extreme solutions and an interesting intermediate solution (T) for the two-member truss design problem are presented

Solution	x_1 (m^2)	x_2 (m^2)	y (m)	f_1 (m^3)	f_2 (kPa)
Min. Volume	4.60(10^{-4})	9.05(10^{-4})	1.935	0.004013	99,937.031
Intermediate (T)	49.30(10^{-4})	99.89(10^{-4})	2.035	0.044779	8,945.610
Min. of max. stress	39.54(10^{-4})	100.00(10^{-4})	3.000	0.051391	8,432.740

1. The inset plot in Figure 5 reveals that for optimal structures, maximum stress (S) developed is inversely proportional to the volume (V) of the structure, that is, $SV = $ constant, as was predicted above. When a straight line is fitted with the logarithm of two objective values, an $SV = 402.2$ relationship is found for NSGA-II solutions. This is close to the true relationship computed above.
2. The inset plot also reveals that the transition occurs at $V = 0.044779$ m^3, which is also close to the exact theoretical value computed above.
3. To achieve a solution with smaller maximum stress (and larger volume) optimally, both cross-sectional areas (AC and BC) need to be increased linearly with volume, as shown in Figure 6. The figure also plots the mathematical relationships (x_1 and x_2 versus V) obtained earlier with solid lines, which can be barely seen, as the obtained NSGA-II solutions fall on top of these lines.
4. A further investigation reveals that the ratio between these two cross-sectional areas is almost 1:2 and the vertical distance (y) takes a value close to 2 m for all solutions.
5. Figure 7 reveals that the stresses developed on both members (AC and BC) are identical for any optimized solution.

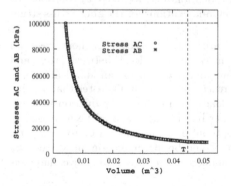

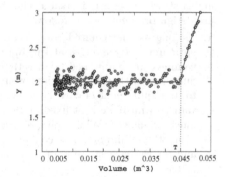

Fig. 7. Variation of stresses in AC and BC of the two-member truss structure problem

Fig. 8. Variation of y for the two-member truss structure problem

These are interesting properties about the design problem which may not be so intuitive to a designer. But the above innovized principles can be explained from the mathematical formulation described. Thus, although these optimality conditions can be derived mathematically from the problem formulation given in Equation 3 in this simple two-membered truss structure design problem, they may be often tedious and difficult to achieve exactly for large-sized and complex problems. Applying a numerical optimization technique and investigating the optimized solutions have the potential of revealing such important innovative principles of design.

5.2 Higher-Level Innovizations

Before we leave this case study, we would like to raise another important aspect of the innovization procedure. Since an analysis is performed on the solutions obtained by solving a particular optimization problem (that is, for fixed values of all problem parameters), one may wonder how the innovized principles will change if different parameter values are used. In the context of the above truss structure design, the parameters kept fixed for the entire analysis were (i) upper limit of developed stress, S_{max}, (ii) upper limit of cross-sectional areas, A_{max}, and (iii) lower and upper bound of y. It would be interesting to investigate whether the innovized principles deciphered above will still be valid parametrically for variations of these parameters. For example, one may think that the reason for the fixed-x_2 solutions (near the smallest stress value) occurred due to the use of a small A_{max}. It may be worthwhile to ponder whether the two-pronged behaviour of the Pareto-optimal front observed above would still remain if the cross-sectional limit A_{max} were increased.

To get a complete idea of the innovized principles, one needs to redo the multi-objective optimization runs for different values of problem parameters and perform further analysis. Figure 9 shows the Pareto-optimal fronts obtained with different A_{max} values and by keeping rest all parameters the same as before. Interestingly, in all simulations the two-pronged behaviour appears, meaning that the property of fixed-y solutions for smaller volume solutions followed by fixed-x_2 solutions for smaller stress values is universal. Higher-level innovizations for the above truss structure design problems are as follows:

1. As long as the required cross-sectional areas can be accepted, there exists an optimum y. By fixing y at this optimum value, a trade-off between stress and volume can be obtained by directly changing x_1 and x_2 by an identical rate.
2. Since x_2 (in this configuration) would reach the upper limit faster than x_1 due to its requirement of carrying a larger load, for any further reduction in stress value, x_2 must be kept fixed at the upper bound and y should be increased as much as allowed. While doing so, the optimal procedure would be to reduce x_1. Thus, the minimum stress configuration would be for the maximum value of x_2 and for the largest value of y or the smallest value of x_1, whichever happens faster.
3. Figure 9 also shows that all fronts produce the same relationship $SV \approx 400$ kN for optimality. Since, $y = 2$ is an optimal solution for any A_{max}, ideally, $SV = 2 \times 20(16 + y^2)/y$ or 400 kN for all cases. Thus, for all optimal trusses having no bounds on cross-sectional size, an optimal truss will have $SV = 400$ kN.

Similarly, further higher-level innovizations can be investigated by varying other fixed parameters.

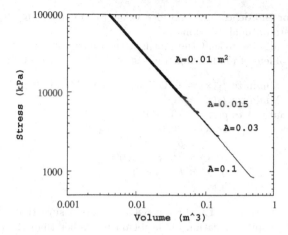

Fig. 9. Pareto-optimal fronts for different values of A_{max} show two-pronged behaviour for the truss structure design problem

6 Welded Beam Design

The welded beam design problem is well studied in the context of single-objective optimization [17]. A beam needs to be welded on another beam and must carry a certain load F (Figure 10). We want to find four design parameters (thickness of the

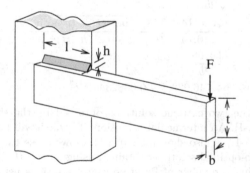

Fig. 10. The welded beam design problem

beam, b, width of the beam t, length of the weld ℓ, and thickness of the weld h) for which the cost of the beam is minimum and, simultaneously, the vertical deflection at the end of the beam is minimum. The overhang portion of the beam has a length of 14 in, and force $F = 6,000$ lb is applied at the end of the beam. It is intuitive that a design which is optimal from the cost consideration is not optimal from rigidity consideration (or end deflection) and vice versa. Such conflicting objectives lead to interesting Pareto-optimal solutions.

In the following, we present the mathematical formulation of the two-objective optimization problem of minimizing the cost and the end deflection [10, 4]:

$$
\begin{aligned}
&\text{Minimize } f_1(\mathbf{x}) = 1.10471h^2\ell + 0.04811tb(14.0 + \ell), \\
&\text{Minimize } f_2(\mathbf{x}) = \frac{2.1952}{t^3 b}, \\
&\text{Subject to } g_1(\mathbf{x}) \equiv 13,600 - \tau(\mathbf{x}) \geq 0, \\
&\qquad\qquad g_2(\mathbf{x}) \equiv 30,000 - \sigma(\mathbf{x}) \geq 0, \\
&\qquad\qquad g_3(\mathbf{x}) \equiv b - h \geq 0, \\
&\qquad\qquad g_4(\mathbf{x}) \equiv P_c(\mathbf{x}) - 6,000 \geq 0, \\
&\qquad\qquad 0.125 \leq h, b \leq 5.0, \\
&\qquad\qquad 0.1 \leq \ell, t \leq 10.0.
\end{aligned}
\tag{8}
$$

There are four constraints. The first constraint makes sure that the shear stress developed at the support location of the beam is smaller than the allowable shear strength of the material (13,600 psi). The second constraint makes sure that normal stress developed at the support location of the beam is smaller than the allowable yield strength of the material (30,000 psi). The third constraint makes sure that thickness of the beam is not smaller than the weld thickness from a practical stand-point. The fourth constraint makes sure that the allowable buckling load (along t direction) of the beam is more than the applied load F. A violation of any of the above four constraints will make the design unacceptable. The stress and buckling terms are highly nonlinear to design variables and are given as follows [17]:

$$
\tau(\mathbf{x}) = \sqrt{(\tau')^2 + (\tau'')^2 + (\ell\tau'\tau'')/\sqrt{0.25(\ell^2 + (h+t)^2)}},
$$

$$
\tau' = \frac{6,000}{\sqrt{2}h\ell},
$$

$$
\tau'' = \frac{6,000(14 + 0.5\ell)\sqrt{0.25(\ell^2 + (h+t)^2)}}{2\{0.707h\ell(\ell^2/12 + 0.25(h+t)^2)\}},
$$

$$
\sigma(\mathbf{x}) = \frac{504,000}{t^2 b},
$$

$$
P_c(\mathbf{x}) = 64,746.022(1 - 0.0282346t)tb^3.
$$

Table 3 presents the two extreme solutions obtained by the single-objective GA and also by NSGA-II. An intermediate solution, T (which will be explained latter), obtained by NSGA-II, is also shown. Figure 11 shows these two extreme solutions and a set of Pareto-optimal solutions obtained using NSGA-II. The obtained front is verified by finding a number of Pareto-optimal solutions using the NC method [15].

6.1 Innovized Principles

Let us now analyse the NSGA-II solutions to decipher *innovized* design principles:

1. Although Figure 11 shows an apparent inverse relationship between the two objectives, the logarithmic plot (inset) shows that there are two distinct behaviours between the objectives. From an intermediate transition solution T (shown in Table 3 and in Figure 11) near the smallest-cost (having comparatively larger

Table 3. The extreme solutions for the welded-beam design problem

Solution	x_1 (h) (in)	x_2 (ℓ) (in)	x_3 (t) (in)	x_4 (b) (in)	f_1	f_2 (in)
Min. Cost	0.2443	6.2151	8.2986	0.2443	2.3815	0.0157
Min. Deflection	1.5574	0.5434	10.0000	5.0000	36.4403	$4.3904(10^{-4})$
Intermediate (T)	0.2326	5.3305	10.0000	0.2356	2.5094	0.0093

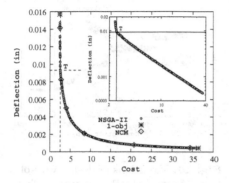

Fig. 11. NSGA-II solutions are shown for the welded beam design problem

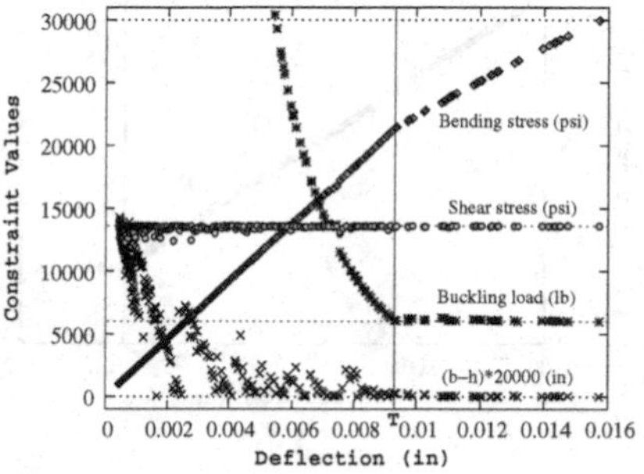

Fig. 12. Constraint values of all Pareto-optimal solutions are shown for the welded beam design problem

deflection) solutions, objectives behave differently than in the rest of the trade-off region. For small-deflection solutions, the relationship is almost polynomial ($f_1 \approx O(f_2^{-0.890})$).

2. Figure 12 plots the constraint values for all trade-off solutions. It is apparent that for all optimal solutions the shear stress constraint is the more critical and active. For small-deflection (or large-cost) solutions, the chosen bending strength (30,000 psi) and allowable buckling load (6,000 lb) are quite large compared to the developed stress and applied load. Any Pareto-optimal solution must achieve the maximum allowable shear stress value (13,600 psi). Thus, in order to improve the design, selection of a material having a larger shear strength capacity would be wise.

3. The transition point (T) between two trade-off behaviours (observed in Figure 11) happens mainly from the buckling consideration. Designs having larger deflection values (or smaller cost values) reduce the buckling load capacity, as shown in Figure 12. When the buckling load capacity becomes equal to the allowable limit (6,000 lb), no further reduction is allowed. This happens at a deflection value close to 0.00932 in (having a cost of 2.509).

4. Interestingly, there are further innovizations with the design variables. For small-deflection solutions, the decision variable b must reduce inversely ($b \propto 1/f_2$) with deflection objective (f_2) to retain optimality. Since for these solutions, only the shear stress constraint is active, and since the shear stress constraint

does not involve the variable b, this variable does not get set by the constraint. On the other hand, b has opposite effects between cost and deflection. Thus, the optimal solutions reflect a similar pattern of variation in b: a reduction in b causes a reduction in cost and an increase in deflection (Figure 12).

5. For small-deflection solutions, the decision variable t remains constant, as shown in Figure 13. This indicates that for most Pareto-optimal solutions, the height

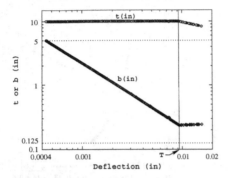
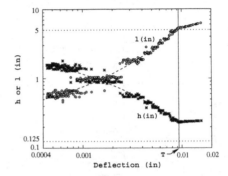

Fig. 13. Variations of design variables t and b across the Pareto-optimal front are shown for the welded beam design problem

Fig. 14. Variations of design variables h and ℓ across the Pareto-optimal front are shown for the welded beam design problem

of the beam must be set to its upper limit. Although t causes an inverse effect on cost and deflection, as apparent from the equations, the active shear stress constraint involves t. Since the shear stress value reduces with an increase in t (apparent from the formulation), it can be argued that fixing t to its upper limit would make a design optimal. Thus, if in practice solutions close to the smallest-cost solution are not desired, a beam of identical height ($t = 10$ in) may be procured, thereby simplifying the inventory.

6. However, an increase in ℓ and a decrease in h with an increase in deflection (or a decrease in cost) are not completely monotonic, as can be seen from Figure 14. These two phenomena are not at all intuitive and are also difficult to explain from the problem formulation. However, the innovized principles for arriving at optimal solutions seem to be as follows: for a reduced cost solution, keep t fixed to its upper limit, increase ℓ and reduce h and b. This 'recipe' of design can be practised only until the applied load is strictly smaller than the allowable buckling load.

7. Thereafter, any reduction in cost optimally must come from (i) reducing t from its upper limit, (ii) increasing b, and (iii) adjusting other two variables so as to make buckling, shear stress, and constraint g_4 active. In these solutions, with decreasing cost the dimensions are reduced in such a manner so as to make the bending stress increase. Finally, the minimum-cost solution occurs when the bending stress equals to the allowable strength (30,000 psi, as in Figure 12). In this solution all four constraints become active, so as to optimally utilize the materials for all four purposes.

8. To achieve very small cost solutions, the innovized principles are different: for a reduced cost solution, reduce t and increase ℓ, h and b. Thus, overall a larger ℓ is needed to achieve a small cost solution.

6.2 Higher-Level Innovizations

Here, we redo the innovization procedure for one different value of three allowable limits: Shear strength in constraint g_1 is increased by 20%, bending strength in constraint g_2 is increased by 20%, and buckling limit load in constraint g_4 is reduced by 50%. We change them one at a time and keep the other parameters identical to their previous values. Figure 15 shows the corresponding Pareto-optimal fronts for these three cases. The following innovizations are obtained:

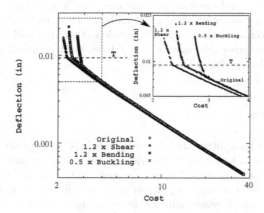

Fig. 15. Effect of material strength and buckling load limit on the Pareto-optimal front for the welded beam design problem

1. It is clear that all three cases produce similar dual behaviour (different characteristics on either side of a transition point) in the Pareto-optimal front, as was also observed in the previous case. All other innovizations (such as t being constant and b being smaller with increasing deflection, etc.) mentioned earlier remains the same in all three cases.
2. The minimum-cost solution depends on all three constraint (g_1, g_2 and g_4) limits, but the minimum-deflection solution only depends on the limit of the shear stress constraint (g_1). However, with this solution, variables t and b take their largest allowable values of 10 in and 5 in, respectively.
3. An increase of shear strength by 20% causes the solutions to change. Recall that the shear stress constraint (g_1) was the most critical constraint in the original case. An increase in shear strength value also makes the constraint active for all new trade-off solutions. Since solutions change, a slightly different trade-off front emerges. Interestingly, the location of the transition point along the deflection axis does not change (since the buckling load limit is not changed).

4. An increase in bending strength of 20% does not change smaller-deflection solutions. Since a higher bending limit is allowed now, better cost solutions are found. A solution with a cost of 2.3545 is now obtained with a deflection value of 0.021 in. The location of the transition point in unaffected by this change in bending strength value.

5. Finally, a decrease in the buckling load limit by 50% changes the location of the transition point (which moves towards a larger-cost solution); however the rest of the original Pareto-optimal front remains identical to the original front.

Thus, we conclude with confidence that (i) shear strength has a major role to play in deciding the optimal variable combinations (the shear stress constraint remains active in all cases), (ii) bending strength has an effect on the smallest-cost solution alone, as only this solution makes the bending constraint active, and (iii) the buckling load limit has the sole effect of locating the transition point on the Pareto-optimal front. These pieces of information provide adequate knowledge about the relative importance of each constraint and the variable interactions for optimally designing a welded beam over an entire gamut of cost-deflection trade-offs. It is unclear how such valuable innovative information could have been achieved otherwise merely from a mathematical problem formulation.

7 Conclusions

In this chapter, we have introduced a new knowledge discovery procedure (which we called 'innovization'), based on multiobjective optimization and a post-optimality analysis of optimized solutions. We have argued that the task of a single-objective optimization results in a single optimum solution which may not provide enough information about useful relationships between design variables, constraints and objectives for achieving different trade-off solutions. On the other hand, consideration of at least two conflicting objectives of design should result in a number of optimal solutions, trading off the two objectives. Thereafter, a post-optimality analysis of these optimal solutions should provide useful information and design principles about the problem, such as relationships among variables and objectives which are common among the optimal solutions and differences which make the optimal solutions different from each other. We have argued that such information should often introduce new principles for optimal designs, thereby allowing designers to uncover innovations for solving the problem at hand.

On a number of engineering design problems having mixed discrete and continuous design variables, many useful innovizations (innovative design principles) are revealed. Interestingly, many such innovizations were not intuitive and not known before. The ease of application of the proposed innovization procedure has also become clear from different applications. It is also clear that the proposed procedure is useful and ready to be used in other more complex design tasks. The procedure will enable designers to perform the innovization task once and for all to the problem at hand and the knowledge thus gained will go a long way in understanding the intricacies of the problems and in solving such future design tasks. On another note, since the Pareto-optimal front obtained using NSGA-II are verified by other single-objective optimization techniques, the reported trade-off solutions also remain as 'benchmark' optimal solutions to these problems.

However, the innovization procedure suggested here must now be made more automatic and problem independent, as far as possible. In this regard, an efficient data mining technique is in order to evolve innovative design relationships from the Pareto-optimal solutions. Although some apparent hurdles of this task have been pointed out in this chapter, we are currently pursuing ways to overcome them.

Finally, it is also worth mentioning that similarly to the expectation of common properties existing among Pareto-optimal solutions (as discovered and demonstrated amply in this chapter), commonality principles may also be expected to exist in other kinds of trade-off solutions, such as among weakly Pareto-optimal solutions, locally Pareto-optimal solutions [5], and robust or reliable Pareto-optimal solutions [8]. It would be interesting then to investigate how the innovized relationships get changed from one type of optimal solution to the other. For example, such an analysis may provide answers to questions such as how *robust* Pareto-optimal solutions differ from Pareto-optimal solutions themselves. Another interesting extension of this study would be to consider three or more conflicting objectives of design, and a resulting post-optimality analysis may yield higher-level innovizations than those that may be obtained with the two-objective procedure. The ease and ability of NSGA-II to handle different vagaries of design variables (discrete, Boolean, real-valued, etc.), nonlinearities in constraint and objective functions, scalability in problem size, and multimodality and multi-objectivity in problem formulations make such an innovization task tractable and worth performing.

References

[1] H. P. Benson. Existence of efficient solutions for vector maximization problems. *Journal of Optimization Theory and Applications*, 26(4):569–580, 1978.

[2] V. Chankong and Y. Y. Haimes. *Multiobjective Decision Making Theory and Methodology*. New York: North-Holland, 1983.

[3] C. A. C. Coello, D. A. VanVeldhuizen, and G. Lamont. *Evolutionary Algorithms for Solving Multi-Objective Problems*. Boston, MA: Kluwer Academic Publishers, 2002.

[4] K. Deb. An efficient constraint handling method for genetic algorithms. *Computer Methods in Applied Mechanics and Engineering*, 186(2–4):311–338, 2000.

[5] K. Deb. *Multi-objective optimization using evolutionary algorithms*. Chichester, UK: Wiley, 2001.

[6] K. Deb. Unveiling innovative design principles by means of multiple conflicting objectives. *Engineering Optimization*, 35(5):445–470, 2003.

[7] K. Deb, S. Agrawal, A. Pratap, and T. Meyarivan. A fast and elitist multi-objective genetic algorithm: NSGA-II. *IEEE Transactions on Evolutionary Computation*, 6(2):182–197, 2002.

[8] K. Deb and H. Gupta. Searching for robust Pareto-optimal solutions in multi-objective optimization. In *Proceedings of the Third Evolutionary Multi-Criteria Optimization (EMO-05) Conference (Also Lecture Notes on Computer Science 3410)*, pages 150–164, 2005.

[9] K. Deb and N. Gupta. Optimal operating conditions for overhead crane maneuvering using multi-objective evolutionary algorithms. In *Proceedings of the Genetic and Evolutionary Computation Conference, (GECCO-2004)*, pages 1042–1053, 2004. Lecture Notes in Computer Science (LNCS) 3102.

[10] K. Deb and A. Kumar. Real-coded genetic algorithms with simulated binary crossover: Studies on multi-modal and multi-objective problems. *Complex Systems*, 9(6):431–454, 1995.

[11] D. E. Goldberg. *The design of innovation: Lessons from and for Competent genetic algorithms*. Kluwer Academic Publishers, 2002.

[12] J. Jahn. *Vector optimization*. Berlin, Germany: Springer-Verlag, 2004.

[13] P. Korhonen, S. Salo, and R. Steuer. A heuristic for estimating nadir criterion values in multiple objective linear programming. *Operations Research*, 45(5): 751–757, 1997.

[14] S. Ltd. TEFC 3 phase squirrel cage induction motor catalogue, http://globaludyog.com/pumps/sl.htm.

[15] A. Messac and C. A. Mattson. Normal constraint method with guarantee of even representation of complete pareto frontier. *AIAA Journal*, in press.

[16] K. Miettinen. *Nonlinear Multiobjective Optimization*. Kluwer, Boston, 1999.

[17] G. V. Reklaitis, A. Ravindran, and K. M. Ragsdell. *Engineering Optimization Methods and Applications*. New York : Wiley, 1983.

User-Centric Evolutionary Computing: Melding Human and Machine Capability to Satisfy Multiple Criteria

Ian C. Parmee[1], Johnson A. R. Abraham[2], Azahar Machwe[1]

[1] ACDDM Lab, Faculty of Computing, Engineering and Mathematical Sciences, University of the West of England, Frenchay Campus, Coldharbour Lane, Bristol, BS16 1QY, UK ian.parmee@uwe.ac.uk; azahar.machwe@uwe.ac.uk

[2] Level E Ltd, ETTC, The King's Buildings, Mayfield Road, Edinburgh EH9 3JL, UK johnson@levelelimited.com

Summary. This chapter centres around the use of interactive evolutionary computation as a search and exploration tool for open-ended contexts in design. Such contexts are characterized by poor initial definition and uncertainty in terms of objectives, constraints and defining variable parameters. The objective of the research presented is the realization of 'user-centric' intelligent systems, i.e., systems which can overcome initial lack of understanding and associated uncertainty, whilst also stimulating innovation and creativity through a high degree of human / machine interaction. Two application areas are used to illustrate how, through the adoption of bespoke visualization techniques, flexible representations, and machine learning agents that 'observe' the evolutionary process, this objective can be achieved.

1 Introduction

Conceptual design and early-stage decision-making processes are largely people-centred activities where human judgement, intuition, experiential knowledge and personal / team preference play major roles. Handling multiple criteria that are so often poorly defined during these stages is extremely difficult. Although some form of machine-based problem representation may be available, there is generally a degree of uncertainty in terms of its fidelity. To attempt to develop a closed, definitive computational approach to multi-criteria satisfaction under these conditions is not only an extremely difficult task but one which could be considered misguided. The definitive inclusion of the many human-centred nuances associated with preference, previous experience, political reasoning, intuition and is beyond current capability and this situation is unlikely to change in the short, medium or even long term. Whether it is necessary or desirable to attempt to replace the rich human capability that can so often introduce innovation and creativity is a pertinent question. A far more natural and potentially powerful approach is to identify the manner in which the complex individual and team-based human behaviour associated with early stage design and decision making can best be supported and enhanced by computational technology. We must not lose this behaviour through the introduction of inflexible computational approaches. Far better to develop systems that capture user knowledge, experience and intuition and utilize these in a mutually beneficial manner whilst also learning from the interaction with regard to future system development.

The chapter centres around the use of evolutionary computation as a search and exploration process from which high-quality, relevant information can be extracted to support a better understanding of poorly defined and uncertain problem areas. The use of multiple criteria and subjective evaluation by users is critical in the problem areas we consider. In these contexts, related work now extends across several fields which present differing levels of complexity. Typical difficulties experienced relate to representation of the problem space and the multiple criteria (qualitative and quantitative) which together provide an evaluation of generated solutions.

An overview of two application areas illustrating the nature of these problems and the manner in which they can be overcome is presented. References direct the reader to associated papers providing far more detail. Differing levels and types of user interaction are apparent, ranging along the explicit-implicit spectrum introduced in Parmee and Abraham (2005), which moves from well-established interactive evolutionary computation (IEC) (Takagi and Ohsaki 1999; Kim and Cho 2005; Carnahan and Doris 2004) to interaction involving the extraction of information and associated continuous development of the problem representation. Other computational intelligence approaches for machine learning and search and exploration are included.

The manner in which user interaction can overcome problems relating to poor representation and poorly defined objectives within EC systems is investigated. The objective is the realization of user-centric intelligent systems that overcome initial lack of understanding and associated uncertainty with regard to multiple criteria and support an improving knowledge base whilst stimulating innovation and creativity.

2 Evolving Appropriate Representations

Problem formulation and reformulation is a well-known research area within the design research community, especially when considering innovative and creative design [Gero 1994, Goel 1997, Su 1990]. This is associated with the development of a designer's understanding of a problem during the early investigative stages that may result in radical changes in problem representation and criteria formulation and ranking. There is also the integration of knowledge through analogical or metaphorical transfer from another problem area which can be of significant benefit to the development of innovative approaches (Gero and Shi 1999; Brown 1998). The following sections, however, concentrate on the utilization and integration of computational intelligence (CI) techniques with these early decision-making processes with the aim of the eventual establishment of user-centred intelligent systems that greatly enhance human capability.

Uncertainty and poor definition are inherent features during early stage design / decision making. An immediate requirement for information to improve understanding can be confounded by many interacting variable parameters and multiple objectives that defy full quantitative representation and require a degree of subjective evaluation. Problem representation may, in the first instance, be based merely upon qualitative mental models arising from experiential knowledge, discussion and sparse available data. Mental representations play a significant role in defining initial direction. Concepts based upon current understanding require both quantitative and qualitative exploration to generate further relevant information that supports and enables meaningful progress.

Generally, the development of computational problem representations supports exploration through the evaluation of solutions against criteria perceived to be relevant *at a particular point in time*. Initial representations based upon current understanding and any available relevant data will likely be relatively basic, and user confidence in the fidelity of model output may therefore be low. However, such representations provide essential problem insight despite their apparent shortfalls. Seemingly high performance solutions identified in terms of quantitative criteria followed by qualitative human evaluation utilizing experiential knowledge and intuition provides an indication of concept viability and model fidelity. An iterative user- / machine-based exploratory process can commence where gradual improvements in understanding contributes to better representations, a developing knowledge-base and the eventual establishment of computational models that support more rigorous analysis. A highly interactive process thus emerges supporting the development of representation through knowledge discovery. Such a human- / machine-based development may run concurrently with, and be enhanced by, other forms of investigation and data / information gathering.

A high degree of assumption relating to objective representation generally provides a starting point for investigation. An initial variable parameter set may be selected with later addition or removal of variables as the sensitivity to various aspects becomes apparent. Constraints may be softened to allow exploration of perceived non-feasible regions. Quantitative objectives may change as significant payback becomes apparent through a reordering of objective preferences. Some non-conflicting objectives may merge whilst conflict between others may stimulate problem reformulation. The initial design space is therefore a moving feast rich in relevant and potentially opinion-changing information (Parmee 2002). It is quite possible that final solutions will be identified from a space that bears little resemblance to the problem space that provided a starting point for investigation.

During early design and decision-making processes we could be considered to be concurrently negotiating two design spaces, i.e.,

1. The machine-based quantitative space that is bounded and inflexible when considered stand-alone (i.e., the space defined by all possible variable combinations of a computational design model). Evolutionary search and exploration utilizing machine-based criteria representations to evaluate solutions can rapidly provide novel information from this space that aids problem understanding at a human level. Such understanding and subsequent search space redefinition can radically alter the initial bounds.
2. The investigators' mental representations of the problem. These representations are bounded only by current knowledge and understanding. The development of this problem space relies upon external stimuli that includes the output from machine-based representation plus human intuition and judgement at both a quantitative and a qualitative level.

The appropriate melding of these two spaces supports a holistic, knowledge-based approach that can result in significant step changes to machine-based objective representation and in overall understanding. This could be considered a general description of how we progress when faced with poorly defined problems that initially seem beyond our perceived analytic capabilities.

Using this description the chapter explores a human-centric utilization of evolutionary computation, machine learning and agent-based approaches integrated with enabling computational technologies to significantly enhance knowledge discovery in terms of variable space / objective space relationships and associated representation development processes. Novel human-centred computational design processes should lead to innovation and competitive product development through continuous knowledge discovery. The continuous development of both qualitative and quantitative objectives plays a primary role in this process.

3 Multiobjective Information Generation, Capture and Visualization

Earlier related research by the authors investigated the visualization of design information within interactive evolutionary design (IED) processes (Parmee et al. 2001; Parmee 2001; Parmee 2002). These user-centric processes have increasingly included the integration of software agent-assisted analysis of GA output that provides further support to the designer in the identification of complex relationships between variables and multiple objectives (Cvetkovic and Parmee 2001). Designer preference has also been extensively addressed (Cvetkovic and Parmee 2003) and novel evolutionary multiobjective approaches have been introduced (Parmee and Watson 1999).

The concurrent further development of the cluster-oriented genetic algorithm (COGA; Parmee 1996) has provided the means to extract wide-ranging information relating to appropriate variable ranges, solution and variable sensitivity and the degree of conflict between included objectives.

Cluster-oriented genetic algorithms (COGAs) provide the means to identify high-performance (HP) regions of complex conceptual design spaces. COGAs identify HP solution regions through the online adaptive filtering of GA-generated solutions. COGA comprises two primary components: the diverse search engine which utilizes a genetic algorithm to search the design space, identifying regions of high performance relating to a particular objective, and the adaptive filter (AF), which extracts and stores information relating to each identified region. The adaptive filter (AF) copies high fitness designs from the evolving population to the final clustering set (FCS). The user can vary the severity of the filtering mechanism in order to identify regions ranging from succinct groupings of very high-performance solutions to larger regions of high- and lower-performance solutions. Sufficient regional set cover (in terms of number of solutions) can be achieved to allow significant qualitative and quantitative design information to be extracted. COGA development and application has been well documented. Many associated papers can be downloaded from http://www.ad-comtech.co.uk/Parmee-Publications.htm

COGAs played a significant role within the Interactive Evolutionary Design System (IEDS), providing two-dimensional (variable) projections of high-performance

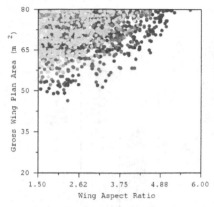

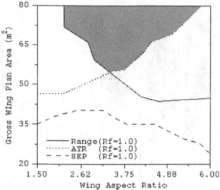

Fig. 1. COGA single-objective output showing projection of high-performance (HP) solutions relating to ATR objective onto a two-variable hyperplane

Fig. 2. COGA air frame design output relating to three objectives again projected onto 2-D variable space. This gives clear indications of degree of conflict between objectives

regions relating to conceptual airframe design of military aircraft (Parmee and Bonham 1999; Bonham and Parmee 1999). The airframe conceptual design model is represented by nine variable parameters mainly in the form of design ratios. There are 11 possible outputs, each of which can be considered an objective. The HP region projections, relate to both single (Fig. 1) and multiple objectives (Fig. 2). The projections provide excellent graphical representations from which much relevant information can be extracted as described in Parmee and Abraham 2004. In particular, objective conflicts are plain to see in those hyperplanes comprising prime variables to which all included objectives are particularly sensitive, as is the case in Fig. 2. Here, the shaded region comprises HP solutions that are common to both ATR and FR objectives, thereby illustrating low conflict whereas the remoteness of the SEP high-performance region indicates a higher degree of conflict. The hyperplane shown in both figures relates to Gross Wing Plan Area and Wing Aspect Ratio variables, whereas the objective projections relate to Attained Turn Rate (ATR), Ferry Range (FR) and Specific Excess Power (SEP) objectives.

However, the designer cannot be expected to search through all 2-D hyperplanes to benefit from such a clear graphical representation hence the development of the parallel coordinate box plot (PCPB) (Fig. 3), which, from an analysis of COGA output, gives an overall perspective of much of the information relating to variable and objective sensitivities. (Parmee and Abraham 2004).

The PCBP acts as a 'one-stop shop' which the user can visit and decide on which 2-D hyperplanes to view. The PCBP shows the HP solution distribution of each objective across all variable dimensions (V_n). The length of the three vertical axes related to each variable indicates to what extent the COGA HP solution output

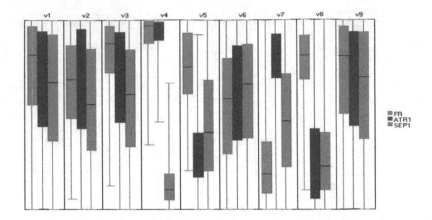

Fig. 3. The Parallel Co-ordinate Box Plot (PCBP)

for each objective covers each variable range. The colour-coded box plots relate to each objective. The median is marked within the box and the box extends between the lower and upper quartile HP solution values within the variable set. The degree of overlap of the three boxes indicates the degree of conflict between the objectives.

For instance, the box plots of variables 4, 5, 7 and 8 suggest high degrees of conflict that could be better expressed graphically by viewing appropriate pair-wise hyperplanes (one of which is shown in Fig. 2). A full description of this representation can be found in Parmee and Abraham 2004. In addition, COGA data can be processed in terms of variable attribute relevance analysis (Inselberg 1985) in addition to standard skewness calculations to verify the visual information available in the PCBP (Abraham and Parmee 2004). The resulting ranking identifies variables 4, 5, 7 and 8 as those variables to which the objective set is most sensitive. Hence, the user can concurrently view the variable / objective interaction from a number of different perspectives.

A further perspective is achieved by mapping COGA output onto objective space (Fig. 4). This mapping supports a far better understanding of the spatial relationships between high-performance solutions that lie on and close to a Pareto frontier (Fig. 5). For comparison, a Pareto front has been generated by SPEA (Zitzler et al. 2002). A nondominated sorting of the COGA high-performance sets results in an approximate Pareto Front (Fig. 7) which is complete in terms of the ATR and FR objectives due to low conflict and the common HP region, but which would be incomplete in terms of the SEP / ATR1 objectives due to high conflict and no common region (see Fig. 6). Reducing the severity of the AF in subsequent SEP and/or ATR COGA runs allows lower-performing solutions into the FCS, thereby closing this 'conflict gap'. If the setting of the AF filter is too high, the missing solutions cannot enter the FCS. Lower the AF setting, and they enter the FCS and the gap is closed.

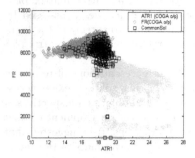

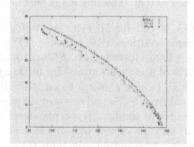

Fig. 4. COGA projection of HP solutions relating to two objectives. The lighter shaded clusters show HP solutions relating to ATR and FR objectives. The common black HP solutions satisfy both

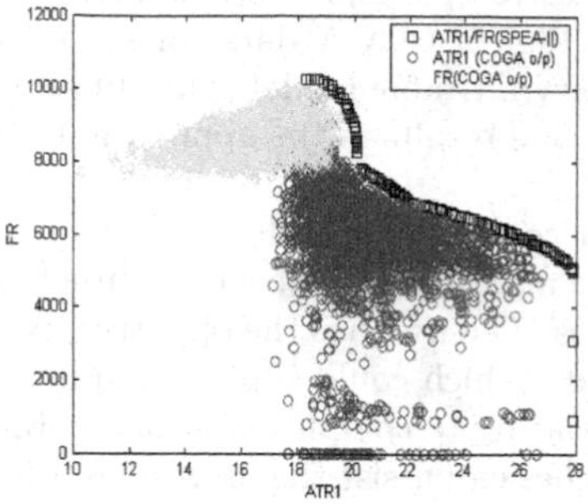

Fig. 5. Similar projection illustrating how the edges of the cluster provide a close approximation to a Pareto frontier generated for the two objectives using the SPEA MOGA approach

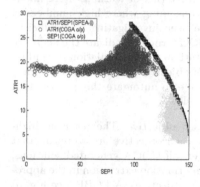

Fig. 6. The distribution of solutions for objective ATR1 and SEP1 against SPEA-II Pareto front. Note the 'conflict gap' that relates to the disjoint HP regions of Fig. 2

Fig. 7. Nondominated sorting of the high-performance sets of the COGA runs relating to ATR1 and FR results in overlapping curves that represent a full approximate Pareto front

The objective space projections can be represented in a scatter matrix plot, as shown in Fig. 8. COGA data for six of a total of 11 objectives is assessed by the preliminary airframe model. The subsequent nondominated sorting of the high-performance sets results in the approximate Pareto curves of the scatter matrix plot in Fig. 9.

This user-centred approach generates highly visual representations of results in both variable and objective space. A direct mapping for each solution exists across the two spaces. The user has the opportunity to view several differing perspectives of generated data which could support implicit learning and the development of tacit knowledge relating to complex variable / objective interactions. It is suggested that such approaches can assist the user in building a better 'intuitive map' of the highly complex relationships.

COGA is inherently user-interactive as the user can explore conflict relationships via the settings of an Adaptive Filter (Parmee 1996). The AF controls the entry of lower performance solutions of selected objectives into the final high-performance solution sets. By lowering the AF threshold it is possible to close the 'conflict gap' evident in Figs. 2 and 6 between ATR1 and SEP and between FR and SEP. The user must first decide which objective should be compromised in terms of accepting lower-performing solutions.

Although this exploration can be beneficial in terms of learning more about the problem space, it could also be very time consuming where high numbers of objectives exist. More recent work has therefore been investigating how sufficient data can be generated through relatively short COGA runs to indicate the degree of conflict between objectives via graphical presentation. The assimilation of conflict characteristics by the user supports the setting of objective preferences which allow the relaxation of the AF setting. It is also possible to automate this process to a greater or lesser extent. The process is outlined below.

Step 1: COGA is run for 50 generations for each objective. The resulting high-performance solutions that pass the AF for each objective are stored in that objective's final clustering set FCS_i. However, all COGA generated solutions, irrespective of fitness, are stored for further information extraction in the appropriate set D_i (where i relates to an individual objective). A PCBP (see Fig. 3) is generated from FCSs and the user determines which variables are causing the highest degree of conflict.. Hyperplanes relating to any two objectives (similarly to Fig. 2) can be viewed to provide a clear impression of the degree of conflict and objective space projections (Fig. 8), and approximate Pareto frontiers (Fig. 9) can also be studied.

Step 2: Where no significant conflict is evident between objectives from the PCBP, hyperplane projections and/or projections onto objective space, these objectives can be grouped and need not be included further in the process as high-performance solutions are evident across all of them. This reduces the overall number of objectives that require attention.

Step 3: Having assimilated available information, the user then ranks the remaining conflicting objectives from 1 to 5 to indicate the degree of importance of each objective. The higher the rank, the less the objective performance should be allowed to be compromised by lower AF settings and the resulting lower performance solutions in the FCS.

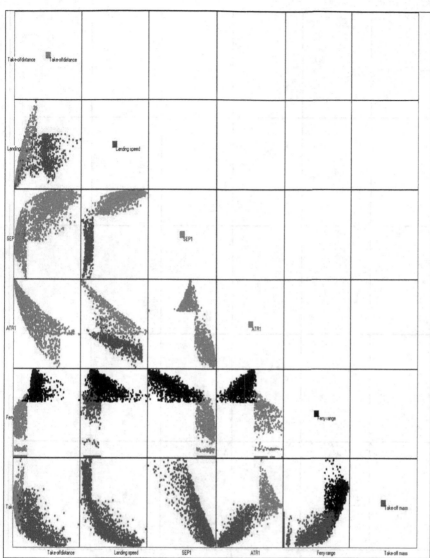

Fig. 8: Scatter Matrix plot of solutions relating to six objectives in the objective space

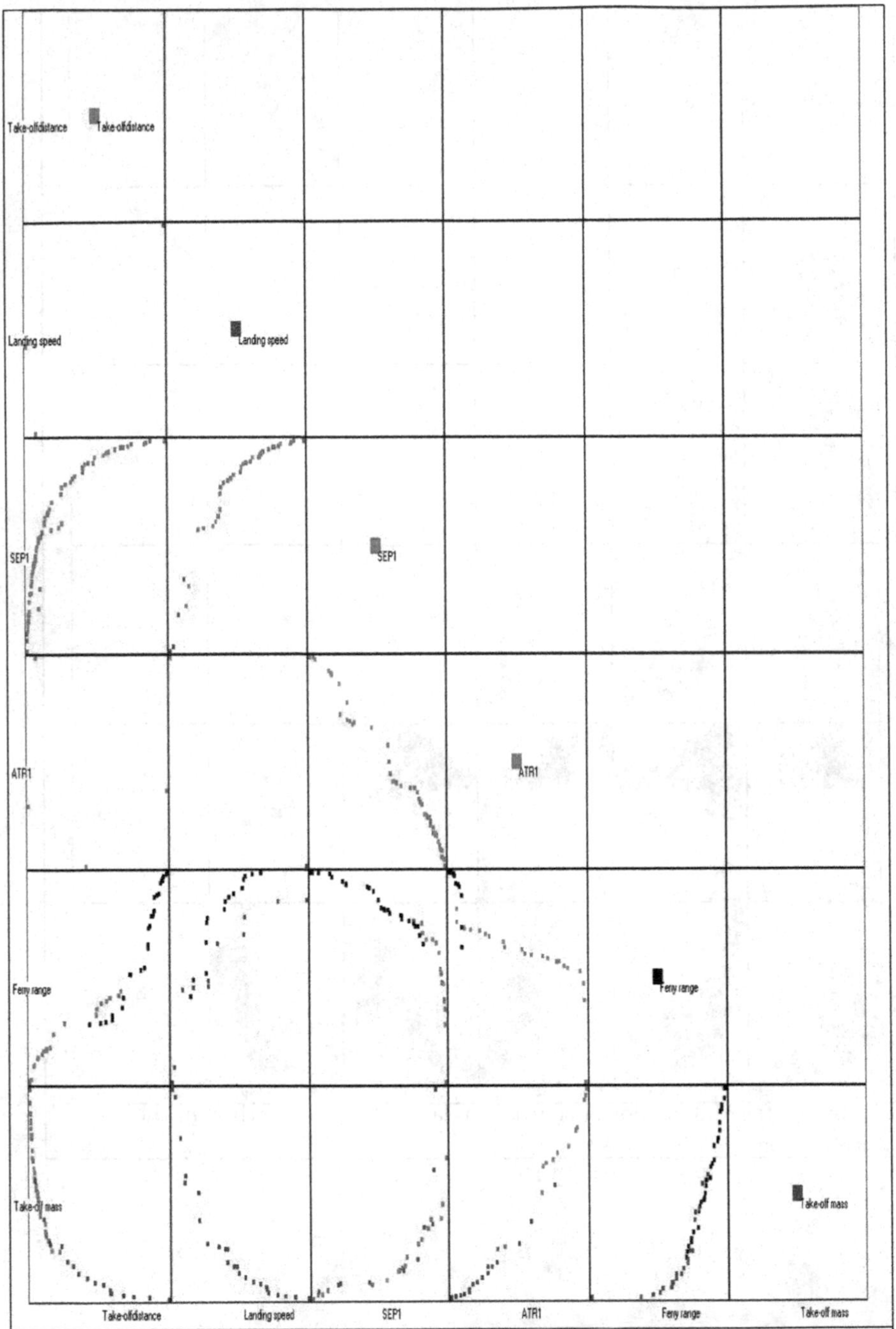

Fig. 9: Distribution of nondominated solutions objective space.

Step 4: Each filter setting is appropriately altered by a preset percentage amount dependant upon rank, and the AFs are then applied to the *already generated* D_i sets.

Step 5: The filtered solutions for each objective are then nondominantly sorted to give approximate Pareto fronts between any pair. Visually, these will be seen either as relatively complete fronts or as fronts with 'conflict gaps'.

Step 6: Max, Min rules are used to check whether the fronts of any two objectives are overlapping or whether a gap exists. This is checked for all the possible pair-wise combinations of the selected objectives. If there is no overlap, steps 4 and 5 are repeated for the objectives involved. If there is overlap, the iteration ceases and the identified fitness thresholds are used to filter the solutions in the remaining stages of COGA runs. During the iterative process the overlap for each objective pair is tracked, and if an objective has overlap with all the remaining objectives during the iterative process, the associated AF setting is not altered in further iterations.

The user can determine the degree of automation to be utilized. The iterative process involving steps 4, 5 and 6 can be interactive, where the user is provided with a matrix scatter plot of the Pareto fronts between different objectives. The user can then choose the fitness threshold for each objective such that she is able to establish an overlap between different objective fronts based on her preferences. The chosen filter threshold can be used to filter the solutions in the later COGA runs. It is quite likely that objective preferences will change during the overall process as understanding improves and overall better solution performance becomes possible by reformulating initial assumptions and representations.

It would become more difficult to interact with the system as the number of objectives increase due to information overload, at which point the automated process can provide assistance. It would be entirely the user's decision as to how interactive the process should be. Generated data and associated graphics can be archived for further study not only by the individual but also by the decision-making team. Traceability should not be an issue, as decision points dependent upon output and subjective judgement can be identified and tagged throughout the process.

It is stressed that the main objective of the above process is to better understand the problem at hand in order to evolve the best representation of the problem in terms of variable and objective definition. It is very likely that the representation will continue to change over several iterations until sufficient understanding of the problem space results in a more definitive model.

4 Concurrent Handling of Quantitative and Qualitative Criteria

The airframe design criteria of the previous section are quantitatively defined within a conceptual parametric model of the system under design. In this case the majority of the variable parameters are ratios that can provide indications of preferred preliminary design, e.g., short, broad wings rather than slender narrow wings. However, the presence of qualitative objectives (which may be considered as important as the quantitative ones) requires an appropriate degree of subjective evaluation and presents the need for a more explicit form of user interaction. Objectives relating to aesthetic qualities of a design, for instance, may be integrated with a parametric model to some extent via accepted and well-founded rules governing relative layout, but to truly evaluate the manner in which an artefact is pleasing to the user, the designer (with, preferably, the end user) needs to be directly involved in the evaluation process.

Such issues have been under investigation for some time within the well-established field of interactive evolutionary computation (IEC) which considers varying degrees of user involvement in the assessment of EC-generated solutions. A positioning of work described in the previous section along an implicit / explicit interactive evolution spectrum can be found in Parmee 2004. Other recent examples of design research involving evolutionary computation, visualization and varying degrees of user-interaction include the concept generation work of Avigad et al. (2004) and Grierson's work relating to the visualization of multi-dimensional Pareto frontiers (2002).

Recent research within the ACDDM Lab has concentrated upon a number of specific issues relating to the manner in which sets of criteria requiring differing degrees of user evaluation can be combined within a user-interactive EC approach. Such issues relate to the development of flexible problem representation, appropriate interface design and the manner in which user-fatigue and cognitive overload caused by the online evaluation of EC-generated solutions can be reduced through the introduction of online machine learning techniques and solution reduction via concept clustering. The overall objective of this work has been to develop an overall, flexible framework that supports a user-centred approach to problem solving where multiple quantitative and qualitative criteria play a major role. This development is running parallel to that of the more implicit form of user interaction described in the previous section, although the intention is to integrate these approaches in future work.

The research relates to user-centric evolutionary design systems which integrate machine-based evaluation of engineering and rule-based aesthetic criteria with the designer's subjective aesthetic evaluation of design solutions. A detailed discussion of factors which make such a system difficult to implement within a real-world context can be found in Machwe et al. (2005a). The research has led to a generic framework for an Interactive Evolutionary Design Environment (IEDE), as shown in Fig. 1.

The initial problem formulation for the IEDE involved the design of simply supported beam bridges against structural engineering criteria and aesthetic criteria (both rule-based and subjective). Several novel concepts have been introduced, including agent-based construction and repair of population members; agent-based evaluation of aesthetic criteria; object-based design representation; and a case-based machine learning subsystem (Machwe and Parmee 2005b and 2006a).

The design of 'urban furniture' in the form of novel and aesthetically pleasing seating arrangements for parks and other public areas followed subsequently to the bridge design. Simple structural analysis of the resulting forms is combined with both rule-based and user-led aesthetic evaluation at a more complex level.

The following sections provide a review of the development of an overall architecture and problem representation that supports such an integration of quantitative and qualitative objective evaluation.

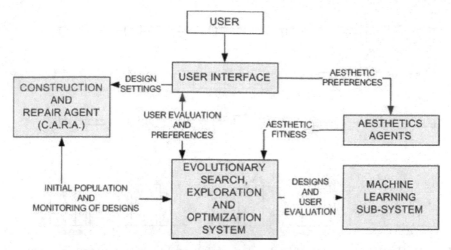

Fig. 10. Interactive Evolutionary Design Environment

4.1 Representation

A highly flexible object-based representation is utilized within the IEDE which allows the representation of diverse designs whilst being sufficiently robust to being manipulated by evolutionary algorithms (Rosenmann 1997; Bentley 2000). The initial bridge design work used a collection of primitive elements to represent a structure. For example, any structure made up of LEGO bricks can be represented as a collection of primitive design objects each with a specific x and y position and a predefined length (along X) and height (along Y). Flexibility also supports the use of different elements with different design properties.

Fig. 11 shows some of the non-optimized 2-D bridge designs generated using the IEDE object-based representation followed by the final designs generated using various engineering, rule-based aesthetic and subjective fitness criteria. The urban furniture (seating arrangement) extends the representation into three dimensions. Elements representing the seat, backrest and legs of a simple bench-type arrangement have been created. This shows the flexible nature of the object-based representation, which can used to represent almost any design form.

4.2 CARA

Without the construction and repair agent (CARA), the object-based representation would not be able to generate meaningful shapes. The construction agent (CA) builds the initial design population using a flexible rule base. Fully free-form designs can be generated by just random placements of elements or a fully defined skeleton can be provided for the placement of elements. The repair agent (RA) ensures that designs remain feasible after undergoing mutation operations, i.e., it maintains the structural integrity of the solutions during the evolutionary design process.

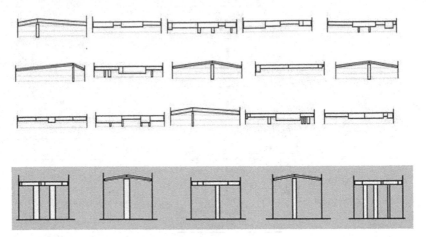

Fig. 11. Initial population of bridge designs followed by best designs generated from the interactive system

4.3 Structural and Aesthetic Criteria

The CARAs can currently create three kinds of bridges: simple beam bridges without support, simple beam bridges with supports and simple beam bridges with angled span sections and supports. Thus, an initial population can consist of a mixture of three designs. These designs are assessed in terms of the designer's aesthetic preferences in addition to structural and cost criteria. Simple length/depth ratio criteria have initially been utilized whilst also minimizing material cost. Column design is assessed via standard buckling criteria.

It is impossible to create a set of aesthetic rules which can be applied universally. However, certain guidelines to provide limited machine-based estimation of aesthetics (Machwe and Parmee 2005a and 2005b) can be utilized. It is difficult therefore to incorporate aesthetic criteria as part of a design system unless some form of designer interaction is utilized. In the present work aesthetics are evaluated at two separate levels. On the machine side, aesthetic fitness is evaluated using a set of rules (or guidelines). On the human side, the designer has the option of ranking the solutions using subjective assessment. This ensures that while certain aesthetic rules are included within the design evaluation, the subjective aspect of aesthetic design is not ignored. In the bridge design system the following rule-based aesthetics have been coded:

1. Symmetry of support placement
2. Slenderness ratio
3. Uniformity in thickness of supports
4. Uniformity in thickness of span sections

Many other quantitative rules exist, but aesthetic evaluation has been kept relatively simple during these formative stages of the study where current design representation does not support detailed aesthetic evaluation. Each aesthetic is evaluated by a separate 'aesthetic agent'.

The 'user-assigned fitness' is the ranking or fitness given to a design by the user on a scale of 0 to 10 (10 being the best). Furthermore, the user can mark solutions for preservation into the next generation. Overall user evaluation operates thus:

1. User stipulates the frequency of user interaction (e.g., once every ten generations).
2. User evaluates a preset number of population members from the initial population (usually the top ten members in terms of stability, material usage and explicitly defined aesthetic criteria).
3. The EP system runs.
4. Population members are evaluated by the user every n generations.
5. Steps 3 and 4 are repeated until user terminates the process.

The overall fitness evaluation therefore comprises structural criteria in terms of stability and material usage plus rule-based 'aesthetic fitness' and human-based 'user-assigned fitness'. It is quite apparent that there are a number of objectives here that could be concurrently manipulated via a standard evolutionary multiobjective technique or by COGA. However, the research has concentrated upon the establishment of the overall system in terms of addressing major issues relating to representation, user-interaction and machine-based assimilation of user preference. In the first instance, in order to establish proof of concept, the various objectives have been combined within a weighted sum representation.

Within the bridge design, CARA system rules relating to the slenderness ratio of spans, the positioning of supports and the thickness of span elements were used. These are also used in the bench design problem with an obvious increase in the number of such rules due to the three-dimensional nature of the problem. A variable number of components (or elements) possessing sets of properties such as style, position and dimensions have also been included, e.g., seat elements, leg elements and back element.

Some initial results from the seating arrangement design system utilizing a relatively well-structured CARA rule-set that produces bench-like designs are shown in Fig. 12. Engineering criteria plus rule-based and user-assigned aesthetic criteria have again been used to determine the fitness of the solutions. Fig. 13 shows some resulting designs when using fully free-form construction of the initial population, i.e., random placement of elements (Machwe and Parmee 2006b). Here, only a basic fitness criterion of minimizing the ground footprint and creating a well-connected structure is used in addition to aesthetic criteria. For further details the reader is directed to [10].

4.4 Machine Learning

The online machine learning subsystem (Fig. 14) reduces the cognitive load on the designer during the interactive evolutionary process. It is an online learning system utilizing case-based reasoning (CBR) to assimilate the subjective aesthetic preferences of the designer. A comparative study showed CBR to be the most promising technique (Machwe and Parmee 2006a) in that the design information can be stored as it stands without transformation to other formats such as fuzzy variables or input

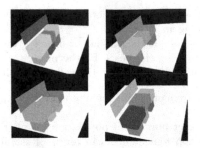

Fig. 12. Generated bench designs

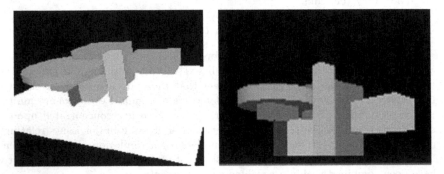

Fig. 13. Interactively evolved free-form seating arrangements

values for neural networks. Such a transformation can destroy essential information. It also proved to be the only technique suitable for online use. The retrieval part of the CBR uses nearest-neighbour distance metrics to measure the difference between the new design and the designs in the Case base. The design closest to the newly generated design has its user-assigned fitness awarded to the new design.

The population size is set at 20 and the user is shown the top ten solutions, with respect to machine-based fitness functions, from each generation. The user also has the freedom to explore and evaluate individual solutions from the population. The user-based experimentation involved identifying 'interesting' solutions in the early generations and then evaluating how well the online machine-based learning system is able to identify the user's preferences as the shapes evolve.

Since CBR learning is online the cases are not carried over from past runs. Initially the case library is empty. After the first generation of user evaluations the case library begins to fill. Solutions not examined by the user are assigned zero fitness. Once there are a number of cases in the case library, the machine learning system starts ranking solutions and the user has the option to change the machine-assigned rank.

With the learning system operating in the background the number of changes made by the user (to the machine assigned rank) decrease with each generation as the machine assimilates user preferences, as can be seen in Fig. 15. For further information on the implementation and detailed results the reader is directed to Machwe and Parmee 2006a.

5 Discussion

This largely speculative paper provides an overview of the research that has illustrated various levels and forms of user interaction when dealing with both qualitative and quantitative criteria. Much greater detail can be found in the cited papers.

The research in Section 3 illustrates the manner in which information relating to complex interactions between variables and objectives can be extracted from evolutionary search and exploration processes. The succinct presentation of this information from several differing perspectives supports the designer in both decision making and in the iterative development of the design representation. Section 4 introduces a highly flexible representation that can readily be changed through online user development of an agent-oriented rule base and then proceeds to illustrate how a user-centred approach can significantly enhance multiple objective satisfaction where some of the included objectives are difficult, if not impossible, to represent computationally. The case-based machine learning approach offers a great deal of potential across all levels and forms of user-centred evolutionary systems in terms of assimilating user preferences and reducing cognitive load.

However, this approach also has limitations. Specifically, a case-based system is only as good as the similarity measure used to retrieve the cases from the library. In the case of numeric representation, data retrieval is relatively easy using the nearest neighbour approach. In the case of nonnumeric representation, retrieval can require conversion of solutions to numeric or pseudo-numeric form. Also, the efficiency of learning in a case-based system depends on the number of stored cases. The larger the number of stored cases, the higher the efficiency. Clearly, the number of stored cases cannot be infinite since there are practical limits to storage and retrieval of solutions from very large case libraries. Other learning methods such as neural networks and fuzzy rule-based systems do not have these extra storage and retrieval efficiency requirements.

Perhaps we can now imagine developing relatively basic machine-based conceptual design representations based upon our current understanding and then rapidly exploring the multi-variate/multiobjective space described by these representations using combinations of local and global search techniques. As the search progresses, the overall system extracts and accumulates information relating to complex characteristics of the design domain whilst also discovering viable solutions. Solutions are initially identified that best satisfy objectives / constraints seemingly relevant in terms of current understanding, whilst background processes extract information from areas of the problem space previously visited, and present this in a succinct manner to the user.

The degree of difficulty of satisfying initial objectives within existing variable bounds or within existing objective preference ranking becomes quantifiable and presentable through background data processing as the search progresses. Online user actions such as constraint softening, objective preference variation and modification of variable ranges may change the nature of the space and search direction, whilst machine-based software agents acting as information collators, processors and presenters provide indications of the effects of such changes. These agents constantly

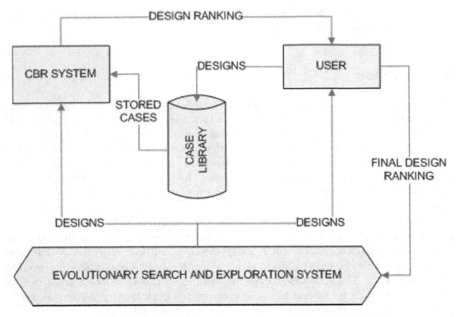

Fig. 14. The CBR system integrated with the IEDE

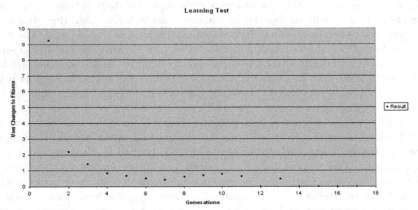

Fig. 15. Number of user-made changes to machine-assigned fitness during different generations

advise the user on interesting solution correlations or redirect her to previously visited areas now possibly of more interest. Concurrent, finer-grained, localized search processes may be spawned to explore specific regions. These actions become semi-autonomous as, through a machine learning capability, the agents become more 'aware' of your requirements in terms of both quantitative and qualitative objectives. The environment becomes more immersive as you react to the information being presented. User online actions become an integral part of the exploration process, reacting to feedback from the system to make iterative changes to the problem landscape.

At any point this relatively continuous exploration process can be paused and relevant information downloaded and presented to the decision-making team for discussion. An easily understood graphic provides a recorded history of user-instigated changes, thereby supporting traceability and allowing analysis of the logical progression of the team's thinking, based upon extracted information. The presentation of such material promotes discussion and allows the perspectives of others to be integrated into further exploratory interactive activity via appropriate problem redefinition and reformulation.

As this iterative interactive process continues, confidence in the developing design models increases, the knowledge base becomes well-founded and uncertainty significantly decreases. A natural result is a reduction in user interaction as we move from a high-risk concept definition phase through an intermediate phase of increasing confidence to the final stages of detailed analysis of a well-defined design space. As illustrated to some extent by the research presented, many of the component parts of the envisaged user-centred intelligent system described above are at a stage of development where their collective utilization is possible. It is suggested that the flexibility of CI technologies is such that specific problems are unlikely to be insurmountable. For instance, although a machine-based representation of an evaluation function may cause problems, the user-centric approach supports complete or partial human evaluation of solutions against any number of criteria, and this can initially play an integral role in evaluation.

Such systems, continuously running as background processes, can support the development of in-house knowledge and expertise whilst reducing lead times to the discovery of innovative products when allied with complementary investigative processes. Current ACDDM collaborators in the pharmaceutical industry are already integrating our user-centric search, exploration and optimization processes using such in-house networked PC resource.

From an academic and industrial point of view, further development and utilization of such systems within a research environment could support significant leaps in understanding related to the characteristics of poorly-defined complex design spaces. The ability to rapidly and efficiently play 'what-if' against multiple qualitative and quantitative objectives whilst concurrently gathering high-quality information that either confirms or contradicts current thinking suggests an environment well suited to the support of knowledge discovery and innovation. The role of human intuition, experience and judgement within such an environment would be paramount, whilst the inherent support of agent-based entities in terms of data processing and presentation would be invaluable.

The development of such people-centred computational environments for conceptual design and early stage decision making is the overall objective of the Institute for People-Centred Computation (http://www.ip-cc.org.uk). This virtual institute established across the UK Universities of Bristol, Cambridge, Cardiff, Newport and West of England is intent on the identification of common languages, methodologies and practices across multiple disciplines in order to facilitate generic system development.

References

[1] Abraham, J. A. R. and Parmee, I. C. (2004) Extraction of emerging multi-objective design information from COGA data, *Procs. of Adaptive Computing in Design and Manufacture VI*, Springer, London.

[2] Avigad, G., Moshaiov, A. and Brauner, N. (2004) Concept-based interactive brainstorming in engineering design. *J. of Advanced Computational Intelligence and Intelligent Informatics, 8(5)*.

[3] Bentley, P. J. (2000) Exploring Component-Based Representations - The Secret of Creativity by Evolution? In *Proc. of the Fourth International Conference on Adaptive Computing in Design and Manufacture (ACDM 2000)*, I. C. Parmee (ed), University of Plymouth, UK. pp. 161–172.

[4] Bonham, C. R. and Parmee, I. C. (1999) An investigation of exploration and exploitation in cluster-oriented genetic algorithms. *Procs. of the Genetic and Evolutionary Computation Conference*, Orlando, Florida, USA: 1491–1497.

[5] Carnahan, B. and Dorris, N. (2004) Identifying relevant symbol design criteria using interactive evolutionary computation, *Procs. Genetic and Evolutionary Computing Conference (GECCO)*, Seattle, USA.

[6] Cvetkovic, D. and Parmee, I. C. (2003) Agent-based support within an interactive evolutionary design system. *J. Artificial Intelligence for Engineering Design, Analysis and Manufacturing*, Cambridge University Press, 16(5): 331–342

[7] Cvetkovic, D. and Parmee, I. C. (2001) Preferences and their Application in Evolutionary Multi-objective Optimisation. *IEEE Transactions on Evolutionary Computation*, 6(1):42–57.

[8] Gero J. S., Louis S. J., Kundu, S. (1994) Evolutionary learning of novel grammars for design improvement. *J. Artificial Intelligence for Design, Analysis and Manufacture*, 8: 83–94.

[9] Gero J. S., Shi X. G., (1999) Design development based on an analogy with developmental biology. *CAADRIA '99*, J. Gu and Z. Wei (eds), Shanghai, China: 253–264.

[10] Gero, J. S. (2002) Computational models of creative designing based on situated cognition, in T Hewett and T Kavanagh (eds), *Creativity and Cognition*, New York, ACM Press, USA.

[11] Goel, A. K. (1997) Design, analogy and creativity. *IEEE Expert, Intelligent Systems and their Applications*, 12(3): 62–70.

[12] Grierson, D. E., Khajehpour, S. (2002) Method for Conceptual Design Applied to Office Buildings. *Journal of Computing in Civil Engineering*, 16 (2) pp. 83–103.

[13] Inselberg, A. (1985) The plane with parallel coordinates, *The Visual Computer*, 1: 69–91.

[14] Kim, H. S., Cho, S. B. (2005) Fashion design using interactive genetic algorithm with knowledge-based encoding. In: Y. Jin (Ed), *Knowledge Incorporation in Evolutionary Computation*, Springer Verlag.

[15] Machwe, A., Parmee, I. C., Miles, J. C. (2005a) Overcoming representation issues when including aesthetic criteria in evolutionary design. *Procs. ASCE Int. Conf. on Computing in Civil Engineering*, Cancun, Mexico.

[16] Machwe, A., Parmee, I. C. and Miles, J. C. (2005b) Integrating Aesthetic Criteria with a user-centric evolutionary system via a component-based design

representation. *Proceedings of International Conference on Engineering Design*, Melbourne, Australia.

[17] Machwe, A. and Parmee, I. C. (2006a) Introducing machine learning within an interactive evolutionary design environment. *Procs. Design 2006*, Croatia.

[18] Machwe, A. and Parmee, I. C. (2006b) Integrating aesthetic criteria with evolutionary processes in complex, free-form design — an initial investigation. *Procs. IEEE Congress on Evolutionary Computation*, Vancouver, Canada.

[19] Parmee, I. C. (1996). The maintenance of search diversity for effective design space decomposition using cluster oriented genetic algorithms (COGAs) and multi-agent strategies (GAANT). *Procs. 2nd International Conference on Adaptive Computing in Engineering Design and Control*, Plymouth, UK; pp 128–138.

[20] Parmee, I. C. (2001) *Evolutionary and adaptive computing in engineering design*. Springer Verlag, London.

[21] Parmee, I. C. (2002) Improving problem definition through interactive evolutionary computation. *J. Artificial Intelligence for Engineering Design, Analysis and Manufacturing*, 16 (3), Cambridge University Press: 185–202.

[22] Parmee, I. C. (2004) *User-centric evolutionary design*. Procs. Design 2004, Dubrovnic.

[23] Parmee, I. C., Abraham, J. A. R. (2005) Interactive Evolutionary Design. In: Y. Jin (Ed), *Knowledge Incorporation in Evolutionary Computation*, Springer Verlag.

[24] Parmee I. C., Abraham J. A. R. (2004) Supporting Implicit Learning via the Visualisation of COGA Multi-objective Data. *Procs. IEEE Congress on Evolutionary Computation 2004*, Portland: 395–402.

[25] Parmee, I. C., Bonham, C. R. (1999) Towards the support of innovative conceptual design through interactive designer/evolutionary computing strategies. *J. Artificial Intelligence for Engineering Design, Analysis and Manufacturing*, Cambridge University Press, 14: 3–16.

[26] Parmee, I. C., Watson, A. W. (1999) Preliminary Airframe Design using Co-evolutionary Multi-objective Genetic Algorithms. *Proceedings of the Genetic and Evolutionary Computation Conference*, Orlando, Florida, USA:1651–1665.

[27] Parmee, I., Watson, A., Cvetkovic, D., Bonham, C. R. (2000) *Multi-objective Satisfaction within an Interactive Evolutionary Design Environment*. Journal of Evolutionary Computation; MIT Press; 8(2): 197–222.

[28] Rosenman, M. A. (1997) An exploration into evolutionary models fornon-routine design,*Artificial Intelligence in Engineering*, 11(3):287–293.

[29] Su N. P. (1990) *The principles of design*. Oxford University Press, New York.

[30] Takagi, H., Ohsaki, M. (1999) IEC-based Hearing Aid Fitting. *Procs. International Conference on System, Man and Cybernetics (SMC '99)* ,IEEE, Vol 3, 657–662.

[31] Zitzler, E., Laumanns, M., Thiele, L. (2002) SPEA2: Improving the Strength Pareto Evolutionary Algorithm for Multiobjective Optimization. *Procs. Evolutionary Methods for Design, Optimisation, and Control*, CIMNE, Barcelona: 95–100.

Multi-competence Cybernetics: The Study of Multiobjective Artificial Systems and Multi-fitness Natural Systems

Amiram Moshaiov

School of Mechanical Engineering
The Iby and Aladar Fleischman Faculty of Engineering
Tel Aviv University, Ramat Aviv, Tel Aviv 69978, Israel
moshaiov@eng.tau.ac.il

> Scientists discover the world that exists; engineers create the world that never was.
>
> Theodore von Kármán

Summary. This chapter provides a comparative discussion on natural and artificial systems. It focuses on multiobjective problems as related to the evolution of systems either naturally or artificially; yet, it should be viewed as relevant to other forms of adaptation. Research developments in areas such as evolutionary design, plant biology, robotics, A-life, biotechnology, and game theory are used to support the comparative discussion. A unified approach, namely multi-competence cybernetics (MCC) is suggested. This is followed by a discussion on the relevance of a Pareto approach to the study of nature. One outcome of the current MCC study is a suggested analogy between species and design concepts. Another resulting suggestion is that multi-fitness dynamic visualization of natural systems should be of a scientific value, and in particular for the pursuit of understanding of natural evolution by way of thought experiments. It is hoped, at best, that MCC would direct thinking into fruitful new observations on the multi-fitness aspects of natural adaptation. Alternatively, it is expected that such studies would allow a better understanding of the similarities and dissimilarities between the creation of natural and artificial systems by adaptive processes.

1 Introduction

Comparing natural and artificial systems has been the focus and drive of the fathers of cybernetics. Such a comparative approach has also served as a major stimulator in the development of the field of Evolutionary Computation (EC). Observing the bio-inspired field of EC we identify its strong link with the field of Mathematical Programming (MP). Many developments in EC could be viewed as advancements in MP as related to both single-objective and multiobjective Optimization (SOO and MOO, respectively). The similarity between SOO, as implemented in EC, and natural evolution is quite apparent.

The similarities between natural evolution and optimality have been extensively discussed in the literature, and the comparison between optimality and adaptation has been a subject of ongoing debate (e.g., [1]).

As outlined in Section 2.3, most of the available discussions on optimality as related to natural evolution can be viewed as referring to SOO rather than MOO. Considering the significance of MOO in the development of artificial systems, the above observation seems striking. Therefore, it is justifiable to explore the relations between MOO and natural evolution. This chapter provides a discussion on this topic using research developments in areas such as evolutionary design, plant biology, robotics, A-life, biotechnology, and game theory. It should be noted that the focus of this chapter is on adaptation as related to evolution; yet some aspects of the discussion should be relevant to other forms of adaptation.

The following contains four sections. Section 2 provides the background needed for the suggested comparison. In Section 3, several observations are made with respect to the suggested comparison. In addition, Section 3 provides a definition of Multi-competence Cybernetics (MCC) and explains the notion of multi-fitness. Section 4 includes a short comparison between natural and artificial design, as well as a recently suggested comparison between design concepts and species as related to MCC. Section 5 provides a short list of MCC questions that might shed some light on future MCC research topics. Finally, Section 6 summarizes and concludes this chapter.

2 Background

This section provides some overview of issues that are relevant to the comparative discussion and suggestions of this chapter.

2.1 Introduction to Cybernetics

Existing Definitions and Scope

The traditional definition of cybernetics, as "the science of communication and control in the animal and the machine," is attributed to Norbert Wiener [40]. The fathers of cybernetics, such as Wiener, studied analogies and metaphors between animals and machines starting at the level of a neuron up to and including the level of societies. It should be pointed out that there are a host of different definitions of cybernetics, as listed by the American Society for Cybernetics (see: http://www.asc-cybernetics.org/foundations/definitions.htm). Of special interest to the current discussion are the nontraditional definitions such as "the art of securing efficient operation" (L. Couffignal), "the mathematical and constructive treatment of general structural relations, functions and systems" (F. von Cube), "the art and science

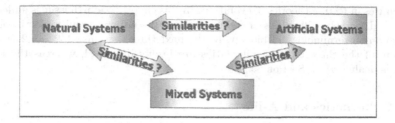

Fig. 1. The scope of modern cybernetics

of manipulating defensible metaphors"(G. Pask), and "the art and science of human understanding" (H. Maturana). It is clear from this collection that cybernetics can be viewed from different and much broader perspectives than that of the original definition. Modern cybernetics involves three types of systems, as schematically depicted in Figure 1.

As suggested in Figure 1, and in accordance with Pask's definition of cybernetics, the scope of cybernetics involves comparisons between the different systems. This issue is briefly described in the following.

The Two Viewpoints of Cybernetics

Cybernetics includes two interesting viewpoints. In fact, Holland [18] implicitly refers to them in his discussion on the role of genetic algorithms by stating that: "It should be emphasized that the plans (algorithms) set forth have a dual role". Referring to the upper part of Figure 1, the first view of cybernetics is aimed at studying natural systems to support the development of better man-made systems (arrow pointing to the right), whereas the second viewpoint involves using new ideas, which are generated as a part of the development and the analysis of artificial systems, to possibly find new explanations to nature (arrow pointing to the left). When considering the first viewpoint of cybernetics, it should be noted that the common engineering design process is substantially different from the way evolution creates its biological products. Yet, the desire to imitate or at least be inspired by nature is strong and has been proved to be fruitful from the engineering standpoint (e.g., soft computing methods, bio-inspired robotics).

Although most of the work in cybernetics could be viewed as focusing on the first viewpoint, the second view should not be ignored. Given the success of bio-inspiration, some sort of similarity must exist, and a major question is whether the similarity is applicable to also the second view of cybernetics. As an example of the second viewpoint, consider the use of EC in explaining natural evolution as in [11]. EC has made an extensive use of metaphors and analogies from its early days, and this has provided a rich vein for its continual development. Yet, there are several difficulties when considering the use of EC to study natural systems. In the more general sense, the difficulties of using the second viewpoint of cybernetics are related to (i) controversies concerning Artificial life (A-life) studies in general, (ii) controversies concerning the association of adaptation with optimality, and (iii) difficulties in testing theories of evolution. In this chapter a fourth issue is added and discussed, namely, difficulties in trying to relate the notion of multiobjectiveness to the common terminology of biology. In the next three sections (2.2, 2.3, 2.4), some aspects of the three types of difficulties are briefly discussed, whereas the fourth issue is dealt with in Section 3.

2.2 Cybernetics and A-life

The scientific exploration of nature and its evolution is an ongoing process that involves observations, theories, and, occasionally, experiments. Acceptable theories, which are based on observations, are always subject to the possibility of being replaced or extended. Darwinism has already been extended into neo-Darwinism as scientific knowledge has expanded. In spite of the fact that the basic ideas of Darwin still prevail, the pursuit for a better and more complete understanding of natural evolution is far from over. The study of evolution has inherent difficulties due to the time scale involved, the lack of complete information about the past, the complexity of natural systems, and the difficulties of performing experiments.

The second viewpoint of cybernetics might help us somewhat compensate for these inherent difficulties. A related approach is that of the Evolution of Artificial Creatures (EAC). EAC is a research topic of relevance to fields such as Robotics, Mechatronics, and Cybernetics. It is an experimental setup for research in A-life; a field that attempts to investigate living systems through the simulation and synthesis of life-like processes in artificial media. In spite of its controversial nature, the A-life research approach has about two decades of recorded research achievements with a growing research community and related conferences. One way to view A-life studies is to consider it as *thought experiments* as suggested in [13]. This means, according to Di Paolo et al. [13], that "although simulations can never substitute for empirical data collection, they are valuable tools for re-organizing and probing the internal consistency of a theoretical position." Such a scientific justification to A-life helps us resolve, to some extent, the controversial aspect of this approach.

2.3 Cybernetics, Adaptation, and Single-Objective Optimization

Securing efficient operation, as stated in the proposed definition of cybernetics by Couffignal (see Section 2.1), suggests a close relationship between cybernetics and optimization of systems. It also hints at close relationships between cybernetics and both control and adaptation. Efficiency is usually associated with some measure with respect to a goal or an objective. Holland [18] lists three major components in the adaptation of a system, namely the environment, the adaptation plan to induce improvements, and a performance measure to be associated with the environment. When referring to artificial systems, Holland [18] states "Here the plans serve as optimization procedures..." In support of his uniform treatment of adaptation, Holland provides illustrations from different fields. In all of his illustrations, when referring to optimization, it appears that the reference is to SOO (including the weighted sum of performances). Holland's suggestion seems logical in view of the similarity between the notion of fitness of an organic individual and performance of an artificial individual. This fundamental aspect of evolving natural (artificial) systems, namely fitness (performance), serves to measure which organic (artificial) individual has a better chance to 'survive' (selected as a candidate in an adaptation step for optimality).

It seems suitable to continue this review on the resemblance between adaptation and optimality with some historical aspects of the use of the famous and confusing term 'survival of the fittest', coined by Spencer [37]. Fittest tends to imply optimization and an optimum, and therefore it requires clarification. The notion of survival of the fittest has been in the centre of ongoing debates, which date back to the time of Darwin. At the extreme, it has been the focus of some theological discussions. In such debates it has been claimed to be a tautology; namely, that it means survival of those better at survival, and hence is meaningless. Gould [16] suggests that the rebuttal by Darwin is most compelling. According to Gould, "Darwin insisted, in principle at least, that fitter organisms could be identified, before any environmental test, by features of presumed biomechanical or ecological advantage." The term survival of the fittest has also played a role in discussions on what is known as social Darwinism, and in particular with respect to the justification of controversial ideologies such as racialism. These kinds of debates may have caused a certain resentment to the term, and probably contributed to the need to better express what adaptation is all about. In fact, although many researchers in the field of EC are still using the term 'survival of the fittest', most contemporary biologists almost exclusively use the alternative term of 'natural selection', and acknowledge its complex nature. Biologists tend to agree that natural selection plays a role in the evolution of traits as an adaptation process, but may fail to agree about the significance of its role with respect to other evolutionary forces (e.g., [30]).

Both fitness and performance are typically considered as scalars. Usually, performance, as applied to optimization, is understood as a value, which is measured with respect to some objective or to a weighted sum of objectives. With respect to the latter it should be noted that the weighted sum of objectives should be considered as an SOO approach. Fitness is aimed at describing the natural capability of an individual of a certain genotype to reproduce, namely to be able to transfer at least a part of its genetic material to the next generation. Fitness of a genotype in biology is commonly measured either in absolute or relative terms. In the former measuring method, fitness is a ratio of the number of individuals after selection to that before selection, as related to a particular genotype. To measure absolute fitness is usually difficult; hence the idea of a relative fitness has emerged. In both methods, fitness is a scalar. Wright [41] suggested studying natural evolution by visualizing the distribution of fitness values as if it were a landscape. For this purpose a distance measure between genotypes is needed. The concept of a fitness landscape, or adaptive landscape, involves the set of all possible genotypes, their degree of similarity, and their related fitness values.

A similar visualization is commonly used in SOO, where the values of the performance of all solution candidates are visualized as a landscape. In a maximization problem the aim is to find the peak or peaks of the landscape. When taking an adaptationist viewpoint, and using the metaphor of landscape as described above, evolution might be viewed as a local optimization rather than a global one. For example, Orzack and Sober [30] defined adaptationism as "the claim that natural selection is the only important cause of the evolution of most nonmolecular traits and that these traits are locally optimal." Although their view of adaptationism is somewhat extreme, the general understanding is that natural selection is similar to local optimization. In fact, as pointed out by Parker and Maynard Smith [31], optimization and game theories have been widely used, particularly by field biologists, to analyze evolutionary adaptation. Yet, as it appears from the description of Parker and Maynard, in such studies a single optimization criterion rather than several criteria, is associated with fitness. One should not confuse the notion of payoffs, which is used in such studies, and that of criteria, since (as stated in [31]) "payoffs are expressed in units of the criterion to be maximized".

Taking the above into account, one should conclude that there is a similarity between adaptation and optimality, in particular with respect to SOO. When viewed closely, it appears that most discussions which deal with adaptation versus optimization and do not explicitly refer to SOO and/or MOO should be considered as implicitly referring to SOO. As already pointed out in the introduction (Section 1), it should be noted that this chapter deals primarily with adaptation in the sense of evolution.

So far the discussion has focused on the similarity between fitness of an organic individual and performance of an artificial individual, as related to adaptation and optimization. When focusing on shape and structure, in nature and in the artificial, it appears valid to further discuss the similarity in physical terms, such as energy, rather than in biological terms (such as the number of individuals after selection). Bejan [6] has investigated such a similarity with respect to tree-like structures, and has generalized his observations into a theory. According to his constructal theory, "For a finite-size system to persist in time (to live), it must evolve in such a way that it provides easier access to the imposed currents that flow through it." The constructal theory, which has emerged from the design of engineered systems, assumes that geometric forms that appear in nature are predictable through optimization under constraints. Furthermore, similarly to studies on adaptation and optimization, the discussion in [6] refers to SOO. The only apparent exception is the citation from the work of Nottale [29] on fractals, which states, "One of the possible ways to understand fractals would be to look at the fractal behaviour as the result of an optimization process...Such a combination...may come from a process of optimization under constraint, or more generally of optimization of several quantities sometimes apparently contradictory..." Interestingly, this citation is left by Bejan [6] without a discussion on the possible role of MOO in the constructal theory. Recently, in [7], a MOO approach to the design of heat exchangers has been discussed in conjunction with the constructal theory. Yet, no reference has been made with respect to heat exchangers in nature.

In summary, there are studies, although of a controversial character, on the similarities between natural and artificial systems as related to optimization. Most of such studies, which use optimization theory to explain evolutionary adaptation, either explicitly or implicitly refer to SOO. There are, however, some exceptions, which are discussed in Section 3.

2.4 Validation of Adaptation Theories

The second viewpoint of cybernetics, which has been described in Section 2.1, may help produce new theories and explanations about nature. Yet, any borrowed idea, from engineering design or alike, needs validation. Recalling the idea of the role of A-life as thought experiments, as discussed in Section 2.2, it is worthwhile to note that Parker and Maynard Smith [31] have used a similar argument. They have justified optimality theory in evolutionary biology by saying that "Optimization models help us test our insight into the biological constraints that influence the outcome of evolution. They serve to improve our understanding about adaptations, rather than to demonstrate that natural selection produces optimal solutions."

In some sense, the use of optimization models in the study of natural adaptation could be viewed as a part of A-life. In any case, if model predictions match the actual observations, then one may hope to have made the right assumptions about the natural process and its modeling. Clearly, models by themselves cannot validate a theory and empirical evidence is a must. Unfortunately, it is well known that empirical research on natural evolution has many limitations, and has not resulted in a well-accepted evolution theory, but rather into a variety of opinions and debates (e.g., [1, 30, 35]). While evolution theories and their extensions are difficult to substantiate by empirical evidence, it is noted that thought experiments, on ideas such as those presented in this chapter, might lead to future planning of evolutionary experiments. As noted by Sarkar [35], with respect to empirical adaptationism, such tests might become increasingly plausible with the advent of large sets of complete genomic sequences.

2.5 Multiobjective Problems in Engineering Design

The following provides background on engineering design in the spirit of the second viewpoint of cybernetics. Namely, ideas from engineering design which are presented here are to be borrowed (in Sections 4 and 5) for the pursuit of understanding nature.

General

Product development commonly involves trade-offs among conflicting objectives (e.g., accuracy vs. cost). The significance of such trade-offs to creative design has been highlighted in the TRIZ method, which resulted from a comprehensive study of patents by Altshuller, as described in [36]. Traditionally, multiobjective problems (MOPs) have been treated by a SOO-like approach using either a weighted sum of the objectives or a goal attainment method. Such problem definitions and solution techniques could be viewed as range-dependent approaches. Modern processing technologies provide a means to consider parallel search methods which are suitable for range-independent MOPs that may involve a search towards a Pareto front and the associated nondominated solutions (see the introduction chapter of this volume).

EC tools are known to be suitable for supporting engineering design (e.g., [8]). Their attractiveness for engineering design has been strengthened by the recent developments of reliable and generic multiobjective evolutionary algorithms (MOEAs), and by the introduction of interactive EC methods for engineering design (see recent reviews by Coello [10] and Parmee [32], respectively). Pareto-based search has also been implemented for engineering design and other applications by non-EC methods (e.g., [21]). Yet, evolutionary multiobjective search and optimization techniques are becoming the most popular methods to solve MOPs in general and for engineering design in particular [10]. The majority of such studies concerns the search of particular Pareto-optimal designs from the set of alternative designs.

Recently, a nontraditional MOP approach, involving set-based concepts rather than particular designs as the focus of the search and selection, has been developed at Tel-Aviv University aiming at the support of engineers. The brief description

of the concept-based approach, which is given below, follows a recent review by
Moshaiov and Avigad [25]. There are two main reasons for the outlining of the
concept-based approach below. First, this background provides a typical spectrum
of engineering considerations that are quite common to multiobjective search and
optimization in design. Second, as pointed out in Moshaiov [23, 24], species and
design concepts might be similar, at least in some metaphorical sense. In fact, this
observation served as a trigger for the work presented here, which summarizes and
continues the suggestions of [23, 24].

An Overview on the Concept-Based Approach

The concept-based approach involves the search and selection of conceptual designs.
The major motivation for the development of the concept-based approach is rooted
in the significance of conceptual design to the survivability of companies (e.g., [38]).
The concept-based approach is not restricted to MOPs. Yet, its development efforts have concentrated on MOPs due to the nature of engineering design, which
commonly involves trade-offs among conflicting objectives [25].

The concept-based approach deviates from the traditional representation in
which each concept has a one-to-one relationship with a point in the objective space.
In general, a conceptual solution should be viewed as a category of solutions. Hence,
in contrast to the traditional approach, in the concept-based approach a conceptual solution is represented by a set of particular solutions. This allows *performance
variability*, which results from the particular solutions that are associated with a
conceptual solution. The set-based concept representation provides a stage for a
synergistic human-computer interaction. In the concept-based approach, computers are utilized to extensively search the decision space at the level of particular
solutions, whereas humans articulate their preferences at the level of conceptual solutions. Such preferences may be articulated not only at the level of concepts, but
also at the level of sub-concepts (e.g., [4]).

In addition to such inherent concept-related preferences, concept-based MOPs
may involve range-related preferences. Both types of preferences could be implemented either a priori or interactively during the search. The recent review paper
by Moshaiov and Avigad [25] lists a variety of EC studies and contributions, which
have been made at Tel Aviv University on the concept-based approach. Among the
studied concept-based topics are a dynamic goal approach, a Pareto approach, a
structured EC approach with sub-concepts, interactivity by preferences of concepts
and sub-concepts, subjective-objective fronts, various concept robustness issues, concept selection by variability and optimality, extension to an epsilon-Pareto approach,
generalization to path planning, application to simultaneous mechanics and control
design, and various computational aspects.

It should be noted that in engineering design the selected solution might not
necessarily be from the Pareto-optimal set (e.g., [4, 32]). Yet, an understanding of
the concepts' relative performances along and in the vicinity of the front is significant
to concept and solution selection (e.g., [26]).

This is illustrated in Figure 2a. Assume that the figure contains the performances
of all solutions of two concepts. Both concepts (designated by stars and circles)
play a role in the front. Yet, when a look beyond the front is taken, the 'star'
concept of Figure 2a might be more robust than the 'circle' one. This may have
happened when the solutions of the first two ranks are to be disregarded due to

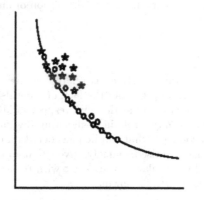

Fig. 2a. Two concepts   **Fig. 2b.** Three concepts

some uncertainties. Alternatively, human preferences might result in the excluding of one or both concepts and the selection of another concept (not shown) that is not on the concept-based front but rather on the subjective-objective front, as described in [4].

Recently, Avigad and Moshaiov [3] argued that concept selection measures should not depend solely on Pareto optimality but should also account of variability in the objectives afforded by a concept. This is illustrated in Figure 2b. Here, the performances, in a bi-objective space, of three concepts are depicted as circles, stars, and black dots. In this min-min problem, the concept-based Pareto-front consists of solutions from the first and second concepts (circles and stars), and yet one should not ignore the third concept because, in comparison with each of the other concepts, it has a better variability with respect to the objectives. In engineering design, such variability might be important due to the variability of market demands (e.g., [5]). The variability and optimality issue adds up to the interactivity and concept robustness issues that motivated the use of an epsilon-Pareto approach for concept-based problems [24, 26].

In summary, the concept-based approach deals with the search and selection of conceptual designs by way of a set-based representation of each concept in a multiobjective space. In this chapter, the concept-based approach is used for a comparison between design concepts and species (see Section 4.3). The comparative discussion of Section 4.3 follows the cybernetic principles and ideas that are described below.

3 Introduction to Multi-competence Cybernetics

This section discusses the notion of Multi-competence Cybernetics (MCC). It starts with general observations concerning MOO as related to nature (Section 3.1). Next, in Section 3.2 the term MCC is introduced and justified as a replacement for the term Multiobjective Cybernetics (MOC), which has been originated and used by the author in [24]. Finally, Section 3.3 provides some insight to the notion of multi-fitness.

3.1 General Observations

Observing the mainstream literature on natural evolution, as related to the comparison of adaptation with optimality, it is striking to note the lack of a consistent and extensive discussion on the similarities and dissimilarities with respect to MOO. This seems astonishing because MOPs play a major role in engineering design (as described in Section 2.5), and nature has produced what can be considered remarkable designs. An intriguing question has to be raised, namely, given that natural adaptation is possibly related to SOO, could a similar relation exist with respect to MOO? Several related observations are made in the following:

- Pareto-related ideas were not available at the time of Darwin's *Origin of Species*. Yet the following point is quite surprising.
- There is no reference to a multiobjective evolutionary theory; optimality theory in evolutionary biology seems to involve the use of a criterion and not a mixture of criteria (e.g., [31]).
- The notion of objectives is controversial with respect to nature.
- A notion equivalent to trade-offs of objectives might be that of trade-offs of functions and forms, or that of trade-offs in behaviour.
- There is no well-known general theory of evolution that relates fitness with trade-offs of functions and forms (or something similar).
- The word 'trade-off' has been used with respect to optimal theory of natural evolution; however, it has referred to the counteracting costs and benefits of strategy changes with respect to a criterion, and not in a multicriterion sense (e.g., [31]).
- There are, however, suggestions to use MOO in studying biological systems, and evidence of its practical results (e.g., [14, 27]).
- Adopting TRIZ, as suggested in [9], to a biological patents' database, might shed light on possible analogies as related to trade-offs.
- There is an increasing evidence of studies that could be viewed as belonging to multiobjective A-life, and/or to the related topic of multiobjective robotics (e.g., [39] and [26], respectively).
- Studies on multiobjective robot path planning, such as that of [26], involve the conflicting objectives of fast versus safe. Such characteristics appear essential for survival in nature.
- Multiobjective optimization is used in bioinformatics and computational biology (see a recent review in [17]), yet much of this could be viewed as engineering-related activities.
- Studies on multiobjective machine learning, such as in [19] and in the introduction chapter of this volume, are strongly related to multiobjective aspects of neural networks. Hence, they might be important to an MCC discussion on learning and control aspects of adaptation.
- There is evidence of the use of multicriterion decision making in ecological planning (e.g., [34]), and of the use of multiobjective optimization in bioprocessing (e.g., [20]). Yet, such activities could be viewed as bioengineering-related activities, and do not necessarily provide evidence of any human-independent natural process in the sense of Dawkins's *Blind Watchmaker*.

- MOEA is useful for control (as revealed in reviews such as that of Coello [10]). Yet, it could be viewed as an engineering-related activity.
- There is evidence of the existence of multiobjective game theory (e.g., [15, 22, 28]). Although practised primarily for operations research, it could be viewed as relevant to biology.
- Multiobjective game theory has been implemented in games that have some metaphorical value with respect to biology (e.g., [22]).
- Human behaviour clearly shows the significance of conflicting objectives and conflict resolution in natural systems (societies). This is apparent in studies on multiobjective game theory (e.g., [28]).
- MOEA is used for understanding nature (e.g., [33]). Yet, such studies do not tell much about natural evolution, but rather on human inference in the process of understanding nature.

The above list of observations includes a compilation of evidence as related to MOO and nature in the spirit of the second viewpoint of cybernetics. It provides some evidence that MOO might support understanding of nature at least in the form of thought experiments (see Section 2.2). This falls within the idea expressed by Parker and Maynard [31], namely that "optimization models... serve to improve our understanding about adaptation." To support the extension from SOO to MOO one must try to understand the possible role of MOO in understanding nature. This is explained in the following.

3.2 Defining Cybernetics and Multi-competence Cybernetics

Modern cybernetics is viewed here as *the study of the competence of natural and artificial systems within the scope of analogies and metaphors*. It follows the definitions of von Cube and of Pask, and constitutes a shift from the terminology of Couffignal (see Section 2.1). That is, the notion of securing efficient operation is replaced with the notion of *competence*. The former terminology appears to be adequate only to the first viewpoint of cybernetics which focuses on the design of the artificial, whereas the latter seems to be more appropriate to both points of view; in other words it does not imply the involvement of a designer. Competence should be understood here, in the context of artificial systems, as the designer's objective that reflects the designer's perception of what type of property of the system is to be used when comparing design alternatives. On the other hand, in the case of natural systems, competence should be viewed as fitness in the sense that no purpose should be implied by it. Yet, as pointed out by Parker and Maynard Smith [31], respect to optimality theory in evolutionary biology, the optimization criterion is often an indirect measure of fitness. The suggested broad view on cybernetics refers to the study of the competence of systems, including natural and artificial ones. Hence, Holland's work [18] on adaptation in natural and artificial systems should be viewed as a study within the field of cybernetics.

The transformation from the traditional definition of cybernetics, as 'the science of communication and control in the animal and the machine', to the above one, appears to have a rationale. Understanding communication and control should not be separated from understanding morphology and mechanics, as pointed out by modern research on the evolution of artificial creatures (see Section 2.2). In fact, evolution appears to suggest a mixed view on the *how* and *what* is governed and governing.

The above definition of cybernetics suggests we define Multi-competence Cybernetics (MCC) as *the study of the multi-competence of natural and artificial systems within the scope of analogies and metaphors.* Here, the focus is not on debates such as optimality versus adaptation, or adaptationism versus pluralism. Rather, a unified view is suggested on adaptation in natural and artificial systems that extends ideas such as those presented in [18] to incorporate the notions of multiobjective adaptation in artificial systems and multi-fitness adaptation in nature. In other words, the MCC suggestions made here do not aim at adding to any controversy, but rather at providing a framework for thinking when comparing natural and artificial systems. The proposed unified view could be substantiated by empirical, logical, and simulation-based arguments, using the accumulation of evidence, which is presented in Section 3.1. The proposed MCC approach is further explained in the following.

3.3 Justifying the Notion of Multi-fitness and Its Visualization

The proposed extension from cybernetics to MCC may look trivial but it requires a justification and clarifications. Where artificial systems are concerned, the notion of multi-competence seems clear as it translates to multiobjective. The natural counterpart of the notion of multi-competence as multi-fitness is however not as trivial to justify. In other words, in spite of the fact that comparing the notion of fitness with the notion of performance, or with that of objective, is not rare, the suggested notion of multi-fitness and the related notion of Pareto front would appear strange, unfamiliar, and even unacceptable to most biologists (for an exception see [14]).

By its definition, fitness is to be measured under the same survival condition. One could argue that there are different types of survival threats and that they can appear in nature either separately or together. In fact, some generic classical threats are well known. For example, shortage of food could be a survival threat, and so could a predator. Certain traits or strategy may fit one type of a threat but not necessarily all types of threats. This means that the notion of fitness cannot be separated from the type of survival threat. In other words there could be different types of fitness related to the different generic threats. To further illustrate the issue of multi-fitness we should note that threats on a particular individual might change from one type to another during the individual's lifetime. The changes may also apply to different individuals of a population in a different order. The time scale of such changes may span over generations and not just over the lifetime of the individual. An individual or a species may also change the environment, which adds another dimension to the above discussion. This can be further illustrated and discussed using the terminology of game theory and winning criterion. The *game of survival* is not just one game; it is a series of games. The rules of winning are not fixed, and they may vary with time and space. The criterion (type of threat) may change from one game to the other, and one could also perceive that even one game may have multiple criteria (e.g., [15, 22, 28]), that is, different threats that are happening simultaneously.

One could therefore think of the multi-competence problem in nature as the study of the *trajectories* of individuals and species in a multi-fitness space. As pointed out by Parker and Maynard Smith [31], fitness can be expressed either directly or indirectly. Taking a form and function approach to the indirect expression of fitness, the above discussion could be compared with that of [14] and [27]. According to [14], the study of form-function relations of branched structures could be advanced by the use of multiobjective optimization. In [27], simulated adaptive walks are used to study the early evolution of the morphologies of ancient vascular plants, in a multi-fitness fashion, using multi-tasks and their related fitness landscapes. Clearly, both direct and indirect expressions of fitness suggest that a multi-fitness (multi-competence) dynamic visualization of natural systems should be of scientific value, in particular for the pursuit of understanding of natural evolution by way of thought experiments and A-life studies. It may also be significant for the analysis of empirical data. Such visualization is perceivable up to 3-D but its extension might pose a difficulty. This is similar to the visualization problem that occurs in multiobjective design (e.g., [21]). While saying all of this, one should realize that it is not so clear to what a degree the notion of Pareto front is significant for the understanding of evolution. This issue is further discussed in Section 4.2 following some further description of the general aspects of MCC.

4 Fundamentals of Multi-competence Cybernetics

As suggested in Section 3, understanding analogies and metaphors between the natural and the artificial, as related to MOPs, seems important. Yet, such an attempt is inherently difficult and often speculative. The prime merit of the following is perhaps in raising some questions and pointing at potential approaches that have resulted from research on the concept-based approach in engineering design. Speculation could be avoided by focusing on possible analogies as a means for possible inspiration and for the production of useful metaphors. This could trigger thought experiments that should not be understood as an attempt to necessarily pose any new theories on nature.

The common process of engineering design differs substantially from evolutionary design. Yet, here the interest is primarily on design by artificial evolution as compared with that of nature. In the following section (4.1), some general aspects of comparing these design processes are discussed. Next, Section 4.2 provides an MCC discussion on the notion of Pareto front. Finally, in Section 4.3, an MCC comparison is carried out with respect to the possible similarities of design concepts and species.

4.1 General Aspects

Many topics that have been mentioned in the background (Section 2), especially those related to the concept-based approach, reflect typical issues in engineering design. In particular they relate to evolutionary multiobjective design. Among such typical issues are:

1. The generic nature of design tools, and in particular EC-based ones.
2. The closeness to A-life aspects.
3. The structured nature of the representations of engineering solutions.
4. The uncertain and subjective nature of design goals and objectives.
5. The interest in the nondominated set and the objective trade-offs.
6. The lack of sufficient modeling of performances.

7. The subjectivity of concept-related preferences.
8. The inherent variability of conceptual solutions.
9. The interest in solutions that are robust and the different types of robustness.
10. As above with respect to robust concepts.
11. The need to extend the Pareto approach for the general concept selection problem.
12. The need for an efficient search.

The above issues are typical to engineering design; yet, one may claim some similarities with nature, at least as related to the possibility of thought experiments. The first three items do not pose any serious dissimilarity problem. Items 4 and 5 appear related to the dynamics and variability of the survival conditions in nature (see Section 3.3 and also the discussion in the next paragraph). Items 6 and 7 are related to the difficulties of modeling, which appear to be a common problem in both natural and artificial systems. Items 8–11 relate to the MCC comparison between concept and species, which is discussed in Section 4.3. Finally item 12 demonstrates a major difference between natural and artificial evolution that is related to the notion of purpose in engineering, which does not exist in nature. Some of the above issues are further discussed below.

Engineering design is a purpose-directed process and not a result of the work of a 'blind watchmaker'. It involves dynamic goals, and the exact preference of objectives is uncertain and may vary during the design and among the designers. In nature, since the environment changes with time, and threats are dynamic, evolution is a dynamic process, and fitness and the multi-fitness problem are dynamic as well (see Section 3.3). In the case of conceptual design, the desire to obtain the full spectrum of nondominated solutions is related to the issue of the uncertainty of objectives (e.g., due to variability of market demands [5]). This may resemble a desire to predict natural evolution under the uncertainty of the trajectories of evolution, or in environments with variable conditions. This issue is further discussed below.

4.2 Is Pareto Relevant to the Study of Nature?

Comparing individuals or species in a multi-competence space does not necessarily mean that the notion of a Pareto front is relevant to the understanding of nature. Yet, as already pointed out, there is some evidence for the significance of a Pareto approach to the understanding of natural systems (e.g., [14, 27]). One should realize that the use of the idea of nondominated solutions in engineering design is a result either of postponing the decision on the objective preferences or of trying to compare performances of different solutions under different situations without a preference for a particular situation. In such cases the efforts of obtaining the front allows a better understanding of the design trade-offs. When dealing with nature one should be careful in making Pareto-related statements. It is arguable that a Pareto front can be useful in the analysis of natural solutions; yet, such an analysis should assume that there is no particular trajectory of scenarios. In spatio-temporal evolution scenarios, a dynamic weighted sum approach, or a dynamic prioritization approach might be more relevant than the Pareto approach. Such alternatives to the Pareto approach do not necessarily mean that the performances of individuals and species are not bounded in some sense by a global Pareto front. Understanding the applicability of the notion of *nondominated sets* in natural evolution might help to shed some light on its possible contribution to natural diversity. Of particular interest might be the use, in the MCC context, of fuzzy and multiobjective game theory (e.g., [28]). This may help account for the fact that the "assignment of fitness values in nature" by way of conflicting competences might be only partially understood by humans.

4.3 Comparing Concepts and Species

Understanding that an analogy between design concepts and species might exist had an important impact on the development of different concept-based MOEAs (see [2])). The following provides some background on this observation. In biology, the term species commonly refers to the most basic biological classification, comprising of individuals that are able to breed with each other but not with others (except in rare cases). In nature, a niche can be viewed as a subspace in the environment with finite resources that must be shared among the population (society) of that niche, while it competes to survive. In evolutionary algorithms, the term speciation (or 'niching') commonly refers to an automatic technique to overcome the tendency of the population to cluster around one optimal solution in a multimodal function optimization. Speciation techniques help maintain diversity to prevent premature convergence while dealing with multimodality. Speciation could be viewed as an automatic process or operator that gradually divides the population into subpopulations (species). Each of these subpopulations deals with a separate part of the problem (*niche* of the search space). Commonly, a niche refers to an optimum of the domain and the fitness represents the resources of that niche. The common process of speciation is also a niching process, as it finds the niches, while dividing the population into them.

Species that are either competing or cooperating are viewed as coevolving. Competitive coevolution has been computationally employed with single as well as multiple populations. In contrast to niching, where species are automatically formatted, in coevolution of competing species they are commonly predefined (although their populations' relative size may be subject to automatic changes). This situation resembles that of the concept-based approach, in which the association of sets of particular solutions with concepts is predefined. The last observation clearly indicates a possible analogy between concepts and species. Both are represented by subsets of the populations. It seems intuitive to view different species as different *design concepts of nature*.

A crucial part of the algorithm, in [4] and in similar studies, is the penalty functions that are used for the fitness. These include a *front-based concept-sharing penalty* and an *in-concept front niching penalty*. The front-based concept sharing is applied to preserve *concept* diversity, and to prevent a good concept from hindering the evolution of other potential concepts within a front. The in-concept front niching preserves the diversity of *particular solutions* within each concept belonging to a particular nondominated front (rank). In a recent investigation [2], the algorithm, such as the one in [4], has been modified to improve the analogy by eliminating crossover operations between concepts. In [2], a crowding approach has been implemented to penalize the fitness. In developing the penalties and the algorithms, the focus has been on engineering design and the wish to find a good representation of the optimal concepts. With the elimination of crossover operations between concepts in [2], it appears that the process of the simultaneous multiobjective concept-based evolution could be viewed as the evolution of species towards and along a Pareto front.

While supporting the development of computational mechanisms to simultaneously evolve species/concepts towards and along a Pareto front by a metaphorical EC approach, a host of questions should be raised about the applicability of such comparisons to improving the understanding of nature. The main question from the second viewpoint of cybernetics is to what a degree it would be possible to advance the potential analogy between design concepts and species to better understand evolution. Furthermore, it is still questionable if new metaphors might arise from taking an MOO view rather than an SOO view on nature. Clearly, the existing concept-based algorithms have been developed for engineering design applications and not as simulators of natural selection. Yet, as described in Section 3, multi-competence situations in the sense of multi-fitness or multifunctionality do exist in nature. Very basic survival situations in nature could involve trade-offs in behaviours such as fast (to obtain food) versus safe (to avoid dangers), which has been the subject of a concept-based robotic-related study in [26]. Incorporating spatio-temporal evolution scenarios into the concept-based approach might create a new way of studying natural evolution in the sense of the second viewpoint of cybernetics. The following is an open question for future research. Would it be possible to say that, regardless of different scenarios, nature evolves species towards optimality in a multiobjective sense, just as humans are trying to create conceptual designs that are satisfying in some Pareto sense?

Engineering design often involves satisfying solutions that are not necessarily Pareto optimal. Similarly, it is expected that natural selection involves "design solutions" that could be viewed as advancing towards a Pareto front but are not optimal in the Pareto sense. It appears logical to try not only an epsilon-Pareto approach but also a fuzzy Pareto approach.

Of a particular interest for future research is investigating potential analogies and metaphors related to current studies on the robustness of concepts (e.g., [5]), which should not be confused with robustness of particular solutions (e.g., [12]). This topic encompasses different types of robustness with respect to different types of uncertainties, and requires the introduction of measures not only for multiobjective optimality of concepts, but also for their robustness. In this regard, methods of comparison, in the multiobjective sense, of particular solutions and concepts (sets), as well as their rationale, might also serve as an MCC research playground where such questions are asked with respect to species. A more questionable idea is to try and compare the interactivity aspects of the concept-based approach with evolutionary issues of mixed systems (see Figure 1). Finally, it should be noted that, due to the fact that the concept-based approach is a set-based approach, analogies might be explored not only with respect to species but also with respect to other biological categories.

5 Hypothetical MCC Questions

The study of multiobjective optimality and robustness of conceptual solutions, which is motivated by engineering, could be carried out using a multiobjective concept-based EAC. In such design studies the EC approach allows evolution that is purpose-directed. Similarly, EAC can be used as an A-life setup to try and explore the role of MOO in the natural evolution of species with a *blind watchmaker approach*. Performing such independent studies might be complemented with related MCC questions. The above discussion in section 4 raises some interesting MCC questions. Among such speculative questions are the following:

1. Is Pareto optimality relevant to natural selection in any sense?
2. As above, with respect to local versus global fronts.
3. Given the dynamic aspects of the survival conditions in nature, could it be possible to compare it as similar to the varying market demands in engineering?
4. Does robustness of concepts have a biological counterpart to robustness of species?
5. As above, in relation to descendants of a biological ancestor.
6. Could evolving Pareto-optimal/robust design concepts be related to game-based theories of evolution?
7. Would it be possible to use ecology and biotechnology multiobjective planning to support MCC-based studies of natural evolution?
8. What are the consequences of a Pareto approach to natural evolution with respect to discussions on natural diversity?
9. What would be the implications of the use of fuzzy multiobjective game theory in MCC studies?

The above list of MCC questions could certainly be extended. Of a particular interest are related questions about other forms of adaptation in nature, as well as questions associated with evolutionary developmental biology. Such issues are left for future research.

6 Summary and Conclusions

This chapter introduces Multi-competence Cybernetics (MCC). The current study focuses on a comparative discussion concerning the multi-competence evolution of systems in nature and the artificial. Research developments in areas such as evolutionary design, plant biology, robotics, A-life, biotechnology, and game theory are used to justify the proposed MCC approach. Several questions are raised, which are related to a long-standing controversy on adaptationism and optimality. Among such questions is that on the relevance of a Pareto approach to the study of nature. At the risk of a controversial position, this chapter suggests a comparison between species and engineering design concepts and hints at possible analogies with respect to their multi-competence. Another resulting suggestion is that multi-fitness dynamic visualization of natural systems should be of a scientific value, in particular for the pursuit of understanding of natural evolution by way of thought experiments. In addition, future MCC research directions are proposed. It is concluded that MCC is a justified framework of thinking that has a ground in past and present findings both in engineering design research and biology. Yet, its scope, as demonstrated here, is bound to be controversial, which makes it both an intriguing and an exciting research area. It is hoped, at best, that MCC would direct thinking into fruitful new observations on the multi-fitness aspects of natural adaptation. Alternatively, it is expected that such studies would allow a better understanding of the similarities and dissimilarities in the creation of natural and artificial systems by adaptive processes.

Acknowledgments

The author is grateful to the many colleagues who have agreed to serve on the IPC of the related IEEE-ICCC-MOC 2005 workshop attempt. Thanks also to the IPC members and coorganizers of the IEEE/RSJ IROS-MOR 2006 Workshop on Multi-Objective Robotics, and the GECCO-ENAS-2007 Workshop on The Evolution of Natural and Artificial Systems: Metaphors and Analogies in Single and Multi-Objective Problems. X. Yao and the University of Birmingham should be acknowledged for an inspiring environment and their support of the author's Sabbatical during 2005. Also to be acknowledged are A. Bejan, J. Niklas, E. Sober and J. Knowles for reading and commenting on the draft. Finally, my student G. Avigad should be acknowledged for his ideas and dedication that created the foundation for the concept-based approach, and for his comments on the draft of this chapter.

References

[1] Abrams, P. (2001) Adaptationism, optimality models, and tests of adaptive scenarios. In: Orzack, S. H., Sober, E. (eds) Adaptationism and optimality. Cambridge University Press, Cambridge, pp. 273–302

[2] Avigad, G., Moshaiov, A. (2006) Simultaneous concept-based EMO. Report at: http://www.eng.tau.ac.il/~moshaiov, also submitted to the IEEE Trans on EC

[3] Avigad, G., Moshaiov, A. (2007) Set-based concept selection in multi-objective problems: optimality and variability approach. Report at: http://www.eng.tau.ac.il/~moshaiov

[4] Avigad, G., Moshaiov, A., Brauner, N. (2005a) Interactive concept-based search using MOEA: The hierarchical preferences case. Int J of Computational Intelligence, 3:182–191

[5] Avigad, G., Moshaiov, A., and Brauner, N. (2005b) MOEA for concept robustness to variability and uncertainty of market's demands. Proc of the 1st EC workshop in the 9th AI*IA conf on AI, Milan, Italy

[6] Bejan, A. (2000) Shape and structure, from engineering to nature. Cambridge University Press, Cambridge

[7] Bejan, A., Lorente, S. (2006) Constructal theory of generation of configuration in nature and engineering. J of Applied Physics 100:041301-27

[8] Bentley, P. J. (1999) (ed) Evolutionary design by computers. Morgan Kaufmann, San Francisco, California

[9] Bogatyreva, O., Pahl A-K., Vincent, J. F. V. (2002) Enriching TRIZ with biology — The biological effects database and implications for teleology and epistemology. Proc of the ETRIA World Conf, Strasbourg, pp. 301–307

[10] Coello, C. A. C. (2005) Recent trends in evolutionary multiobjective optimization. In A. Abraham, L. Jain and R. Goldberg (eds) Evolutionary multiobjective optimization: Theoretical advances and applications, Springer-Verlag, London, pp. 7–32

[11] Dawkins, R. (1986) The blind watchmaker. Longman Scientific and Technical, Harlow

[12] Deb, K., Gupta, H. (2005) Searching for robust Pareto-optimal solutions in multi-objective optimization. In: Evolutionary Multi-Criterion Optimization, volume 3410 of LNCS, Springer, pp. 150–164

[13] Di Paolo, E. A., Noble, J., Bullock, S. (2000) Simulation models as opaque though experiments. In: Bedau, M. A., McCaskill, J. S., Packard, N. H., Rasmussen, S. (eds) Artificial Life VII: the 7th Int Conf on the Simulation and Synthesis of Living Systems. Reed College, Portland, Oregon, MIT Press/Bradford Books, Cambridge MA, pp. 497–506

[14] Farnsworth, K. D., Niklas, K. J. (1995) Theories of optimization, form and function in branching architecture in plants. Functional Ecology, 9:355–363

[15] Fernandez, F. R., Hinojosab, M. A., and Puertoa, J. (2004) Set-valued TU-games. European J of Operational Research 159:181–195

[16] Gould, S.J. (2002) The structure of evolutionary theory. The Belknap Press of Harvard University Press, Cambridge and London

[17] Handl, J., Kell, D. B., Knowles, J. (2006) Multiobjective optimization in bioinformatics and computational Biology. IEEE/ACM Transactions on Computational Biology and Bioinformatics, 4 (2):279–292

[18] Holland, J. H. (1975) Adaptation in natural and artificial systems. The University of Michigan Press, Michigan

[19] Jin, Y. (ed) (2005) Multi-objective machine learning. Springer, Berlin

[20] Mandal, C., Gudi, R. D., Suraishkumar G. K. (2005) Multi-objective optimization in aspergillus niger fermentation for selective product enhancement. Bioprocess Biosyst Eng, 28:149–164

[21] Mattson, C. A., Messac, A. (2005) Pareto frontier based concept selection under uncertainty with visualization. Optimization and Engineering, 6:85–115

[22] Meijer and Koppelaar, (2003) Towards multi-objective game theory. GAME-ON conference, available at: http://mmi.tudelft.nl/~meijer/files/meijer-gameon03.pdf

[23] Moshaiov, A. (2006a) Multi-objective design in nature and in the artificial. Invited keynote paper, Proc of the 5th Int Conf on Mechanics and Materials in Design, Porto, Portugal

[24] Moshaiov, A. (2006b) Multi-objective cybernetics and the concept-based approach: Will they ever meet? The PPSN 2006 Workshop on Multiobjective Problem Solving from Nature, (PPSN 2006), available at: http://dbkgroup.org/knowles/MPSN3/Moshaiov-MO-cybernetics.pdf

[25] Moshaiov, A., Avigad, G. (2007a) Concept-based multi-objective problems and their solution by EC. Proc of the User-centric EC Workshop of the GECCO 2007 Conf, London, UK

[26] Moshaiov, A., and Avigad, G. (2007b) The extended concept-based multi-objective path planning and its A-life implications. Proc the 1st IEEE Symposium on A-life, in 2007 IEEE Symposium Series on Computational Intelligence, Honolulu, Hawaii, USA

[27] Niklas, K. J. (2004) Computer models of early land plant evolution. Annu. Rev. Earth Planet. Sci. 32:47–66

[28] Nishazaki, I., Sakawa, M. (2001) Fuzzy and multiobjective games for conflict resolution. Studies in Fuzziness and Soft Computing 64, Physica-Verlag, Heidelberg.

[29] Nottale, L. (1993) Fractal space-time and microphysics, World Scientific, Singapore

[30] Orzack, S. H., Sober, E. (2001) Introduction, in Orzack SH, Sober E (Eds.) Adaptationism and optimality, Cambridge University Press, Cambridge

[31] Parker, G. A., Maynard Smith, J. (1990) Optimality theory in evolutionary biology. Nature, 348:27–33

[32] Parmee, I. C. (2005) Human centric intelligent systems for design exploration and knowledge discovery. Proc of ASCE 2005 Int Conf on Computing in Civil Eng, Cancun, Mexico

[33] Poladian, L., Jermlin, L. S. (2006) Multi-objective evolutionary algorithms and phylogenetic inference with multiple data sets. Soft Comp, 10:359–368

[34] Pukkala, T. (2002) (ed) Multi-objective Forest Planning, Kluwer Academic Publishers, Durdrecht

[35] Sarkar, S. (2005) Maynard Smith, optimization, and evolution. Biology and Philosophy

[36] Savransky, S. D. (2000) Engineering of creativity: Introduction to TRIZ methodology of inventive problem solving. CRC Press LLC, Boca Raton, Florida

[37] Spencer, H. (1864) Principles of Biology, Williams and Norgate

[38] Sobek, D. K., Ward, A. C. (1996) Principles from TOYOTA'S set-based concurrent engineering process. Proc of the 1996 ASME Design Engineering Technical Conferences and Computers in Engineering Conference, Irvine, California, USA

[39] Teo, J., Abbass, H. A. (2005) Multiobjectivity and complexity in embodied cognition. IEEE Trans. on Evolutionary Computation, 9 (2):337–360

[40] Wiener, N. (1948) Cybernetics or control and communication in the animal and the machine. MIT Press, Cambridge

[41] Wright, S. (1932) The roles of mutation, inbreeding, cross-breeding and selection in evolution. Proc of the 6th Int Congress of Genetics, pp. 356–366

Part IV

Scaling up Multiobjective Optimization

Fitness Assignment Methods for Many-Objective Problems

Evan J. Hughes[1]

Dept. Aerospace, Power and Sensors
Cranfield University
DCMT, Shrivenham, Swindon, UK.
ejhughes@theiet.org

Summary. This chapter considers a number of alternative methods for fitness assignment in evolutionary algorithms for multiobjective optimization. Most of the fitness assignment methods in the literature were designed to work for any number of objectives, in principle; but, in practice, some of the more popular methods (e.g. those in NSGA-II, IBEA and SPEA) do not perform well on problems with four or more objectives. We investigate why this is the case, considering two aspects of performance: convergence towards the Pareto front and drive towards a set of well spread solutions. The visualization of induced fitness surfaces is used to understand the effects of the different fitness assignment methods, and both Pareto- and non-Pareto-based methods are analysed.

1 Introduction

1.1 Background

All optimization algorithms, whether of conventional design or based on evolutionary methods, rely on being able to perform a direct comparison between two competing solutions. In order to derive a selective pressure (or gradient) towards an optimum, the comparison should yield that either solution A is superior to B, or vice versa. For progress towards the global optimum, then the comparison must also report that the true superior solution is indeed superior. If the solutions are equivalent, then there is no information as to which may be genotypically closer to the optimum and no progress towards the optimum can be made unless a superior solution to either A or B exists elsewhere, or can be generated somehow.

With single-objective problems, the assignment of a degree of *fitness* that is used to compare two solutions is often straightforward. Complexities are introduced, however, when constraints are also considered. With more than one objective, it is likely that there no longer exists a single solution, but, rather, the best objective values are described by the Pareto front. The concept of nondomination applies and infers that two solutions lying on the Pareto front are therefore equivalent *until some additional external preferences are applied*. In order to derive a gradient or selective pressure, however, the optimization algorithm will still require a single-dimensional fitness assignment method that allows solution A to be compared directly with solution B, even though the algorithm may be maintaining an entire Pareto front in a single run.

Pareto ranking methods alone, as described in the Introduction chapter of this volume (pp. 1–26), will create a selective bias towards solutions on the Pareto front, but will not necessarily produce solutions that are spread across the front or at the edges. Additional elements in the fitness assignment are required to aid the Pareto ranking in order to create a diverse solution set.

Various forms of Pareto ranking and sharing/clustering have been exploited in recent years to develop a large number of multiobjective optimization algorithms which can solve bi-objective optimization problems effectively and reliably, for example, NSGA-II and SPEA-II. However, it is known that many of the methods which are efficient on bi-objective problems do not scale well to problems with large numbers of objectives (four or more typically cause issues) [7, 5].

As the number of objectives increases, typically the proportion of nondominated solutions within a search population increases [3]. The result is that if all of the solutions are nondominated, then all of the solutions will have the same Pareto rank, and the search towards the Pareto front reaches a plateau. In practice, the selective pressure is low even in the early phases of the search when dominated solutions exist, as only few Pareto ranks are needed to classify the population. The secondary elements of the fitness assignment function now dominate. These secondary elements are often sharing or clustering methods and serve to distribute the solutions across the nondominated front. Thus, with many-objectives, the initial optimization progression is weakly towards the Pareto front; then, in the later stages of the optimization, the solutions are spread out evenly. As the dimensionality increases, the spreading actions dominate rapidly giving nondominated solutions distributed evenly, but not near the true Pareto front.

Real engineering problems are often characterized by many objectives, many constraints, or both. Often, problems have constraints where information on the degree of constraint is available, and the constraints can be converted to objectives (see this volume, pp. 53–75). The constraint conversion however increases the dimensionality of the objective space (primarily in the early phases of the search until the constraints are satisfied).

The problem is how to design a many-objective fitness assignment method that will allow an optimization algorithm to produce nondominated fronts that are both well spread and a good approximation of the true Pareto front. Currently, there are few algorithms that are designed specifically to tackle many-objective problems.

1.2 Many-Objective Fitness Assignment Methods

Two alternative approaches have been employed to date to derive useful fitness assignment processes for many objective problems: either augment the Pareto ranking concept with functions that can aid the progression towards the Pareto front, or use approaches that do not use Pareto ranking. A general observation is that methods based on Pareto ranking still perform well on bi-objective problems but may have computational performance issues when scaling to very large numbers of objectives, whereas many non-Pareto ranking methods scale well computationally but are not necessarily so efficient at approximating the Pareto front with low numbers of objectives. Why do these methods behave so differently?

Fundamentally, the fitness assignment process maps the multiobjective space into a single dimension to allow the solutions to be ranked. There are some idealized optimization behaviours we would like to promote through the fitness assignment process:

1. A solution that is dominated should not be assigned a fitness superior to its dominating solution;
2. If two solutions are nondominated, the solution with the closest neighbours should be inferior;
3. If two solutions are nondominated, the solution closest to the true Pareto front should be superior;
4. Constrained solutions should be inferior to feasible solutions.

These four idealized optimization behaviours all assume that the objective functions (and constraint satisfaction) improve (in a minimization or maximization sense) as the true solution quality improves. It must be remembered that it is possible for the objective functions to be good indicators of true solution performance only in specific regions of the search space, leading to multimodalities and, therefore, 'local' Pareto optimal fronts. However, the fitness assignment process is by definition applied *after* the objectives have been defined and therefore must assume that the objective functions do indeed provide a true reflection of the solution quality. The ability of any optimization algorithm to escape local optima is a property not of the fitness assignment process but of how the new trial solutions are generated; generally, in evolutionary methods, mutation-based techniques are employed to help search for global optima.

Whenever a high-dimensional space is mapped to a lower dimension, information has to be discarded and a compromise is often made, leading to a nonideal fitness assignment and, therefore, potentially inappropriate ranking of the solutions. There are also examples where the idealized behaviour may not always produce the best performing algorithm: the handling of constraints may well be improved by compromising on item 4 [8]. Constraint handling techniques are considered further in this volume, pp. 53–75.

Many approaches to multiobjective fitness assignment exploit Pareto ranking methods, which treat item 1 above as the dominant requirement, with item 2 as a secondary ranking element. Interestingly, many of these methods (for example, NSGA, NSGA-II, MOGA, SPEA, SPEA-II) have no mechanism for addressing item 3 directly. Instead the algorithms rely on the concept that, in bi-objective spaces, driving away from dominated solutions (item 1) is a very good approximation of item 3 (moving towards the Pareto front). However, for many-objective spaces where the majority of solutions are nondominated, the approximation breaks down as item 1 becomes ineffective. Thus, we can now see the mechanism which describes the variation in behaviour between the Pareto ranking and non-Pareto methods: whether they address item 3 indirectly through Pareto concepts, or directly through other means.

Unfortunately, item 3 above is difficult to convert to a fitness metric, and if we knew the Pareto front from the design of the fitness assignment process, we would have solved a large part of the problem, with only the corresponding decision space region to be identified! All of the methods that function well for many-objective problems have item 3 as the primary fitness assignment mechanism, with item 2 and (sometimes) item 1 as secondary processes.

1.3 Fitness Assignment Visualisation

When designing and evaluating alternative fitness assignment mechanisms, it is useful to be able to visualize the behaviour of the assignment process in the objective space. A simple mechanism is to create a sample set of solution points in objective space, and then evaluate the fitness that would be given to each solution. One of the solution points can then be moved across the entire objective region, usually by gridding the space to some convenient resolution, and then graphing the variation of fitness of each solution point (including a graph for the point that is moving) as the single point is moved. The result for a bi-objective problem is a set of 3D surfaces that describes how the fitness of the moving and fixed points vary as the one objective point moves. If we draw contour lines of constant fitness on these surfaces, then we can create maps of *isofitness contours* that can be used to visualize the behaviour of the fitness assignment method.

Although the isofitness contour concept can be extended to many objectives, high-dimensional visualisation becomes an issue. However, in conjunction with the four idealized optimization behaviours described in Section 1.2, the key characteristics of the fitness assignment process can often be determined from visualisation in two dimensions, and the expected behaviour in many dimensions predicted accurately. For simple aggregation functions, where the fitness of a point is independent of the location of other objective vectors, the isofitness contours alone suffice; a detailed example is provided in Section 3.1.

For more complex ranking methods, we need to visualize how the rank order would be influenced by the geometry of the points. For these methods (such as NSGA) it is more useful to consider a contour of *relative isofitness*. A map of relative fitness is calculated by subtracting the fitness map of a fixed point from the fitness map of the moving point (assuming that the assigned fitness value is to be minimized in the ranking process). For example, Figure 1 shows the locations of five fixed points. As a sixth point is moved through the objective region, the fitness of the point of interest (at [0.7, 0.85] in the figure) is calculated and subtracted from the fitness that the sixth moving point would have at the current location of the sixth point on the graph. The result is that points which lie on a zero-valued contour are directly equivalent to the fixed point being studied, when considered for ranking. All regions which dominate the fixed point of interest should have a value less than 0, and all regions that are dominated should have values greater than 0. Thus, item 1 in Section 1.2 would be satisfied. To satisfy item 2, nondominated solutions that are more crowded than the point of interest should have a positive relative fitness, and nondominated solutions that are less crowded should have a negative relative fitness. To satisfy item 3, the local gradient of the fitness contour should always be to improve all objectives, i.e., falling towards the origin of the graph (in a minimization sense), or, as a worst case, to leave an objective unchanged. Any regions where the gradient is towards degrading an objective value may allow a solution nearer to the true Pareto front to appear inferior to one further away. Figure 1 demonstrates these regions of relative fitness and the direction of the gradients on the contour where points are directly equivalent to the point of interest at location [0.7, 0.85]. Section 2.2 provides a detailed example of the isofitness contour visualisation process in action.

The use of isofitness contours to visualize many-objective fitness assignment methods provides a simple mechanism to analyse *typical* fitness behaviours. The

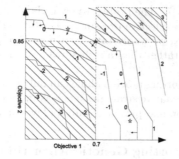

Fig. 1. Example of a relative isofitness map showing an idealized relative fitness behaviour in the dominated, dominating and nondominated regions of the fixed point at [0.7, 0.85] being examined

process can be automated in high dimensions where the dominated and dominating regions can be assessed for correct relative performance, and the local gradients can be calculated in the nondominated region and tested for any reverse gradient conditions. However, although automation can identify fitness assignment methods that are unlikely to work well in many-objective spaces, the ability of the fitness assignment methods to create well-spread solution sets is difficult to ascertain. In addition, the shape of the isofitness contours is often conditioned on the distribution of the trial points in the objective region. Thus, for accurate automated analysis a Monte Carlo process is advised where many different example sets of objective vectors are tested in order to explore the potential for adverse fitness behaviours.

1.4 Chapter Structure

This chapter discusses the behaviour of existing fitness assignment methods designed for bi-objective problems, and methods that can function with many objectives. Enhancements to Pareto ranking are discussed in Section 2, and non-Pareto methods are discussed in Section 3. Section 3.3 discusses how some of the fitness methods may be used to aid the visualisation of the Pareto front with many objectives, and Section 4 concludes.

2 Pareto Ranking Extensions

2.1 Introduction

It is known that as the dimensionality of the objective space increases, the proportion of solutions which are nondominated in the initial random population tends to increase rapidly [3].

Generally, with many-objective problems, and, therefore, with very few dominated individuals, the selective pressure on the remaining population is very low, as nondominated individuals are considered equivalent. The spreading mechanisms dominate the selection process and the solutions are spread, rather than progressing towards the Pareto front [7]. The problem now is: how can Pareto ranking methods be extended to restore the selective pressure towards the Pareto front (i.e., item 3 in Section 1.2)? Realistically, we would rather have a set of points close to the Pareto front but poorly spread, rather than a well-spread set of solutions that are far from the true optimal surface.

The Nondominated Sorting process alone can only separate a population into individual rank layers. Alternative strategies, such as those used in SPEA and MOGA, count levels of domination and provide similarly layered structures.

2.2 Nondominated Sorting Genetic Algorithm

The following example is based on the classic NSGA algorithm (see this volume, p. 19) that consists of a nondominated sorting step, followed by sharing within the sorted layers. The weakness in the original sharing method was that a priori knowledge was often needed in order to set the share radius. For demonstration, here a large fixed share radius of $\sigma = 0.4$ has been used.

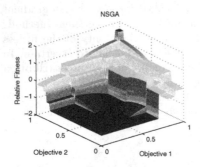
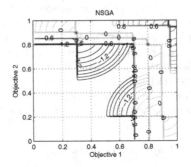

Fig. 2. 3D relative fitness surface for point [0.7, 0.85] and the NSGA method with one moving and five fixed objective vectors. Spreading factor is $\sigma = 0.4$

Fig. 3. Relative isofitness contours for point [0.7, 0.85] and the NSGA method with one moving and five fixed objective vectors. Spreading factor is $\sigma = 0.4$

Figure 2 shows the relative fitness surface for point [0.7, 0.85] in a six point set, five of which are in fixed locations and the sixth is moved through the objective region in order to generate the fitness surface. The values on the graph are the differences in fitness value between the moving and the fixed point at [0.7, 0.85]. Figure 3 shows the corresponding relative isofitness contour map. Thus, if the moving point were at location [0.4, 0.63], then the figures show that the fitness of the moving point would be less by a value of 1.2 and therefore more likely to be selected after ranking.

The key feature of Figure 3 is that although the nondominated front is visible clearly (as a line connecting all the nondominated points), many of the contours traverse *across* the objective space, rather than focusing towards the origin (which would be an ideal solution). If we consider point [0.2, 0.8] for example, the relative fitness is very low, demonstrating that the point is highly attractive. If we now consider point [0.75, 0.5], it is nondominated with respect to the test point at [0.7, 0.85]; however, the local gradient is focused towards improving objective 1, but *degrading* objective 2! Therefore the requirement of item 3 in Section 1.2 is compromised as we are moving away from the Pareto front. The problem with the classic NSGA algorithm is that as the dimensionality increases, regions with adverse fitness gradient structures become more common and the optimization process is compromised.

With the contour alignment in the nondominated region approximately normal to the nondominated front, the selective focus is on spreading the solutions, rather than on driving them towards the Pareto front. With many objectives, the problems are exacerbated.

2.3 Nondominated Sorting Genetic Algorithm II

Figures 4 and 5 show the relative fitness surface and relative isofitness contours for the NSGA-II fitness assignment process with point [0.7, 0.85] as a fixed reference. NSGA-II uses crowding distance rather than fitness sharing (the use of $-\infty$ at the edges has been modified to provide a consistent and representative fitness landscape). It is clear that the fitness gradient is changed significantly over the original NSGA algorithm. The crowding operator is calculated based on the location of the neighbour solutions relative to the point (in the same nondominated rank), and the result is that the fitness in the local region remains constant as long as the local neighbour solutions are the same. The fitness surface is now dominated by plateaus, rather than by continuous gradients. The optimization performance is improved over NSGA as there are fewer regions where the gradient is away from the Pareto surface; however, they still exist and point [0.75, 0.5] is a good example. Unfortunately, within the plateaus, there is no selective pressure to either converge towards the Pareto surface, or to spread evenly, however a plateau is preferable to a reverse-gradient in the nondominated region. In a practical algorithm, a moderate or large population size would be desirable in order to reduce the scale of each plateau region (i.e., for smaller distances to neighbours). As the NSGA-II algorithm maintains the elite solutions within the working population, in practice the population sizes are often sufficiently large to make the ranking process perform well in low dimensions. With many-objective problems, however, still having areas in the nondominated region of reverse gradient degrades the algorithm performance.

2.4 Multiobjective Genetic Algorithm

The Multiobjective Genetic Algorithm (MOGA) [3] counts the number of solutions that a point is dominated by, and then uses a sharing mechanism to spread the solutions. The fitness landscape that is obtained is very similar to NSGA; however, not all implementations of the algorithm confine the sharing mechanism to individual rank layers, of the same domination count, as in the NSGA algorithm. Figures 6 and 7 show the relative fitness surface generated from the MOGA algorithm *when the sharing was not confined to rank layers*.

In the figures, a share distance of $\sigma = 0.1$ has been used. Issues with reverse gradient in the nondominated region are visible clearly in many regions (point [0.85, 0.3]

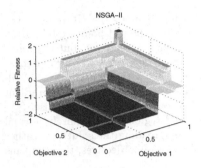

Fig. 4. 3D relative fitness surface for point [0.7, 0.85] and the NSGA-II method with one moving and five fixed objective vectors

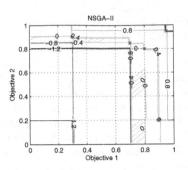

Fig. 5. Relative isofitness contours for point [0.7, 0.85] and the NSGA-II method with one moving and five fixed objective vectors

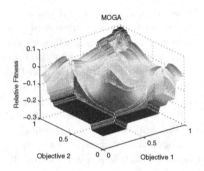

Fig. 6. 3D relative fitness surface for point [0.7, 0.85] and the MOGA method with one moving and 5 fixed objective vectors. Spreading factor is $\sigma = 0.1$

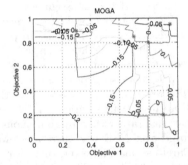

Fig. 7. Relative isofitness contours for point [0.7, 0.85] and the MOGA method with one moving and 5 fixed objective vectors. Spreading factor is $\sigma = 0.1$

for example). The point [0.85, 0.85] is interesting, as the gradient towards it converges as a local attractor. This is a direct result of using global sharing, rather than restricting sharing to within rank/domination layers, and is caused by empty regions within the dominated space being emphasized by the sharing process. Additionally, the point [0.85, 0.85] shows that item 1 in Section 1.2 is compromised, as there are solutions in the dominated region that are superior to the test point at [0.7, 0.85]. If rank/domination layer sharing is applied, then the fitness surface is very similar to the surface obtained with NSGA. The MOGA algorithm does not scale well to many-objective problems either.

2.5 Hypervolume Selection

Fundamentally, the hypervolume metric [10] assesses the total volume that lies between a chosen reference point that acts as a corner to a hypercube and the nondominated front, described by a set of points, which intersects the hypercube. The closer the nondominated front is to the Pareto front, the larger is the hypervolume.

A simple way to use the hypervolume metric to augment the basic Pareto ranking process is to first use nondominated sorting to establish the front that a particular point belongs to, and to then calculate the hypervolume of the set of points which includes all points on that front and all points that are worse. The point of interest is then removed from the set and the hypervolume recalculated, allowing a change in hypervolume, ΔS, to be established. The change ΔS is then normalized by the maximum possible hypervolume to give an indicator of *local* worth $\Delta S/V_{max}$. The fitness is the Pareto rank layer index number minus $\Delta S/V_{max}$. As long as $\Delta S/V_{max} < 1$ then the Pareto ranking structure will be preserved, with the local shaping of the fitness surface being provided by the hypervolume metric.

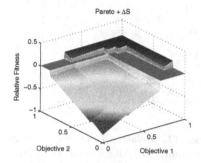

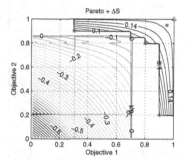

Fig. 8. 3D relative fitness surface for point [0.7, 0.85] and the hybrid Nondominated Sorting/Hypervolume method with one moving and five existing objective vectors

Fig. 9. Relative isofitness contours for point [0.7, 0.85] and the hybrid Nondominated Sorting/Hypervolume method with one moving and five existing objective vectors

For minimization, the hypervolume reference point R is placed in such a way as to be at least weakly dominated by every member of the set to be investigated. Often, the reference point is chosen as the maximum value observed in all objectives so far.

Figures 8 and 9 show the resultant relative isofitness contours for a reference point at $\mathbf{R} = [1, 1]$. It is clear that the fitness gradient is always well defined in the dominated/dominating regions; however in the nondominated region, there is a significant plateau before the nondominated front, which can reduce overall algorithm efficiency. There are no regions of reverse relative fitness gradient, however.

The hypervolume metric produces a weak 'spreading' effect; however, it is clear that points are being driven slowly towards [0.3, 0.9] by a gradient that is primarily independent of solution locations: although it may cause solutions to crowd locally in the area, the focus will aid the discovery of better extreme solutions. The main drive is towards the Pareto front and the distribution of solutions is a weak secondary process with this simple implementation. Many of the practical implementations of the hypervolume metric [6, 2] use extra processes to improve the distribution of the solutions further. The lack of any nondominated regions with reverse gradient characteristics makes the method suitable for many-objective optimization; however, the processing complexity of the hypervolume metric often limits its applicability.

2.6 Indicator-Based Evolutionary Algorithm

The *Indicator-Based Evolutionary Algorithm* (IBEA) process [11] does not use Pareto ranking directly, but instead uses *indicator functions* that allow the fitness of a solution in a population to be determined. The indicators, however, are specifically designed to preserve Pareto rank.

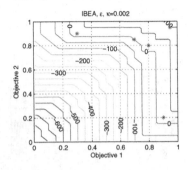

Fig. 10. Relative isofitness contours for point [0.7, 0.85] and the IBEA method with one moving and five existing objective vectors. Scaling factor $\kappa = 0.002$

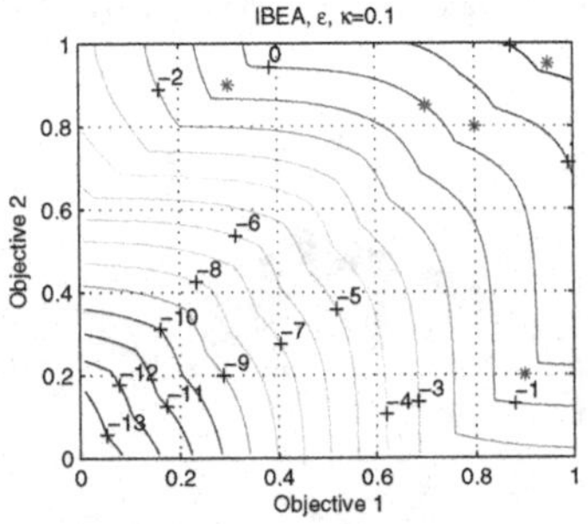

Fig. 11. Relative isofitness contours for point [0.7, 0.85] and the IBEA method with one moving and five existing objective vectors. Scaling factor $\kappa = 0.1$

Figure 10 shows the isofitness contours for the IBEA ϵ indicator in equation 1, with a shape parameter of $\kappa = 0.002$. In (1), O_{iA} is the value of the ith objective of population member A, k is the number of objectives and P is a set of objective vector points that describe the population. For visualisation purposes, the difference in the logarithm of the fitness has been plotted.

$$I(A, B) = \min_{\epsilon}\{O_{iA} - \epsilon \leq O_{iB} \text{ for } i \in \{1, \ldots, k\}\}$$

$$f = \sum_{O_A \in P \neq O_B} -exp(-I(A, B)/\kappa) \tag{1}$$

It is clear that the relative isofitness contours of the ϵ indicator are structured with only a weak relationship to the shape of the nondominated surface. The relative fitness values in the dominated and dominating regions are correct. The fitness gradient is well structured with no reverse gradients, and in regions beyond the edges of the nondominated front will promote good edge exploration. The fitness method does lack any intrinsic directionality to aid a uniform distribution of solutions across the Pareto front, however. The lack of gradient for forming uniform distributions in the population is evidenced in Figure 10 by solutions which are well spread (e.g., [0.9, 0.2]) not being promoted as superior to the point under test at [0.7, 0.85].

Figure 11 shows how an alternative choice of the scaling factor κ can change the structure of the fitness surface. With the larger value of $\kappa = 0.1$, the algorithm will not perform so well on highly concave Pareto fronts as the isocontours do not form sharp 'corners', and will penalize Pareto solutions within a deep concavity. In Figure 11, the points [0.3, 0.9] and [0.9, 0.2] both lie on a fitness contour that is superior to the test point [0.7, 0.85], which is in turn superior to point [0.8, 0.8]. All of the points are nondominated, but the modified fitness function is promoting solutions which are isolated over those which are crowded, and will achieve a superior spread of solutions than with the smaller value for κ.

The configuration is likely to give good performance on convex and mildly concave Pareto fronts. The fitness contours also demonstrate that although a strict Pareto relationship is maintained when comparing two solutions in isolation, the act of combining the results for a population of solutions can produce isofitness contours that do not follow the nondominated front, and yet not compromise any of the ideal requirements itemized in Section 1.2. The IBEA method will work well on many-objective functions; however, an external diversity mechanism, such as clustering of an archive, is recommended to achieve a controlled spread of solutions.

2.7 Summary of Pareto Methods

If the Pareto rank is enforced in the fitness process, with inferior ranks guaranteed to have a worse fitness than true nondominated solutions, it is possible to create optimizers that perform very well in bi-objective problems. However, for many-objective problems, the gradient of the fitness within the nondominated regions must also always focus on improving all objectives to some degree.

The IBEA and hypervolume methods have both demonstrated advantageous fitness gradient structures, but at the expense of limited (if any) solution distribution characteristics intrinsic within the fitness assignment. The desires to move away from existing solutions, yet not degrade performance on any objectives, are conflicting in many circumstances. Realistically, the degree of solution spreading that can be generated by the fitness assignment function alone is limited, but an external archive can be used to help impose a uniform spread of solutions across the Pareto front, for example, through clustering.

3 Non-Pareto Ranking Methods

An alternative many-objective fitness assignment process is to use a method that does not rely on Pareto ranking to sort the population. The simplest of these non-Pareto methods is to use a conventional aggregation approach such as weighted sum (Section 3.1) and perform many single objective optimizations, changing the weight vector set a little each time to enable the entire Pareto surface to be sampled.

A natural extension is to attempt to satisfy all the weight vectors simultaneously in a single run of the optimizer. Multiple Single Objective Pareto Sampling (MSOPS) [4] is one method that develops this concept into a practical algorithm.

Many early multiobjective EAs, such as VEGA and Weighted Average Ranking [1], do not use Pareto ranking methods. Many of these early approaches have some merit in many-objective spaces and are worthy of investigation.

3.1 Repeated Single Objective

In the Repeated Single Objective (RSO) [5] approach, a conventional single objective EA is used, based on an aggregation function, and repeat runs are performed for different target search directions, allowing a Pareto front to be constructed from a sequence of spot solutions. To allow direct comparison with true multiobjective EAs, each run of RSO uses a correspondingly smaller population size and number of generations to keep the total number of evaluations to identify a Pareto front consistent.

The RSO method is very simple but does require an a priori specification of the directions to search, in order to populate the Pareto front, which can be difficult with previously unseen problems. The RSO method is known to be effective in high-dimensional many-objective optimization problems [5]. The performance and applicability of RSO to different objective structures is determined primarily by the choice of the aggregation functions used to identify optimal solutions.

A key benefit of prespecification of search directions is that designer preferences can be incorporated very easily and the search focused only on regions of interest. Additionally, Pareto front analysis may be performed as described in Section 3.3.

Aggregation Functions

Aggregation functions have been used for many years in classical gradient-based optimization. Generally, a single aggregation function will yield a single Pareto point; however, all of the aggregation functions described here are effective with both low- and high-dimensional objective spaces. In practice, once the structure of the Pareto surface has been approximated, a decision has to be made about the particular Pareto point to choose. An aggregation function can then be used in a single-objective optimization framework to identify a single near-Pareto solution.

The following common aggregation functions have been plotted with their fitness functions arranged for objective minimization. The fitness contours have been drawn in the same context as the population-based methods, with the five example population points plotted to allow direct comparison. It should be remembered, however, that the following aggregation methods are conditioned only by the weight vector and control parameters, and not by the location of the other population members. Thus, the isofitness contour plots are contours of true fitness values, but should still allow dominated individuals to be inferior to dominating solutions.

Weighted Sum

The weighted sum score of k objectives is calculated using (2), where w_i is part of the weight vector $\mathbf{W} = [w_1, w_2, \ldots]$ and is the weight of the ith objective O_i.

$$f = \sum_{i=1}^{k}(w_i O_i) \tag{2}$$

Weighted sum will not introduce discontinuities into the gradient of the aggregated function but is able to generate points *only on convex* Pareto fronts. The location of the point on the Pareto front is highly dependent on the shape of the front itself; however, the search 'direction vector' may be described as $\mathbf{V} = \mathbf{W}$.

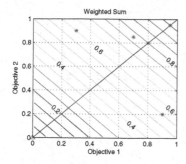
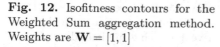

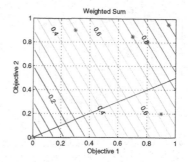

Fig. 12. Isofitness contours for the Weighted Sum aggregation method. Weights are $\mathbf{W} = [1, 1]$

Fig. 13. Isofitness contours for the Weighted Sum aggregation method. Weights are $\mathbf{W} = [2, 1]$

Figures 12 and 13 show the *isofitness contours* for the weighted sum aggregation function with two different weighting vectors. The diagonal line radiating from the origin shows the search direction vector $\mathbf{V}$. The isofitness contours form hyperplanes normal to the search direction $\mathbf{V}$. The weighted sum method will return Pareto points where the normal to the Pareto front is parallel to $\mathbf{V}$, and therefore normal to the isofitness contours. Thus, the aggregation method is unsuitable if points within a concave region of a Pareto front are to be identified. The primary benefit of the weighted sum is that if the gradients of the individual objective functions are continuous, then the gradient of the resulting fitness value, f, will also be continuous. No additional constraints are created, allowing the weighted sum to be used as an aggregation method with *all* optimization algorithms.

Goal Attainment

The goal attainment score of k objectives is calculated by transforming all of the objectives into objective-space constraints using (3),where w_i is the weight of the ith objective O_i, and Z_i is the ith dimension of an idealized reference point $\mathbf{Z}$.

Minimize γ subject to:

$$O_i - w_i\gamma \leq Z_i \ \forall i \in [1, k] \tag{3}$$

The control parameter γ is reduced until the constrained region consists of a single feasible point. This point lies on the Pareto front. If the final value of γ is negative, the reference point $\mathbf{Z}$ has been dominated. Figure 14 shows the isofitness contours for the goal attainment aggregation function. As the isofitness contours always form a 'corner' which has its sides aligned parallel to the objective axes, goal attainment is able to generate points on both convex and concave Pareto sets.

If the optimization process converges to a solution that exactly 'matches' the weight vector, then $C = (O_1 - Z_1)/w_1 = (O_2 - Z_2)/w_2 = \ldots$, where C is a constant, allowing the convergence of the solution with respect to the weights to be assessed. The weight vector corresponds to a point on the Pareto set in the true direction given by the vector $\mathbf{V} = [w_1, w_2, \ldots]$ (after offsetting by the reference point $\mathbf{Z}$). Thus, the

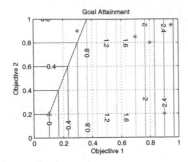

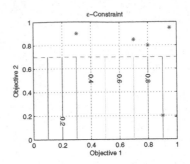

Fig. 14. Isofitness contours for the Goal Attainment aggregation method. Reference point is $\mathbf{Z} = [0.1, 0.2]$, weights are $\mathbf{W} = [1, 3]$

Fig. 15. Isofitness contours for the ϵ-constraint aggregation method. Objective 1 is being minimized, while constraining objective 2 to ≤ 0.7

angle between the vectors $\mathbf{V}$ and $\mathbf{O} - \mathbf{Z}$ indicates whether the solution lies where it was expected or not. If the vector $\mathbf{V}$ lies within a discontinuity of the Pareto set, or is outside of the entire objective space, then the angle between the two vectors will be significant. By observing the distribution of the final angular errors across the total weight set, the limits of the objective space and discontinuities within the Pareto set can be identified. This active probing of regions of interest can only be performed if the weight vectors are defined prior to the optimization run. Section 3.3 provides examples of the process.

Unfortunately, goal attainment relies heavily on the optimization algorithm being able to implement constraints efficiently. Constraint handling in evolutionary processes is possible, but not often efficient.

ϵ-Constraint

The ϵ-constraint metric converts all but one of the objectives into objective space constraints. The optimization process operates on the one remaining objective and the Pareto front point chosen is usually a point that best satisfies the objective, and is just within the feasible region defined by the constraints on the remaining objectives. As the constraint locations are moved, other Pareto points may be identified. Equation (4) describes the process, where C_i is the constraint location of the ith objective O_i.

Minimize $f = O_j$ subject to:
$$O_i \leq C_i \; \forall i \neq j \in [1, k] \tag{4}$$

Figure 15 shows the isofitness contours for the ϵ-constraint aggregation function. Like goal attainment, the isofitness contours always form a 'corner' which has its sides aligned parallel to the objective axes; thus ϵ-constraint is able to generate points on both convex and concave Pareto sets. If the objective to be minimized is chosen carefully, the gradient of the optimization surface can be very favourable; for example if the first three objectives are highly multimodal, but the fourth is unimodal, it makes sense to constrain the first three and optimize the fourth.

Weighted Min-Max

The weighted min-max score of k objectives is calculated using (5),where w_i is the weight of the ith objective O_i.

$$f = \max_{i=1}^{k}(w_i O_i) \tag{5}$$

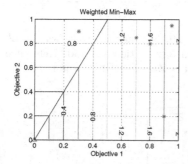

Fig. 16. Isofitness contours for the Weighted Min-Max aggregation method. Weights are $\mathbf{W} = [2\ 1]$

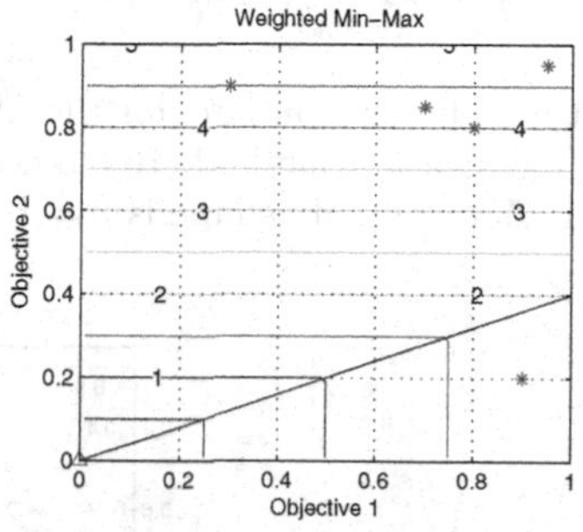

Fig. 17. Isofitness contours for the Weighted Min-Max aggregation method. Weights are $\mathbf{W} = [2\ 5]$

Figures 16 and 17 show the isofitness contours for the weighted min-max aggregation function with two different weighting vectors. Like goal attainment, weighted min-max isofitness lines form 'corners', and the method is able to generate points on both convex and concave Pareto sets. If the optimization process converges to a solution that exactly 'matches' the weight vector, then $w_1 O_1 = w_2 O_2 = \ldots$, allowing the convergence of the solution with respect to the weights to be assessed. The weight vector corresponds to a point on the Pareto set in the true direction given by the vector $\mathbf{V} = [1/w_1, 1/w_2, \ldots]$.

Weighted Min-Max is sometimes also referred to as a Weighted Tchebychev Norm (spelling of Tchebychev may vary). It is a variant of the L^p norm method with $p = \infty$. It can be considered as a weighted L^{∞} metric but with a reference point at the origin.

L^p Norm

The L^p Norm score of k objectives is calculated by using (6),where O_i is the ith objective of the vector $\mathbf{O}$, Z_i is the ith dimension of an idealized reference point $\mathbf{Z}$, W_i is a weighting component and p is a scalar factor that determines the shape of the isofitness contours. For the classic L^p norm methods, unity weighting factors $W_i = 1$ are usually assumed.

$$f = \left[\sum_{i=1}^{k} W_i |O_i - Z_i|^p \right]^{\frac{1}{p}} \tag{6}$$

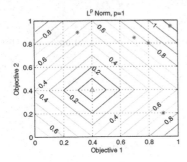

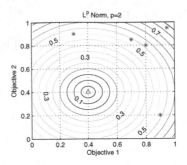

Fig. 18. Isofitness contours for the L^p Norm aggregation method. Reference point is $\mathbf{Z} = [0.4, 0.4]$, shape is $p = 1$

Fig. 19. Isofitness contours for the L^p Norm aggregation method. Reference point is $\mathbf{Z} = [0.4, 0.4]$, shape is $p = 2$

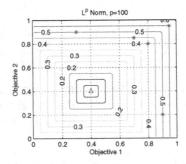

Fig. 20. Isofitness contours for the L^p Norm aggregation method. Reference point is $\mathbf{Z} = [0.4, 0.4]$, shape is $p = 100$

Figures 18, 19 and 20 show the isofitness contours for the L^p norm aggregation function with three different shape parameters. The classic L^p norm does not use a weight vector; rather, the Pareto point that is closest to the reference point $\mathbf{Z}$ will give the lowest aggregated value. Thus, as $\mathbf{Z}$ is moved, points on the Pareto front can be generated. The reference point $\mathbf{Z}$ must dominate the nearest points on the Pareto front; otherwise the optimizer will simply converge to the point $\mathbf{Z}$ if it lies within the objective region.

Figure 18 uses a shape of $p = 1$ and is therefore the L^1 norm, or the *Manhattan* distance. The isofitness contours form lines that have similar properties to the weighted sum; however, if the reference point is placed within a concave region of the Pareto front, points within the concavity can be found, although the reference must be placed very close to the Pareto front in order to identify regions of sharp concavities.

Figure 19 uses a shape of $p = 2$ and is therefore the L^2 norm, or the *Euclidean* distance. The isofitness contours form circular contour lines, allowing shallow concave regions to be identified easily. Sharp concavities will still require the reference point to be placed very close to the Pareto front. With low values of p, the gradient of the objective functions, and therefore the gradient of the aggregated fitness, is maintained.

Figure 20 uses a shape of $p = 100$ and is therefore the L^{100} norm. The isofitness contours approximate corners now, similarly to the corners displayed by the Weighted Min-Max method, allowing even quite sharp concave regions to be identified easily. With these high values of p, the gradient of the aggregated function can be subjected to numerical errors and appear discontinuous. A metric with $p = \infty$ is the Tchebychev Norm, and the infinite power is approximated by a max() operation.

Vector Angle Distance Scaling (VADS)

Vector Angle Distance Scaling (VADS) is a new metric first introduced in [4]. The metric is designed specifically for identifying the *Objective Front*, rather than just the Pareto front. The objective front is the entire leading edge of the feasible objective space region. The Pareto front is therefore a subset of the objective front. If the objective front is identified, then areas where 'gaps' appear in the Pareto set can be analysed: if there are objective front solutions that lie within the gap, then the break in the Pareto front is a discontinuity due to a very deep or reentrant concavity. If there are no objective front solutions in the region, then it is likely that the feasible objective region is comprised of disconnected subregions. In bi-objective problems, it is not difficult to identify regions of discontinuity in the Pareto front alone. However, even with three objectives, a discontinuity may present as a 'hole' and is not simple to identify without knowing the shape of the objective front. All of the metrics so far that are capable of identifying the Pareto surface in the presence of concavities use an isofitness contour that forms a 'corner'. To identify regions of the objective front, a mechanism is needed to form the isofitness contours into acute angles in order to allow deep probing into highly concave regions. The use of these metrics for surface analysis is discussed further in Section 3.3.

The VADS score is the magnitude of the vector of objectives ($|\mathbf{O}|$) divided by the cosine of the angle between the vector of objectives and a target vector, where the resulting angle cosine is then raised to a high power. Thus, an objective vector that forms a point lying on the target vector is assigned a fitness which is the distance along the target vector. As the objective vector strays from the target vector, the fitness is increased rapidly with increasing offset angle.

The cosine of the angle can be calculated conveniently by a dot product operation. The score equation for k objectives is calculated using (7), where $\mathbf{V}$ is the k-dimensional unit-length target vector which describes the point on the objective front to search for, $\mathbf{O}$ is the k-dimensional objective vector, $|\cdot|$ indicates vector magnitude and q is a constant factor for scaling the cosine result (typically $q = 100$). The vector $\mathbf{V}$ may also be described in terms of the weight vector used in the other metrics as the normalisation $\mathbf{V} = \mathbf{W}/|\mathbf{W}|$.

$$f = \frac{|\mathbf{O}|}{\left(\mathbf{V} \cdot \frac{\mathbf{O}}{|\mathbf{O}|}\right)^q} \tag{7}$$

Low values for q may lead to difficulty in identifying very sharp concavities in the objective front. The dot product of the vector $\mathbf{V}$ with the objective vector $\mathbf{O}$ must

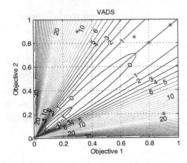

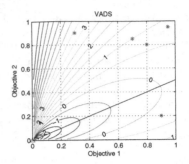

Fig. 21. Isofitness contours for the Vector-Angle Distance Scaling aggregation method. Weights are $\mathbf{W} = [1,1]$, $q = 100$. Logarithm of fitness plotted for clarity

Fig. 22. Isofitness contours for the Vector-Angle Distance Scaling aggregation method. Weights are $\mathbf{W} = [2,1]$, $q = 10$. Logarithm of fitness plotted for clarity

remain positive for the basic VADS metric to function correctly, and consequently objective offsets may be necessary for proper operation.

Figure 21 shows the isofitness contour for a weight vector $\mathbf{W} = [1,1]$ and shaping parameter $q = 100$. The 'tear drop'-shaped isofitness contour is made thinner by increasing q, allowing sharper concavities to be probed. With very high values of q, care must be taken to prevent numerical instability. In the figure, the logarithm has been taken to reduce the dynamic range of the metric values experienced in the optimization process. The use of logarithms allows (7) to be reformulated as shown in (8) and reduces the impact of numerical imprecision.

$$f = \exp((q+1)\log(|\mathbf{O}|) - q\log(\mathbf{V} \cdot \mathbf{O})) \qquad (8)$$

Figure 22 shows the isofitness contour for a weight vector of $\mathbf{W} = [2,1]$ and shaping parameter $q = 10$. With the lower value for q, the 'tear-drop' shaped isofitness contour is fatter and therefore less able to probe deep folds in the objective surface. It is also clear that as the weight vector is changed, the isofitness contour follows the vector, rather than being aligned to the objective axes.

The final solution identified by an optimizer using the VADS metric should have the objective vector $\mathbf{O}$ lying parallel to the target vector $\mathbf{V}$. Thus the angle between the two vectors can be used to assess final convergence. As VADS is tolerant of 'folds' in the objective surface that cause discontinuities in the Pareto front, angular errors between $\mathbf{V}$ and $\mathbf{O}$ indicate non-obtainable sections in the objective region.

3.2 Multiple Single Objective Pareto Sampling

Multiple Single Objective Pareto Sampling (MSOPS) [4] is a technique that allows multiple single objective optimization searches to be run in parallel and therefore exploit a larger effective working population. Each of the aggregated optimizations is directed by its own vector of weights, or target vectors. Thus the algorithm uses a matrix of target vectors to search in parallel. It is also possible to combine searches in different directions and with different reference points, and searches using different aggregation functions all within a single optimization run. The key advantage is that the algorithm does not rely on Pareto ranking to provide selective pressure. As the target vectors are generally decided a priori, MSOPS provides an active probing of the Pareto set, rather than passive discovery.

The operation of MSOPS is to generate a set of target vectors, T, and evaluate the performance of every individual in the population, P, for every target vector, based on a conventional aggregation method. As aggregation methods (e.g., weighted min-max, ϵ-constraint, goal attainment) are very simple to process, the calculation of each of the performance metrics is fast.

Thus, each of the members of the population set P has a set of scores, one for each member of T, that indicate how well the population member satisfied the range of target conditions. The scores are held in a score matrix, S, which has dimensions $\|P\| \times \|T\|$, where $\| \cdot \|$ indicates set cardinality. Each *column* of the matrix S corresponds to one target vector (across the population P). The aggregate fitness, f_i, of the ith member of P is calculated using equation 9, where $f_n(\mathbf{O}_i, \mathbf{V}_n, \mathbf{Z}_n)$ is the aggregation function n with target vector $\mathbf{V}_n$ and reference point $\mathbf{Z}_n$ for objective vector $\mathbf{O}_i$ (which is the ith member of P).

$$f_{i \in P} = \min_{\forall n \in T} \left(\frac{f_n(\mathbf{O}_i, \mathbf{V}_n, \mathbf{Z}_n)}{\min_{\forall j \neq i \in P}(f_n(\mathbf{O}_j, \mathbf{V}_n, \mathbf{Z}_n))} \right) \tag{9}$$

The flexibility of the approach is such that the target vectors can be arbitrary, either generated to give full coverage of the objective space if no a priori domain knowledge exists, or with some structure to target key elements of the search volume. As the fitness combination method employed is based on the set of fixed target vectors, the target vector set determines the final spread of the obtained solutions. As a consequence, the efficiency of the algorithm is reduced in relation to the number of unobtainable target vectors that do not pass through the feasible objective region.

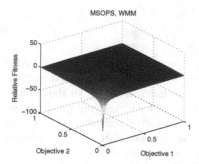

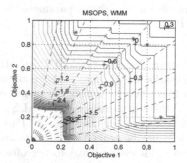

Fig. 23. 3D relative fitness surface for point [0.7, 0.85] and the MSOPS method with one moving and five existing objective vectors and using ten Weighted Min-Max target vectors

Fig. 24. Relative isofitness contours for point [0.7, 0.85] and the MSOPS method with one moving and five existing objective vectors and using ten Weighted Min-Max target vectors

Figures 23 and 24 show the relative fitness surface and contours for the MSOPS algorithm using ten target vectors, referenced at the origin, and the Weighted Min-

Max aggregation function. It is clear in Figure 24 that the relative fitness gradient in the nondominated region is never counter to the objective directions and so will provide rapid progress towards the Pareto front; however, the isofitness contours are also not aligned to the true nondominated surface of the other population members; rather, they are aligned to a combination of the target vectors and nondominated surface. For example, point [0.8, 0.8] is nondominated and yet does not lie on the same fitness contour as the other nondominated solutions as it is far from a target vector. In contrast, point [0.9, 0.2] is closer to a target vector line and so has a better fitness (however, it is still inferior to the test point [0.7, 0.85]). The result is that the final population will cluster around the points where the target vectors cut the Pareto front (or the nearest feasible point in a weighted min-max sense).

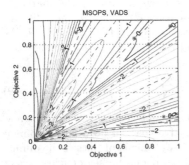

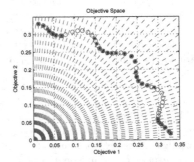

Fig. 25. Relative isofitness contours for point [0.7, 0.85] and the MSOPS method with one moving and five existing objective vectors, and using ten VADS target vectors (plotted as logarithm of fitness). Shape parameter is $q = 100$

Fig. 26. Plot of objective/decision space for function (10). Circles are VADS solutions, stars are Weighted Min-Max solutions

Figure 25 shows the MSOPS relative isofitness contours when using the VADS Aggregation metric. The VADS contours in Figure 25 are very complicated and it is clear that the directions of the target vectors are a dominating factor in the description of the fitness surface. The VADS metric is designed for identifying the *Objective* front profile, rather than just the Pareto front. Thus, highly concave and reentrant surfaces may be probed with this metric. Unfortunately, the relative fitness gradient in the nondominated regions are often not ideal, and optimization performance is compromised in both bi-objective and many-objective problems.

Empirical studies have shown that running the MSOPS algorithm with both Weighted Min-Max and VADS Aggregation will provide superior optimization performance to VADS alone. When both metrics are combined, the weighted min-max process dominates initially and minimizes the effects of reverse gradients in the nondominated regions. As the algorithm converges, the VADS metric can help to provide a more balanced search in difficult regions such as extreme convexities.

3.3 Pareto Front Analysis

To demonstrate the use of RSO or MSOPS for analysis of the objective and Pareto front, the Tanaka two-objective test function has been studied [9]:

$$O_1 = x$$
$$O_2 = y$$
$$0 \geq -(x)^2 - (y)^2 + 1 + 0.1 \cos\left(16 \arctan\left(\frac{x}{y}\right)\right)$$
$$0.5 \geq (x - 0.5)^2 + (y - 0.5)^2$$
$$0 \leq x, y \leq 1 \tag{10}$$

Figure 26 shows the result of MSOPS using combined Weighted Min-Max and VADS, applied to (10) with 51 target vectors (shown as dashed) and a population of 50 (run for 100 generations). The stars indicate the best set of solutions found with the weighted min-max and the circles are the best VADS solutions. The 51 weight vectors were generated a priori using the origin as a reference point and designed to cover the objective space with equal angles between neighbouring vectors.

It is clear that points on the boundary of the objective front have been identified. The 'leading edge' of the objective space is identified by VADS, while Weighted Min-Max finds the Pareto front. The use of two aggregation functions is very useful for analysing the behaviour of the objectives, rather than just the Pareto front. The area around [0.1, 0.3] is a discontinuity in the Pareto front and as such has only been identified in the VADS search. The corresponding plots of angular errors between each target vector and the 'best performing' objective vector for VADS and min-max respectively are shown in Figures 27 and 28, sorted according to the weights with V_1 (the first element of the target vectors) increasing. It is clear that many of the target vectors were satisfied with an error less than 2° to their nearest objective vector for VADS; but there are areas with high errors for weighted min-max, indicating that some target vectors could not be obtained exactly. These errors correspond to the limits of the Pareto set in the VADS plot (the first and the last two target vectors in Figure 27) and also to the discontinuities in the function in the weighted min-max plot (around vectors 15 and 37 in Figure 28).

This example demonstrates that because we know a priori the regions of the Pareto set that are being investigated, based on the set of target vectors, we can quantify how close the optimization result came. With large numbers of objectives though, large numbers of target vectors may be required if a detailed search is to be performed across the entire objective space in one pass. It is simple, however, with RSO or MSOPS to target a range of smaller areas with each run. The areas can be of varying size and diverse in each run if necessary, providing extreme flexibility in the optimization process and incorporation of designer preferences and interactive decision-making. Strategies may be used to yield extra information about the Pareto front, such as generating a set of target vectors that lie on a plane, allowing 'slices' through the Pareto front to be visualized to test for continuity.

3.4 Summary of Non-Pareto Methods

Both RSO and MSOPS are capable of generating objective fronts and Pareto fronts in low- and high-dimensional objective spaces. However, the MSOPS process is more

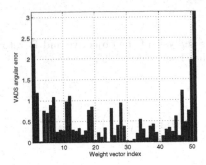

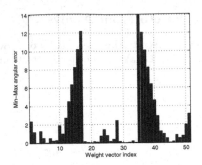

Fig. 27. Plot of angular errors (in degrees) of target vectors for their best performing objective vector using the VADS metric and function (10)

Fig. 28. Plot of angular errors (in degrees) of target vectors for their best performing objective vector using the weighted min-max metric and function (10)

efficient in practice and is recommended if multiple target vectors are to be considered. The RSO algorithm is best when a single final optimal solution is to be generated. Interestingly, although the IBEA method has been described as using Pareto concepts, the relationship to MSOPS is very strong and it could be argued that MSOPS is an indicator-based algorithm that has not been restricted to identifying the Pareto front alone.

As both RSO and MSOPS utilize aggregation functions, the wide variety of functions available allow a comprehensive analysis of the objective and Pareto surface to be performed.

4 Conclusions and Recommendations

Most optimization algorithms to date have focused on bi-objective problems, and many, unfortunately, do not extend well to many-objective problems with four or more objectives. This chapter has shown that the properties of the fitness assignment process can be visualized and analysed to assess the suitability of a method for many-objective optimization. Some of the fitness assignment methods may also be used to aid analysis and visualisation of the objective and Pareto fronts.

It is unlikely that a simple fitness assignment function will provide selective pressure towards the Pareto front while also providing an effective drive towards a set of well-spread solutions. It is more likely that an influence or mechanism external to the fitness assignment process may be needed to ensure that a satisfactory distribution of solutions is obtained, such as clustering or automatic target vector generation.

References

[1] P. J. Bentley and J. P. Wakefield. An analysis of multiobjective optimization within genetic algorithms. Technical Report ENGPJB96, University of Hud-

dersfield, UK, 1996. Online: http://citeseer.ist.psu.edu/62443.html.

[2] L. Bradstreet, L. Barone, and L. While. Maximising hypervolume for selection in multi-objective evolutionary algorithms. In *IEEE Congress on Evolutionary Computation, CEC 2006*, pages 1744–1751, Vancouver, Canada, July 2006. IEEE.

[3] K. Deb. *Multi-objective optimization using evolutionary algorithms*. John Wiley & Sons, 2001. ISBN 0-471-87339-X.

[4] E. J. Hughes. Multiple single objective pareto sampling. In *Congress on Evolutionary Computation 2003*, pages 2678–2684, Canberra, Australia, 8–12 December 2003. IEEE.

[5] E. J. Hughes. Evolutionary many-objective optimisation: Many once or one many? In *IEEE Congress on Evolutionary Computation, 2005*, volume 1, pages 222–227, Sept. 2005.

[6] B. Naujoks, N. Beume, and M. Emmerich. Multi-objective optimisation using S-metric selection: application to three-dimensional solution spaces. In *IEEE Congress on Evolutionary Computation, CEC 2005*, volume 2, pages 1282 – 1289, Edinburgh, UK, Sept. 2005. IEEE.

[7] R. C. Purshouse. Evolutionary many-objective optimisation: An exploratory analysis. In *The 2003 Congress on Evolutionary Computation (CEC 2003)*, volume 3, pages 2066–2073, Canberra, Australia, 8–12 December 2003. IEEE.

[8] T. P. Runarsson and X. Yao. Stochastic ranking for constrained evolutionary optimisation. *IEEE Transactions on Evolutionary Computation*, 4(3):284–294, Sept. 2000.

[9] M. Tanaka, H. Watanabe, Y. Furukawa, and T. Tanino. Ga-based decision support system for multicriteria optimization. In *Conference on Systems, Man and Cybernetics: Intelligent Systems for the 21st Century*, volume 2, pages 1556–1561. IEEE, 22–25 October 1995.

[10] E. Zitzler. *Evolutionary algorithms for Multiobjective Optimisation: Methods and Applications*. PhD thesis, Swiss Federal Institute of Technology (ETH), Zurich, Switzerland, Nov. 1999.

[11] E. Zitzler and S. Knzli. Indicator-based selection in multiobjective search. In *Parallel Problem Solving from Nature - PPSN VIII*, volume LNCS 3242/2004, pages 832–842, Birmingham, UK, 2004. Springer.

Modeling Regularity to Improve Scalability of Model-Based Multiobjective Optimization Algorithms

Yaochu Jin[1], Aimin Zhou[2], Qingfu Zhang[2], Bernhard Sendhoff[1], and Edward Tsang[2]

[1] Honda Research Institute Europe
   Carl-Legien-Str. 30
   63073 Offenbach, Germany
   {yaochu.jin,bernhard.sendhoff}@honda-ri.de
[2] Department of Computer Science
   University of Essex
   Wivenhoe Park, Colchester, CO4 3QS, UK
   {azhou,qzhang,edward}@essex.ac.uk

Summary. Model-based multiobjective optimization is one class of metaheuristics for solving multiobjective optimization problems, where a probabilistic model is built from the current distribution of the solutions and new candidate solutions are generated from the model. One main difficulty in model-based optimization is constructing a probabilistic model that is able to effectively capture the structure of the problems to enable efficient search. This chapter advocates a new type of probabilistic model that takes the regularity in the distribution of Pareto-optimal solutions into account. We compare our model to two other model-based multiobjective algorithms on a number of test problems to demonstrate that it is scalable to high-dimensional optimization problems with or without linkage among the design variables.

1 Introduction

The last decade has witnessed a great success of evolutionary algorithms and other population-based meta-heuristic search methods in solving multiobjective optimization problems [8]. Nevertheless, several challenges still remain to be addressed for population-based search methods to deal with hard, real-world optimization problems. One of these challenges is algorithms' ability to efficiently solve optimization problems of a high search dimension, which is often known as the scalability of optimization algorithms.

For evolutionary multiobjective algorithms to be scalable to high search dimensions, they must be able to effectively take advantage of domain knowledge of the problem at hand during the search. Unfortunately, major search operators of conventional evolutionary algorithms, such as crossover and mutation, are not efficient in taking problem-specific knowledge into account in search. To address this weakness, several approaches have been suggested for incorporating domain knowledge into evolutionary algorithms to guide the sampling process [13], among which model-based optimization methods, such as the estimation of distribution algorithms (EDAs) [6, 18], have widely been studied. It should be noticed that existing EDAs have mainly been developed to solve scalar optimization problems, which are not necessarily suited for solving multiobjective problems.

Another weakness of evolutionary algorithms that use crossover and mutation for generating new candidate solutions is that they do not explicitly exploit the correlation between design variables (also known as variable linkage) [10]. Model-based algorithms are believed to be able to learn the linkage among variables. However, the ability of learning linkage can be at the cost of scalability if the probabilistic model is not chosen appropriately.

This chapter presents a methodology for incorporating additional knowledge into building probabilistic models for solving continuous multiobjective problems. The domain knowledge we use here is the regularity in the distribution of Pareto-optimal solutions, which has largely been overlooked in developing evolutionary multiobjective optimization algorithms. Since regularity is a general property for a large class of multiobjective problems, the proposed framework is applicable to a wide range of real-world problems.

The remainder of the chapter is organized as follows. A brief introduction to solving multiobjective optimization problems using a probabilistic model, together with a short discussion on the main difficulties of model-based algorithms, is presented in Section 2. Three model-based multiobjective optimization algorithms, including the one that takes regularity into account, are described in detail in Section 3. Section 4 provides the experimental setup, such as parameter settings of algorithms, the test functions, and the performance indicators, for quantitatively evaluating the performance of the algorithms. Comparison results of the three models with respect to algorithms' scalability and ability to handle variable linkages are presented in Section 5. A summary and conclusions of the chapter are provided in Section 6.

2 Probabilistic Modeling for Multiobjective Optimization

The basic idea of population-based search using a probabilistic model is to first estimate the probability distribution of the solutions previously generated and to then generate new candidate solutions by sampling the probabilistic model, as shown in Fig. 1.

A large family of probabilistic models can be used to estimate the distribution of continuous and discrete functions [18]. In this chapter, we limit our discussion to continuous optimization problems, where Gaussians or a mixture of Gaussians are employed to model the distribution of the function to be optimized.

It would be ideal if we could use a full joint probability distribution model, i.e., a probabilistic model that considers dependency between all variables. Unfortunately,

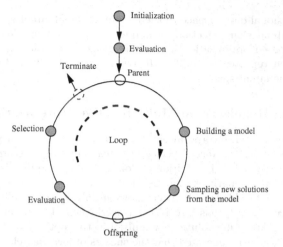

Fig. 1. A generic framework for model-based optimization algorithms using a population

accurate estimation of a full joint distribution model in a high-dimensional space remains an open problem. In order to estimate the distribution accurately, a huge number of data samples are needed, which is impractical in solving real-world problems due to the fact that calculation of the function value for a given design (often known as fitness evaluation in evolutionary optimization) is computationally very expensive. In this context, EDAs that require a huge population size are of very limited practical importance.

Several techniques have been adopted to address the curse of dimensionality. The simplest way to cope with high dimensionality is to neglect the linkage between variables and build a univariate distribution model for each variable [22, 30]. Unfortunately, such models are not able to capture the dependency between variables and they are not recommendable if there are strong correlations between the variables. One popular approach is to use factorized univariate or multivariate distributions, which are able to capture the independence between the variables. A multivariate factorized probability distribution is a probabilistic model in the form of a product of probability density functions. Both univariate factorization [2, 19] and multivariate factorization [2, 19, 3] have been employed for model-based optimization.

A natural extension to models consisting of a single factorized probability distribution is to use a weighted sum of single factorized distributions, which is usually known as a mixture of Gaussians. Such models can often be obtained by dividing the search space into a number of subspaces and then constructing a single factorized distribution for each cluster [3]. This method is of particular interest for multimodal scalar optimization and multiobjective optimization, where more than one solution needs to be achieved.

Although multivariate factorization is able to capture the dependency among at least two variables, it is not straightforward to select a model that is optimal for a given problem [4]. Another approach to factorization is to map the high-dimensional search space onto a latent space of a lower dimensionality, and then a univariate or multivariate factorized distribution can be built. The mapping from the high-dimensional design space to the low-dimensional latent space can often be realized using dimension reduction techniques such as principal component analysis [1]. Model-based optimization algorithms using a distribution model in latent space have been reported in [7, 27, 24]. One main difficulty is to determine the dimension of the latent space.

2.1 Modeling Regularity in Multiobjective Optimization

As previously discussed, incorporation of knowledge into a search process helps to improve the search performance, especially the scalability of the search algorithms to high search dimensionality. In addition to domain knowledge that is specific to each particular problem, regularity in the distribution of the Pareto-optimal solutions is a nice property that holds for a large class of multiobjective optimization problems. So far, this nice property has largely been overlooked. The importance of taking advantage of regularity in evolutionary multiobjective optimization was first advocated in [14], where it is suggested that the success of local search in multiobjective optimization can most probably be attributed to the fact that local search is able to implicitly exploit the regular distribution of Pareto-optimal solutions. In that work, piecewise linear models are constructed in the design space using the nondominated solutions achieved by an evolutionary algorithm. It has been demonstrated that the quality of the solutions generated from the linear models are better than the original solutions.

The regularity property can be induced from the Karush-Kuhn-Tucker condition [21, 26], which indicates that under certain smoothness conditions, the Pareto-optimal set in the design space of a continuous multiobjective optimization problem is an $(m-1)$-dimensional piecewise continuous manifold, where m is the number of the objectives.

The question now is how to efficiently exploit the regularity property using model-based multiobjective optimization. Although it is believed that model-based optimization is able to learn the problem structure, it must be pointed out that the model's ability to capture the problem structure heavily depends on the model in use. This is particularly true for multiobjective optimization, where the final solution is a Pareto front consisting of multiple solutions rather than a single optimum.

Most existing model-based multiobjective optimization algorithms for solving continuous problems employ Gaussian distributions with few exceptions, e.g., in [24], where a Voronoi mesh has been adopted. The most important a priori knowledge that can be derived from the regularity condition is that the Pareto front in the original n-dimensional search space can be modeled in an $(m-1)$-dimensional space without any information loss, where n is the dimensionality of the search space; and, in most cases, $m \ll n$. This knowledge removes exactly the main obstacle in latent-variable-based models, where the dimension of the latent space must be specified. Besides, knowing that the Pareto front is a principal curve or surface, we believe that first-order or second-order polynomials might be more efficient than Gaussian models in modeling the regular distribution of the Pareto-optimal solutions.

Take bi-objective optimization problems as an example. The regularity property has two implications. First, the Pareto front can be described by one or a few sections of a one-dimensional model, regardless how large the design space is. Second, a linear curve is more efficient in leading the population to the final Pareto front, as illustrated in Fig. 2.

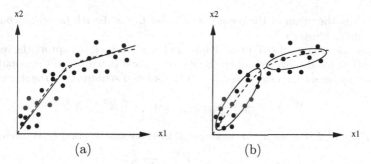

Fig. 2. Modeling Pareto set using (a) linear models; (b) Gaussian models

The idea of modeling regularity in model-based multiobjective optimization has most recently been exploited by the authors, and very competitive results have been achieved [31, 32, 33, 34] compared to some of the state-of-the-art evolutionary multiobjective optimization methods such as NSGA-II [9], GDE3 [17], and MIDEA [5]. In the following, we are going to compare one model-based multiobjective algorithm that exploits regularity to two other model-based multiobjective optimization algorithms with respect to scalability with search dimension, ability to handle variable linkage, and sensitivity to population size.

3 Three Model-Based Algorithms

3.1 Regularity-Based Latent Principal Curve Model (LPCM)

Modeling in a latent space is an attractive idea because the dimension of the latent space is usually much lower than that of the design space. According to the regularity condition, the Pareto front of an m-objective optimization problem can be modeled in an $(m-1)$-dimensional space. For this purpose, the local principal curve analysis (LPCA) algorithm [15] has been employed. One elegant property of LPCA is that it simultaneously groups the population into a number of clusters while mapping it from the n-dimensional design space to the $(m-1)$-dimensional latent space.

The points in the kth cluster (denoted by C^k) can be described by a uniform distribution on an $(m-1)$-dimensional manifold M^k:

$$P^k(S) = \begin{cases} \frac{1}{V^k}, & \text{if } S \in M^k, \\ 0, & \text{else} \end{cases} \tag{1}$$

where S is an $(m-1)$-dimensional random vector in the latent space, V^k is the volume of M^k bounded by

$$a_i^k \le s_i \le b_i^k, i = 1, \cdots, (m-1), \tag{2}$$

and

$$a_i^k = \min{}_{X \in C^k} (X - \bar{X}^k)^T U_i^k, \tag{3}$$

$$b_i^k = \max{}_{X \in C^k} (X - \bar{X}^k)^T U_i^k, \tag{4}$$

where $\bar{X}^k$ is the mean of the points in C^k, and U_i^k is the ith principal component of the data in cluster C^k.

While the uniform distribution defined in Eq. (1) is used to capture the regularity (centroid) in the distribution of the population, local dynamics of the population is described by an n-dimensional zero-mean Gaussian distribution in the design space:

$$N^k(X) = \frac{1}{(2\pi)^{n/2}|\Sigma^k|^{n/2}} \exp\left\{-\frac{1}{2}X^T\Sigma^k X\right\}, \tag{5}$$

where $\Sigma^k = \delta^k I$, I is an $n \times n$ dimensional identity matrix, and δ^k is calculated by

$$\delta^k = \frac{1}{n-m+1}\sum_{i=m}^{n}\lambda_i^k, \tag{6}$$

where λ_i^k is the ith largest eigenvalue of the covariance matrix of the points in cluster k. Here, we assume that the inequality $n > m - 1$ always holds. An illustration of the principal curve and the Gaussian models in a one-dimensional latent space is provided in Fig. 3, where the population is divided into two clusters.

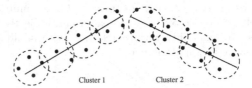

Fig. 3. Modeling the distribution of the population using a uniformly distributed principal curve and a Gaussian perturbation

During the sampling process, the probability at which the model of cluster k is chosen is determined by

$$p(k) = \frac{V^k}{\sum_{k=1}^{K}V^k}, \tag{7}$$

where V^k is the volume of the $(m-1)$-dimensional manifold. In the case of a curve, it is the length of the curve.

The sampling process consists of three steps. In the first step, a point is generated on the $(m-1)$-dimensional manifold M^k according to Eq. (1), and is then mapped onto the n-dimensional design space. Assume S is an $(m-1)$-dimensional random vector generated in M^k for the kth cluster; it is mapped onto the n-dimensional design space in the following way, if the manifold M^k is a first-order principal curve:

$$X_1 = \Theta_0^k + \Theta_1^k S, \tag{8}$$

where X_1 is an n-dimensional random vector, Θ_0^k is the mean of the data in cluster $C(X)^k$, and Θ_1^k is an $n \times (m-1)$-dimensional matrix, composed of the eigenvectors corresponding to the $(m-1)$ largest eigenvalues.

In the second step, an n-dimensional random vector X_2 is generated from the Gaussian distribution defined in Eq. (5). Finally, the following new candidate solution is generated:

$$X = X_1 + X_2. \tag{9}$$

This process continues until all offspring are generated.

3.2 Univariate Factorized Gaussian Model (UGM)

The basic idea for using a univariate factorized normal distribution for modeling the population has been considered in [5]. Before constructing the models, the population is divided into K clusters. For this purpose, the leader clustering algorithm [12] is employed, as suggested in [5]. In this clustering algorithm, it is not necessary to define the number of clusters; however, a threshold that defines the radius of the clusters must be given, which implicitly determines the number of clusters. One major drawback of the leader algorithm is that the clustering result is sensitive to the choice of the initial cluster centre (leader). For cluster $k, k = 1, 2, ..., K$, a Gaussian model is then constructed for each search dimension:

$$p_i^k(x_i) = \frac{1}{\delta_i^k \sqrt{2\pi}} \exp\left\{ -\frac{(x_i - \mu_i^k)^2}{2(\delta_i^k)^2} \right\}, \tag{10}$$

where μ_i^k and δ_i^k are the mean and the standard deviation of the Gaussian model for variable $i = 1, ..., n$. The mean and standard deviation of the univariate Gaussian distribution for cluster k can be calculated according to the individuals that are assigned to the cluster.

During the sampling process, one of the K clusters is chosen randomly with probability $1/K$. For the chosen cluster, one new candidate solution is generated using the n Gaussian models for each design variable. This procedure repeats until all the offspring solutions are generated.

3.3 Marginalized Multivariate Gaussian Model (MGM)

Univariate factorized Gaussian models neglect any correlation between the variables. As a result, the model cannot effectively learn the problem structure if there is dependency among the variables. To address this problem, a joint Gaussian distribution model is considered in this model. However, building an accurate full joint distribution model in a high-dimensional space is almost intractable. For this reason, the population is first grouped into a number clusters and then a joint distribution model is built for each cluster. To cluster the population, the k-means clustering algorithm [12] is adopted. Therefore, the number (K) of clusters needs to be predefined by the user.

For the kth cluster, the following joint distribution model is constructed:

$$p^k(X) = \frac{1}{(2\pi)^{n/2}|\Sigma^k|^{n/2}} \exp\left\{ -\frac{1}{2}(X - \Lambda^k)^T (\Sigma^k)^{-1}(X - \Lambda^k) \right\}, \tag{11}$$

where X is an n-dimensional design vector, Λ^k is an n-dimensional vector of the mean value, and Σ^k is an $n \times n$ covariance matrix estimated by the individuals in the kth cluster.

Different from the univariate factorized model, the probability of sampling the model of the kth cluster is calculated as follows:

$$p(k) = \frac{N^k}{\sum_{k=1}^{K} N^k}, \tag{12}$$

where N^k is the number of individuals in the kth cluster.

3.4 The General Algorithm Framework

For a fair comparison, all three algorithms use the same selection strategy, i.e., the MaxiMin sorting selection algorithm suggested in [28], which is a variant of the crowded nondominated sorting selection proposed in [9]. The first steps in the MaxiMin sorting are the same as those in the crowded nondominated sorting. First, the parent and offspring populations are combined. Second, the combined population is sorted according to the nondominance ranks. During the ranking, nondominated solutions in the combined population are assigned with a rank 1, which belongs to the first nondominated front. These individuals are removed temporarily from the population and the nondominated individuals in the rest of the population, which consists of the second nondominated front of the population, are identified, and assigned a rank 2. This procedure repeats until all individuals in the combined population are assigned a rank from 1 to R, assuming that R nondominated fronts can be identified in total. Instead of calculating the crowding distance as done in NSGA-II, selection starts directly after nondominated sorting. During selection, solutions on the first nondominated front are passed to the parent population of the next generation. If the number of solutions on the first nondominated front is smaller than the population size, those on the second nondominated front are moved to the parent population. However, it can happen that only some of the solutions on a nondominated front can be selected.

Let us assume there are L solutions on the jth nondominated front, and only M solutions are to be selected, where $M < L$. In NSGA-II, M solutions with the largest crowding distances are selected. In the MaxiMin selection method, the extreme solutions on the concerned nondominated front are selected. Then, the solutions that have the maximal distance to the selected solutions from the same nondominated front are selected first. This process is continued until the parent population is filled up. An illustrative example is provided in Fig. 4.

Assume that we need to select ten solutions from 20 in the combined population. We first select the six solutions on the first nondominated front. On the second nondominated front, there are six solutions, from which four will be selected. According to the MaxiMin method, the two extreme solutions A and B are first selected. Then, solution C is selected because the minimal distance from solution C to those selected from the second nondominated front (A and B) is the largest. Afterwards, solution D is selected because its minimal distance to the selected solutions (A, B, and C) is the maximal. It has been shown that the MaxiMin approach can lead to a more diverse population with a lower computational complexity compared to the crowded nondominated sorting selection method [28].

It should be noted that the reproduction strategy in our work is also different from that in MIDEA [5]. In this work, all solutions in the parent population are used to construct the probabilistic model, whereas in MIDEA only a portion of individuals in the parent population is used. In addition, the number of new candidate solutions (offspring) generated from the model equals the number of parents, while in MIDEA only those solutions that are not used in model building are replaced by newly generated offspring.

A generic diagram of the model-based multiobjective optimization algorithms studied in this work is presented in Fig. 5.

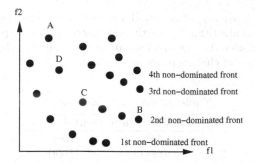

Fig. 4. MaxiMin nondominated selection

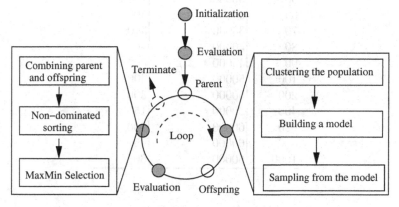

Fig. 5. A generic diagram of the model-based optimization algorithms

4 Experimental Setup

4.1 Parameter Settings

To investigate the scalability of the algorithms' performance with the search dimension, we have performed simulations on the test problems with dimensions of $10, 20, 30, 40, 50, 60, 70, 80, 90, 100$. The sensitivity of the search performance to the size of the population is also studied. To this end, we have used population sizes of $20, 30, 40, 50, 60, 70, 80, 90, 100, 200, 400, 600, 800$, and $1,000$. The baseline for comparison is the maximum number of fitness evaluations, which is listed in Table 1.

In UGM, the clustering of the population is conducted in the objective space using the leader algorithm, as recommended in [5]. The threshold used in clustering is set to 0.2, and a maximum of ten clusters is allowed. For MGM, the population is clustered using the k-means algorithm, and the cluster number is set to 3 if the population size is smaller than or equal to 50; otherwise, the cluster number is set to 5. Note, however, that clustering for MGM is done in the design space. The reason why population clustering is carried out in different spaces is that the leader algorithm produces better results in the objective space while the k-means algorithm shows more stable results in the parameter space.

In LPCM, the clustering of the population is conducted using the local principal component analysis algorithm, where the number of clusters is predefined to 3 for a population size smaller than or equal to 50. In the case of a population size larger than 50, the number of clusters is defined to be 5.

Table 1. Maximum Evaluations

Pop Size	Max Evaluation	Max Gen
20	20000	1000
30	30000	1000
40	40000	1000
50	50000	1000
60	30000	500
70	35000	500
80	40000	500
90	45000	500
100	50000	500
200	60000	300
400	60000	150
600	60000	100
800	60000	75
1000	60000	60

4.2 Test Functions

The performance of the algorithms is studied on six test functions. Three of them are taken directly from the widely used ZDT test functions [35], namely, ZDT1, ZDT2, and ZDT3, whose Pareto fronts are convex, concave and discontinuous, respectively. Note that the ZDT test functions are slightly modified so that the Pareto front in the design space is shifted to

$$x_2 = \cdots = x_n = 0.2,\ x_1 \in [0, 1].$$

In the ZDT test functions, there is no dependency among the design variables. To investigate how the algorithms can deal with variable linkage, three additional test functions derived from the ZDT test functions are also considered, termed ZDT1.2, ZDT2.2, and ZDT3.2. The Pareto front of these three test functions in the objective space is completely the same as the corresponding ZDT functions. However, there is a nonlinear dependency between the design variables. The mathematical descriptions of the six test functions are presented in Table 2.

4.3 Performance Indicators

To evaluate the performance of the algorithms, we adopted two performance indicators(PIs). The first PI is the inverted generational distance (IGD) [25], which

Table 2. Test Instances

Test function	Search space	Objectives
ZDT1	$[0,1]^n$	$f_1(x) = x_1$ $f_2(x) = g(x)[1 - \sqrt{f_1(x)/g(x)}]$ $g(x) = 1 + 9(\sum_{i=2}^{n}(x_i - 0.2)^2)/(n-1)$
ZDT2	$[0,1]^n$	$f_1(x) = x_1$ $f_2(x) = g(x)[1 - (f_1(x)/g(x))^2]$ $g(x) = 1 + 9(\sum_{i=2}^{n}(x_i - 0.2)^2)/(n-1)$
ZDT3	$[0,1]^n$	$f_1(x) = x_1$ $f_2(x) = g(x)[1 - \sqrt{f_1(x)/g(x)} - \frac{x_1}{g(x)}sin(10\pi x_1)]$ $g(x) = 1 + 9(\sum_{i=2}^{n}(x_i - 0.2)^2)/(n-1)$
ZDT1.2	$[0,1]^n$	$f_1(x) = x_1$ $f_2(x) = g(x)[1 - \sqrt{x_1/g(x)}]$ $g(x) = 1 + 9(\sum_{i=2}^{n}(x_i^2 - x_1)^2)/(n-1)$
ZDT2.2	$[0,1]^n$	$f_1(x) = \sqrt{x_1}$ $f_2(x) = g(x)[1 - (f_1(x)/g(x))^2]$ $g(x) = 1 + 9(\sum_{i=2}^{n}(x_i^2 - x_1)^2)/(n-1)$
ZDT3.2	$[0,1]^n$	$f_1(x) = x_1$ $f_2(x) = g(x)[1 - \sqrt{x_1/g(x)} - \frac{x_1}{g(x)}sin(10\pi x_1)]$ $g(x) = 1 + 9(\sum_{i=2}^{n}(x_i^2 - x_1)^2)/(n-1)$

is derived from the generational distance (GD) suggested in [29, 9]. IGD can be expressed as follows:

$$D(P,P^*) = \frac{1}{|P^*|} \sum_{x \in P^*} ||x - x'||_2, \tag{13}$$

where P is the nondominated set achieved by the optimization algorithm, P^* is a reference Pareto-optimal set uniformly sampled from the true Pareto front, x is a solution in reference set P^*, and x' is a solution in set P that has the minimal distance to x. If the reference set represents the true Pareto front adequately well, IGD can effectively measure the accuracy as well as the diversity of the achieved set P. The inverted generational distance is called D-metric hereafter.

The second PI we adopted in the comparison is the difference of the hypervolume (I-Metric for short) between the reference set P^* and the achieved set P [16]:

$$I_H^-(P) = I_H(P^*) - I_H(P), \tag{14}$$

where $I_H(P^*)$ and $I_H(P)$ are the hypervolumes of P and P^*, respectively.

5 Simulation Results

5.1 Scalability with Search Dimension: Without Dependency

The first set of simulations has been performed to study the scalability of the three algorithms with search dimension for a given population size (100) on the three test functions without variable linkage. The simulation results on ZDT1, ZDT2, and ZDT3 are provided in Figs. 6, 7, and 8, respectively, where the best and worst Pareto fronts from 30 independent runs according to the D-metrics are plotted. It can be seen from the figures that the results from LPCM, UGM, and MGM, which are presented on the left, the middle, and the right panels of the figures, are quite similar when the dimension changes from 20 to 100, though degradation in the performance of the MGM is a little more serious than that of LPCM and UGM. This observation can be confirmed by the D-metric and the I-metric of the results listed in Tables 3 and 4, respectively, in which the mean and standard deviation of 30 runs are listed.

The results indicate that both LPCM and UGM have very good scalability with search dimension for problems without variable linkage. It is worth noting that the performance of MGM is also quite good, probably due to the fact that the distribution of the Pareto fronts in ZDT1, ZDT2, and ZDT3 is quite easy to model.

Table 3. Mean and standard deviation of the D-metric for test functions without variable linkage

Instance	Method	Search Dimension				
		20	40	60	80	100
ZDT1	LPCM	0.0043±0.0001	0.0044±0.0001	0.0047±0.0001	0.0051±0.0001	0.0057±0.0002
	UGM	0.0043±0.0002	0.0044±0.0002	0.0046±0.0002	0.0049±0.0002	0.0053±0.0002
	MGM	0.0046±0.0002	0.0046±0.0002	0.0052±0.0004	0.0069±0.0011	0.0091±0.0013
ZDT2	LPCM	0.0040±0.0000	0.0042±0.0001	0.0045±0.0001	0.0048±0.0001	0.0054±0.0002
	UGM	0.0044±0.0001	0.0045±0.0001	0.0047±0.0002	0.0051±0.0002	0.0056±0.0002
	MGM	0.0045±0.0006	0.0043±0.0002	0.0048±0.0005	0.0061±0.0010	0.0096±0.0021
ZDT3	LPCM	0.0051±0.0000	0.0053±0.0001	0.0056±0.0001	0.0060±0.0001	0.0069±0.0003
	UGM	0.0054±0.0003	0.0058±0.0002	0.0069±0.0006	0.0080±0.0008	0.0098±0.0012
	MGM	0.0056±0.0003	0.0055±0.0003	0.0058±0.0004	0.0064±0.0005	0.0081±0.0011

5.2 Scalability with Search Dimension: With Dependency

Simulations are also conducted on the three test functions with nonlinear linkage among the design variables. The best and worst Pareto fronts from 30 independent runs according to the D-metric for the three test functions, ZDT1.2, ZDT2.2, and ZDT3.2 are presented in Figs. 9, 10, and 11, respectively. From the figures (left panel), we find that there is a slight performance decrease of LPCM on ZDT1.2 and

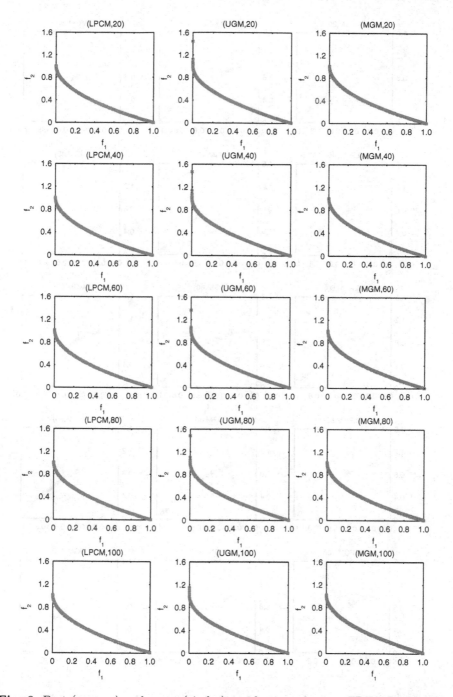

Fig. 6. Best (squares) and worst (circles) nondominated set on ZDT1. Population size 100. The number of design variables ranges from 20 (top row) to 100 (bottom row)

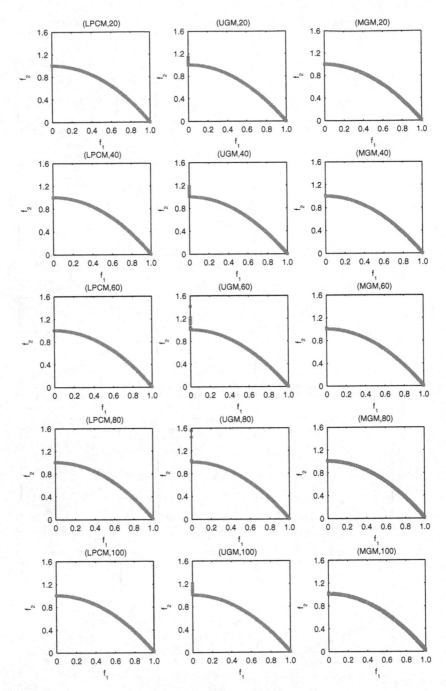

Fig. 7. Best (squares) and worst (circles) nondominated set on ZDT2. Population size 100. The number of design variables ranges from 20 (top row) to 100 (bottom row)

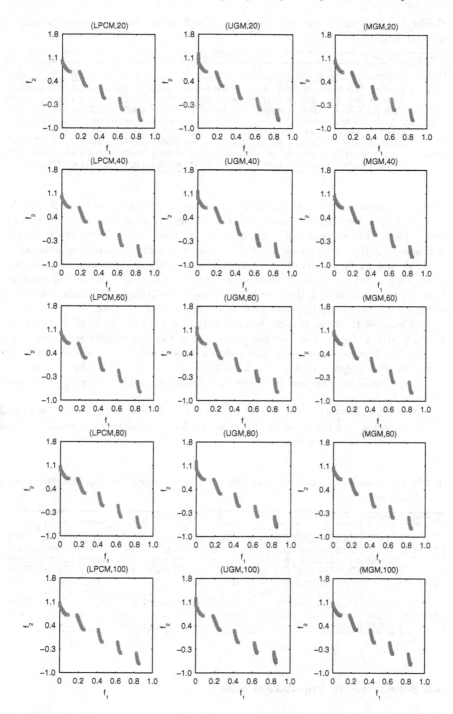

Fig. 8. Best (squares) and worst (circles) nondominated set on ZDT3. Population size 100. The number of design variables ranges from 20 (top row) to 100 (bottom row)

Table 4. Mean and standard deviation of the I-metric for test functions without variable linkage

Instance	Method	Search Dimension				
		20	40	60	80	100
ZDT1	LPCM	0.0048±0.0001	0.0058±0.0001	0.0069±0.0002	0.0083±0.0004	0.0098±0.0005
	UGM	0.0048±0.0002	0.0056±0.0003	0.0064±0.0003	0.0076±0.0004	0.0087±0.0004
	MGM	0.0061±0.0008	0.0061±0.0006	0.0081±0.0011	0.0120±0.0022	0.0163±0.0023
ZDT2	LPCM	0.0049±0.0001	0.0061±0.0002	0.0075±0.0003	0.0090±0.0004	0.0108±0.0006
	UGM	0.0050±0.0001	0.0059±0.0002	0.0071±0.0003	0.0083±0.0004	0.0099±0.0005
	MGM	0.0075±0.0021	0.0063±0.0010	0.0083±0.0014	0.0120±0.0025	0.0202±0.0046
ZDT3	LPCM	0.0043±0.0003	0.0080±0.0008	0.0129±0.0014	0.0185±0.0018	0.0265±0.0033
	UGM	0.0093±0.0061	0.0186±0.0040	0.0311±0.0058	0.0428±0.0079	0.0565±0.0096
	MGM	0.0151±0.0041	0.0107±0.0033	0.0147±0.0034	0.0210±0.0044	0.0328±0.0075

ZDT2.2, comparing its performance on the three ZDT functions without variable linkage. The performance decrease on ZDT3.2 seems more obvious; nevertheless, the best achieved Pareto front still approximates the true Pareto front very well. This indicates that the performance of LPCM scales well with the search dimension.

If we look at the results of UGM (middle panel), the performance becomes very poor. For all the three test functions, no single run is able to achieve a complete Pareto front, regardless of the search dimension. This strongly indicates that UGM is not suited for solving problems in which variable linkage exists.

We can see from the figures (right panel) that for test functions ZDT1.2 and ZDT2.2, MGM is able to achieve the entire Pareto front when the search dimension is low (20). However, the performance degrades seriously as the search dimension increases. This suggests that although MGM is able to capture the linkage between the design variables, the modeling accuracy decreases rapidly when the dimension becomes high.

The above observations made from the plot of the Pareto fronts can be confirmed by the D-metric and the I-metric of the results form 30 independent runs, as listed in Tables 5 and 6, respectively.

Table 5. Mean and standard deviation of the D-metric for test functions with variable linkage

Instance	Method	Search Dimension				
		20	40	60	80	100
ZDT1.2	LPCM	0.0045±0.0001	0.0049±0.0001	0.0058±0.0011	0.0091±0.0044	0.0120±0.0072
	UGM	0.1494±0.0266	0.2165±0.0213	0.2443±0.0224	0.2552±0.0134	0.2676±0.0151
	MGM	0.0138±0.0179	0.2032±0.0540	0.3526±0.0130	0.3624±0.0133	0.3637±0.0125
ZDT2.2	LPCM	0.0042±0.0001	0.0046±0.0001	0.0051±0.0002	0.0063±0.0033	0.0070±0.0020
	UGM	0.1845±0.0232	0.2222±0.0176	0.2408±0.0106	0.2512±0.0114	0.2508±0.0119
	MGM	0.0339±0.0269	0.1916±0.0236	0.2626±0.0125	0.2690±0.0100	0.2718±0.0119
ZDT3.2	LPCM	0.0051±0.0001	0.0109±0.0211	0.0077±0.0027	0.0103±0.0032	0.0126±0.0037
	UGM	0.0554±0.0296	0.0975±0.0284	0.1186±0.0090	0.1221±0.0014	0.1232±0.0008
	MGM	0.0868±0.0370	0.1059±0.0700	0.1428±0.0603	0.1690±0.0288	0.1993±0.0329

5.3 Sensitivity to Population Size

In solving hard real-world problems, the computational cost is often very high. One common approach is to parallelize the fitness evaluation using a computer cluster

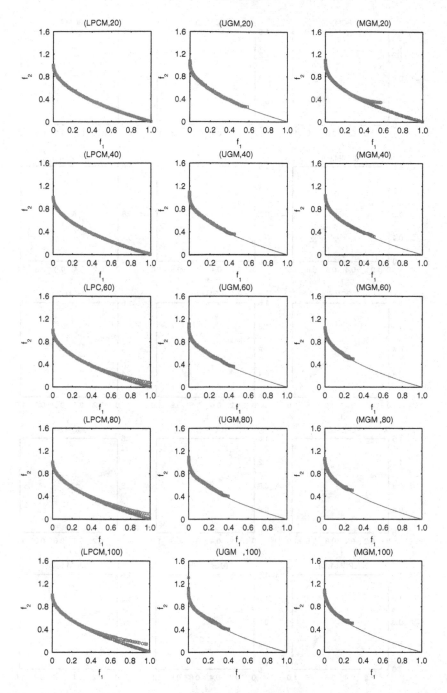

Fig. 9. Best (squares) and worst (circles) nondominated set on ZDT1.2. Population size 100. The number of design variables ranges from 20 (top row) to 100 (bottom row)

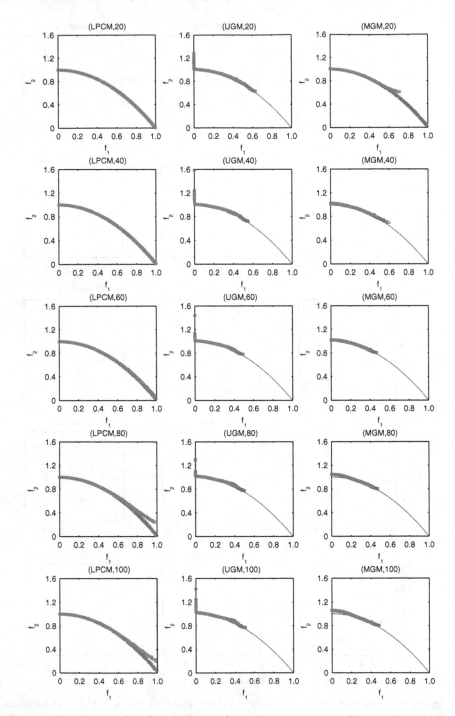

Fig. 10. Best (squares) and worst (circles) nondominated set on ZDT2.2. Population size 100. The number of design variables ranges from 20 (top row) to 100 (bottom row)

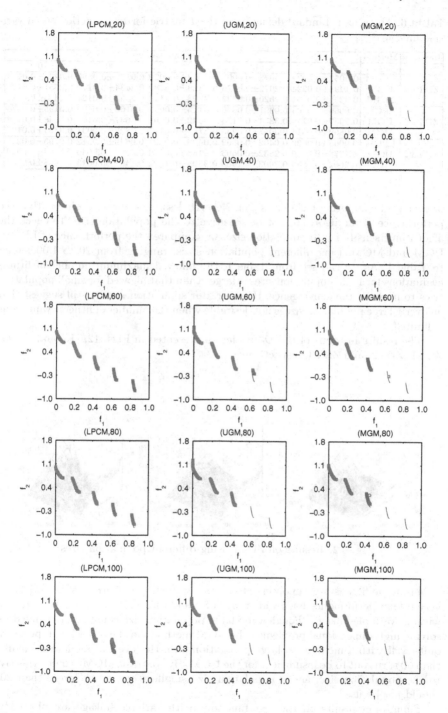

Fig. 11. Best (squares) and worst (circles) nondominated set on ZDT3.2. Population size 100. The number of design variables ranges from 20 (top row) to 100 (bottom row)

Table 6. Mean and standard deviation of the I-metric for test functions with variable linkage

Instance	Method	Search Dimension				
		20	40	60	80	100
ZDT1.2	LPCM	0.0057±0.0001	0.0073±0.0005	0.0102±0.0029	0.0173±0.0090	0.0223±0.0134
	UGM	0.1494±0.0224	0.2072±0.0176	0.2323±0.0182	0.2434±0.0117	0.2551±0.0126
	MGM	0.0241±0.0291	0.2020±0.0422	0.3293±0.0117	0.3391±0.0121	0.3425±0.0114
ZDT2.2	LPCM	0.0066±0.0004	0.0091±0.0006	0.0115±0.0011	0.0160±0.0106	0.0195±0.0078
	UGM	0.2957±0.0249	0.3354±0.0173	0.3542±0.0101	0.3647±0.0107	0.3648±0.0114
	MGM	0.0869±0.0614	0.3053±0.0233	0.3751±0.0117	0.3822±0.0091	0.3854±0.0105
ZDT3.2	LPCM	0.0065±0.0029	0.0402±0.0562	0.0423±0.0221	0.0649±0.0212	0.0795±0.0239
	UGM	0.1778±0.0689	0.2832±0.0503	0.3188±0.0124	0.3282±0.0049	0.3319±0.0029
	MGM	0.2602±0.0834	0.3003±0.1485	0.3895±0.1072	0.4487±0.0613	0.5098±0.0552

or even grid computing techniques [20]. Nevertheless, it is always desirable that the performance of an algorithm not be sensitive to the population size. To check the algorithms' sensitivity to population size, we compared the performance of LPCM, UGM and MGM using different population sizes, ranging from 20 to 1,000; refer to Table 1. As we can see from Table 1, the allowed maximum number of fitness evaluations for large population sizes is larger than that allowed for small population sizes to improve the convergence. However, our simulation results still suggest that an overly large population size is not desirable when the number of fitness evaluations is limited.

The results in terms of the D-metric are presented in Figs. 12, 13, and 14 for ZDT1, ZDT2, and ZDT3, respectively.

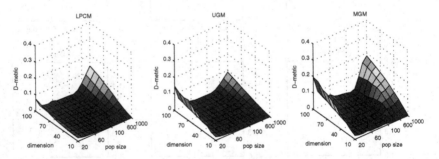

Fig. 12. Results on ZDT1 using different population sizes

From the figures, we can observe that among the three algorithms, LPCM shows very robust performance for a wide range of population sizes on the ZDT functions without variable linkage, though a too large population size is not recommended for solving high-dimensional problems. The UGM method, on the other hand, performs quite well with a medium to large population size. However, a population smaller than 60 turns out to be insufficient for the UGM. By contrast, MGM is quite sensitive to the population size, and a population size smaller than 60 or larger than 200 should not be used.

Simulation results on the test functions with variable linkage are plotted in Figs. 15, 16, and 17, respectively.

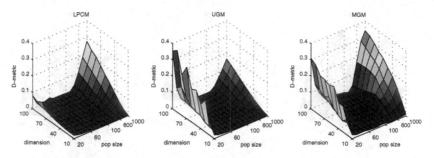

Fig. 13. Results on ZDT2 using different population sizes

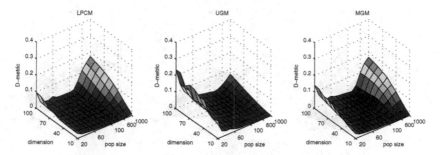

Fig. 14. Results on ZDT3 using different population sizes

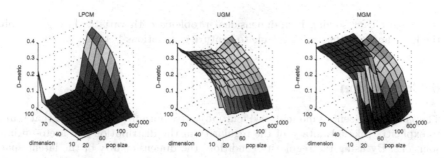

Fig. 15. Results on ZDT1.2 using different population sizes

LPCM distinguishes itself from the other two algorithms more on the test functions with variable linkage. In solving problems with variable linkage, LPCM still performs very well with different population sizes for high-dimensional problems. On ZDT3.2, the performance is not very satisfactory for small population sizes. Contrary to that, UGM performs poorly in most cases, except for the case in which a very large population size is used for a low-dimensional problem. The bad performance of UGM can obviously be attributed to the fact that UGM is not able to efficiently solve problems with variable linkage. From the figures, we can see that MGM works well on low-dimensional problems, though it is quite clear that MGM

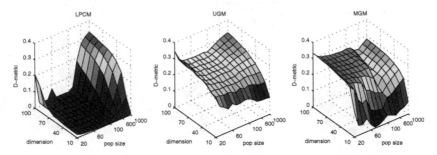

Fig. 16. Results on ZDT2.2 using different population sizes

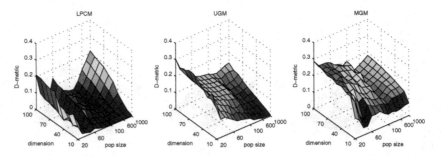

Fig. 17. Results on ZDT3.2 using different population sizes

is not suited for solving high-dimensional problems with variable linkage, despite the fact that it is theoretically able to capture correlations between variables.

6 Conclusion

This chapter presents a model-based multiobjective optimization method that is able to explicitly take advantage of the regularity in the distribution of Pareto-optimal solutions. By using the regularity condition, the dimensionality of the latent space in which the model is constructed is greatly reduced. In addition, a principal curve or surface model is used instead of a joint Gaussian distribution or a factorized Gaussian distribution model.

Simulation studies on comparing the scalability of the three multiobjective optimization algorithms, i.e., LPCM, UGM, and MGM, are conducted on six test problems with or without linkage among the design variables. From the simulation results, we demonstrate that LPCM exhibits excellent scalability to the increase in search dimension for problems with or without variable linkage. We also show that LPCM is in principle insensitive to population size ranging from 20 to 1,000. We show that UGM is also scalable to the search dimension for problems without linkage among design variables. However, the performance of UGM deteriorates drastically when linkage exists among the design variables.

It is somehow surprising that MGM also shows quite good performance on high-dimensional test problems without variable linkage, though it is more sensitive to the population size than LPCM and UGM. While MGM shows better performance than UGM on low-dimensional problems with variable linkage, its performance is as poor as that of UGM for high-dimensional problems, regardless of the population size used.

From our comparative studies, we conclude that explicitly taking the regularity in the distribution of Pareto-optimal solutions into account is very helpful in improving the scalability of model-based multiobjective optimization algorithms.

Our future work is to compare the performance of the algorithms on more complex test problems where stronger correlations exist, such as those suggested in [23, 11]. In addition, the scalability of the algorithms to the number of objectives (see pp. 307–329 and pp. 1–5, this volume) is also an interesting issue to further investigate.

References

[1] C. M. Bishop. *Neural Networks for Pattern Recognition*. Oxford University Press, 1996.

[2] P. A. N. Bosman and D. Thierens. Expanding from discrete to continuous estimation of distribution algorithms: The IDEA. In *Parallel Problem Solving from Nature*, pages 767–776, 2000.

[3] P. A. N. Bosman and D. Thierens. Advancing continuous IDEAs with mixture of distributions and facorization selection metrics. In *Genetic and Evolutionary Computation Workshop on Optimization by Building and Using Probabilistic Models*, pages 208–212, 2001.

[4] P. A. N. Bosman and D. Thierens. Learning probabilistic models for enhanced evolutionary computation. In Y. Jin, editor, *Knowledge Incorporation in Evolutionary Computation*, pages 147–176. Springer, 2005.

[5] P. A. N. Bosman and D. Thierens. The naive MIDEA: A baseline multiobjective EA. In *Third International Conference on Evolutionary Multi-Criterion Optimization*, LNCS 3410, pages 428–442. Springer, 2005.

[6] P. A. N. Bosman and D. Thierens. Numerical optimization with real-valued estimation of distribution algorithms. In M. Pelikan, K. Sastry, and E. Cantu-Paz, editors, *Scalable Optimization via Probabilistic Modeling*. Springer, 2006.

[7] D.-Y. Cho and B.-T. Zhang. Continuous estimation of distrubution algorithms with probabilistic component analysis. In *Congress on Evolutionary Computation*, pages 521–526. IEEE, 2001.

[8] K. Deb. *Multi-Objective Optimization using Evolutionary Algorithms*. John Wiley & Sons, Ltd, Chichester, 2001.

[9] K. Deb, A. Pratap, S. Agarwal, and T. Meyarivan. A fast and elitist multiobjective genetic algorithm: NSGA-II. *IEEE Transactions on Evolutionary Computation*, 6(2):182–197, 2002.

[10] G. R. Harik and D. G. Goldberg. Learning linkage. In *Foundations of Genetic Algorithms*, pages 247–262, 1996.

[11] S. Huband, L. Barone, L. While, and P. Hingston. A scalable multi-objective test problem toolkit. In *Evolutionary Multi-Criterion Optimization*, LNCS 3410, pages 280–295. Springer, 2005.

[12] A. K. Jain, M. N. Murty, and P. J. Flynn. Data clustering: A review. *ACM Computing Surveys*, 31(3):264–323, 1999.

[13] Y. Jin, editor. *Knowledge Incorporation in Evolutionary Computation.* Springer, Berlin, 2005.

[14] Y. Jin and B. Sendhoff. Connectedness, regularity and the success of local search in evolutionary multi-objective optimization. In *Proceedings of the Congress on Evolutionary Computation (CEC 2003)*, pages 1910–1917, Canberra, Australia, 2003. IEEE.

[15] N. Kambhatla and T. K. Leen. Dimension reduction by local principal component analysis. *Neural Computation*, 9(7):1493–1516, 1997.

[16] J. D. Knowles, L. Thiele, and E. Zitzler. A tutorial on the performance assessment of stochastic multiobjective optimizers. Technical Report 214, Computer Engineering and Networks Laboratory, ETH Zurich, Zurich, Switzerland, 2006.

[17] S. Kukkonen and J. Lampinen. GDE3: The third evolution step of generalized differential evolution. In *Proceedings of the Congress on Evolutionary Computation (CEC 2005)*, pages 443–450, Edinburgh, September 2005. IEEE.

[18] P. Larrañaga and J. A. Lozano, editors. *Estimation of Distribution Algorithms: A New Tool for Evolutionary Computation.* Kluwer Academic Publishers, Norwell, MA, 2001.

[19] P. Larranaga, R. Etxeberria, J. A. Lozano, and J. M. Pena. Optimization in continuous domains by learning and simulation of Gaussian networks. In *Genetic and Evolutionary Computation Workshop on Optimization by Building and Using Probabilistic Models*, pages 201–204, 2000.

[20] D. Lim, Y.-S. Ong, Y. Jin, B. Sendhoff, and B.-S. Lee. Efficient hierarchical parallel genetic algorithms using grid computing. *Future Generation Computer Systems - The International Journal of Grid Computing: Theory, Methods, and Applications*, 2007. Accepted.

[21] K. Miettinen. *Nonlinear Multiobjective Optimization*, volume 12 of *Kluwer's International Series in Operations Research & Management Science.* Kluwer Academic Publishers, 1999.

[22] H. Mühlenbein and G. Paass. From recombination of genes to the estimation of distribution I. Binary parameters. In *Parallel Problem Solving from Nature*, LNCS 1141, pages 178–187, 1996.

[23] T. Okabe, Y. Jin, M. Olhofer, and B. Sendhoff. On test functions for evolutionary multi-objective optimization. In *Parallel Problem Solving from Nature*, LNCS 3242, pages 792–802. Springer, 2004.

[24] T. Okabe, Y. Jin, B. Sendhoff, and M. Olhofer. Voronoi-based estimation of distribution algorithm for multi-objective optimization. In *Congress on Evolutionary Computation*, pages 1594–1601, Portland, Oregon, 2004. IEEE.

[25] M. Reyes Sierra and C. A. Coello Coello. A study of fitness inheritance and approximation techniques for multi-objective particle swarm optimization. In *Congress on Evolutionary Computation*, pages 65–72, Edinburgh, 2005. IEEE.

[26] O. Schütze, S. Mostaghim, M. Dellnitz, and J. Teich. Covering Pareto sets by multilevel evolutionary subdivision techniques. In *Second International Conference on Evolutionary Multi-Criterion Optimization (EMO 2003)*, pages 118–132, Faro, Portugal, 2003. Springer, LNCS 2632.

[27] S.-Y. Shin, D.-Y. Cho, and B.-T. Zhang. Function optimization with latent variable models. In *The Third International Symposium on Adaptive Systems*, pages 145–152, 2001.

[28] E. J. Solteiro Pires, P. B. de Moura Oliveira, and J. A. Tenreiro Machado. Multi-objective maxMin sorting scheme. In *The Third International Conference on Multi-Criterion Optimization*, LNCS 3410, pages 165–175. Springer, 2005.

[29] D. A. van Veldhuizen and G. B. Lamont. Evolutionary computation and convergence to a Pareto front. In *Late Breaking Papers at the Genetic Programming Conference*, pages 221–228, Madison, Wisconsin, 1998. Stanford University Bookstore.

[30] Q. Zhang. On stability of fixed points of limit models of univariate marginal distribution algorithm and factorized distribution algorithm. *IEEE Transactions on Evolutionary Computation*, 8(1):80–93, 2004.

[31] Q. Zhang, A. Zhou, and Y. Jin. Modelling the regularity in estimation of distribution algorithm for continuous multi-objective evolutionary optimization with variable linkages. *IEEE Transactions on Evolutionary Computation*, 2007. Accepted.

[32] A. Zhou, Y. Jin, Q. Zhang, B. Sendhoff, and E. Tsang. Combining model-based and genetics-based offspring generation for multi-objective optimization using a convergence criterion. In *Congress on Evolutionary Computation*, pages 3234–3241, Vancouver, BC, July 2006. IEEE.

[33] A. Zhou, Q. Zhang, Y. Jin, B. Sendhoff, and E. Tsang. Modelling the population distribution in multi-objective optimization by generative topographic mapping. In *Parallel Problem Solving From Nature - PPSN IX*, pages 443–452. Spinger, 2006.

[34] A. Zhou, Q. Zhang, Y. Jin, E. Tsang, and T. Okabe. A model-based evolutionary algorithm for bi-objective optimization. In *Congress on Evolutionary Computation*, pages 2568–2575, Edinburgh, September 2005. IEEE.

[35] E. Zitzler, K. Deb, and L. Thiele. Comparison of multiobjective evolution algorithms: empirical results. *Evolutionary Computation*, 8(2):173–195, 2000.

Objective Set Compression
Test-Based Problems and Multiobjective Optimization

Edwin D. de Jong[1] and Anthony Bucci[2]

[1] Institute of Information and Computing Sciences
Utrecht University
PO Box 80.089
3508 TB Utrecht, The Netherlands
dejong@cs.uu.nl

[2] DEMO Laboratory
Michtom School of Computer Science
Brandeis University
415 South St.
Waltham, MA 02454
abucci@cs.brandeis.edu

Summary. We consider a class of optimization problems wherein the quality of candidate solutions is estimated by their performance on a number of tests. Classifier induction, function regression, and certain types of reinforcement learning, including problems often attacked with coevolutionary algorithms, can all be seen as members of this class. Traditional approaches to such *test-based problems* use a single objective function that aggregates the scores obtained on the tests. Recent work, by contrast, argues that useful finer-grained distinctions between candidate solutions are obtained when each test is treated as a separate objective, and that algorithms employing such multiobjective comparisons show favourable behaviour relative to those which do not. Unfortunately, the number of tests can be very large. Since it is well-known that high-dimensional multiobjective optimization problems are difficult to handle in practice, the question arises whether the multiobjective treatment of test-based problems is feasible. To begin addressing this question, we examine a method for reducing the number of dimensions without sacrificing the favorable properties of the multiobjective approach. Our method, which is a form of dimension extraction, finds *underlying objectives* implicit in test-based problems. Essentially, the method proceeds by placing the tests along the minimal number of coordinate axes that still preserve ordering information among the candidate solutions. Application of the method to the strategy set for several instances of the game of Nim suggest the technique has significant practical benefits: a type of compression of the set of objectives is observed in all tested instances. Surprisingly, we also find that the information contained in the arrangement of tests on the coordinate axes reveals important information about the structure of the underlying problem.

1 Introduction

Certain problem domains encountered in machine learning and computational intelligence applications involve an evaluation of candidate solutions that is derived from a set of *tests*. The outcomes of a candidate on these tests are integrated into a scalar or vector which reflects different aspects of the quality of the individual, and which is used to make decisions about keeping, discarding, or modifying that candidate. These domains are called *test-based problems* [6].

Before introducing the ideas that follow, it will be useful to carry forward several illustrative examples of test-based problems:

- **Classifier Induction**: We are given a set of labeled data points and asked to produce a model, a neural network for example, that classifies them as well as possible. We fix a network topology and consider the task as a search through M, the space of possible weights for that topology. In other words, we seek a particular set of weights $m^* \in M$ which minimizes some classification error over the given data set. Each data point can be thought of as a test which an $m \in M$ either passes (classifies correctly) or fails (classifies incorrectly).[3] The error measure is an integration of this test information into a final evaluation of m.

- **Function Regression**: We are given a set of coordinates (x_i, y_i) representing the inputs and outputs of an unknown function and are tasked with finding a function that produces those pairs. We fix a space F of functions (for instance, genetic programs) and construe the task as a search through F for an f^* which minimizes a measure like the RMS error. As with the classifier induction example, we can view each pair (x_i, y_i) as a test of a candidate function $f \in F$. f's outcome on test x_i is its error $|f(x_i) - y_i|$. These individual errors are then integrated over all pairs to form an evaluation of f, e.g., its RMS error.

- **Learning Games of Strategy**: We aim to learn a competent strategy for an instance of the game of Nim. Let P_1 be the set of first-player strategies and P_2 the set of second-player strategies. In order to see how good a particular strategy $r \in P_1$ is, we play it against a variety of second-player opponents $s \in P_2$. Each opponent s can be thought of as a test of r: r plays s with the outcome being a win or loss for r. These outcomes across many P_2 players can then be integrated into an evaluation of r, for instance its worst-case outcome.

- **Coevolutionary Algorithms**: Roughly speaking, coevolutionary algorithms search through instances of two or more distinct roles, utilizing individuals playing one role to evaluate individuals playing the other. A seminal example is Hillis' coevolution of sorting networks (one role) against sets of unsorted lists (a second role which Hillis called *parasites*) [11]. Each possible sorting network is a candidate solution whose sorting abilities can be partially tested by a parasite. To put it differently, a sorting network's evaluation is derived from its outcomes against a number of parasites acting as tests.

The connections between coevolution and multiobjective optimization are worth exploring in more detail. Recent work examining these connections has produced a theory which suggests that any test-based problem can be viewed as a multiobjective

---

[3] In practice we could further differentiate between false positive and false negative outcomes, but for the sake of the example we are simplifying matters.

optimization problem. In this chapter, we aim to explore some of the outcomes of this point of view. In the remainder of this section we will explore the intellectual source of these ideas and then give an overview of the rest of the chapter.

It is worth pointing out that the method described here can be viewed as performing dimension *extraction* in the sense of [4]. The notion of dimension reduction has recently been applied in other work in Evolutionary Multiobjective Optimization; see, for instance, [7] and [15]. Also see the discussion of feature selection and feature extraction in the chapter by Brockhoff et al. in this volume.

1.1 Coevolution, Test-Based Problems and MOO

Traditionally, coevolutionary algorithms have integrated outcomes against multiple tests into a single fitness value, often by averaging or maximizing over all values.[4] That is, a candidate solution interacts with several test individuals and is then given a fitness that is an average or the maximum of its results against these tests (for a discussion on treating interactions in Cooperative Coevolution as tests, see [3]).

Both [8] and [13] argue that a finer-grained comparison can be made by viewing each test as its own objective. Rather than averaging or maximizing over all outcomes, the idea is to treat each outcome as a separate component of a vector of outcomes. Then, to compare two candidate solutions, the same Pareto dominance or Pareto covering relations utilized in multiobjective optimization are employed.

The process of transforming single-objective problems into multiobjective problems by separating the different criteria contributing to the quality of individuals has been named multiobjectivization [12]. The application of this idea within coevolution is called *Pareto coevolution*. In Pareto-coevolution, each test in the population is treated as if it were an objective in a massive multiobjective optimization problem.[5] For instance, if there were a population of 100 parasites in the sorting network example, a Pareto coevolutionary approach might evaluate a sorting network with a 100-dimensional vector of numbers, each number an outcome against a different parasite. Initial approaches to Pareto coevolution bought finer-grained comparison information at the cost of large outcome vectors like this; increasing the dimensionality of the objective space generally complicates the search problem. In other words, while Pareto coevolution has advantages, for instance its mitigation of cycling dynamics, large outcome vectors introduce new problems; see also the chapter by Ficici in this volume. As a result, there has been a drive to reduce the size of these vectors without losing too much of what was gained by using them in the first place.

Along these lines, [6] presents empirical results suggesting that a Pareto coevolutionary algorithm could find what were dubbed the *underlying objectives* of a problem. These are hypothetical objectives that determine the performance of candidate solutions without the need to test candidates against all possible tests. [6] applies a two-population Pareto coevolutionary algorithm, DELPHI, to instances of

---

[4] Simple fitness proportional coevolutionary algorithms typically use a (weighted) average of outcomes as fitness. Cooperative Coevolutionary algorithms sometimes use the maximum outcome as fitness value [14].

[5] A critical difference from evolutionary multiobjective optimization being that in coevolution not all objectives are in hand in advance, but rather are discovered during search.

a class of abstract test games. Figures 13 and 15 of that work suggest that evaluator individuals[6] evolve in a way which tracks the underlying objectives of the problem domain. The results suggest that the algorithm is sensitive to the presence of underlying objectives even though it is not given explicit information about those objectives. [2] makes a similar observation, also empirical though using a different algorithm; Fig. 5 of that work suggests a similar sensitivity to underlying objectives. In both cases, clusters of individuals, rather than single individuals, move along or collect around the objectives of the problem domain. The problem domains considered, namely numbers games [18], were designed to have a known and controllable number of objectives, but the algorithms used in these two studies did not rely on that fact. The work therefore raises the question of whether underlying objectives exist in all problem domains, and whether search learning algorithms can discover them.

A partial answer to this question is found in the notion of *coordinate system* [4]. Coordinate systems, which were defined for a class of test-based problems,[7] can be viewed as a formalization of the empirically observed underlying objectives of [6]. To elaborate, a coordinate system consists of several *axes*. Each axis is a list of tests ordered in such a way that any candidate can be placed somewhere in the list. A candidate's placement is such that it passes all tests before that spot and fails all tests after it. For this reason, an axis can be viewed as measuring some aspect of a candidate's performance: a candidate that places high on an axis is "better than" one which sits lower in the sense that it passes more tests (more of the tests present in the axis, not more tests of the problem as a whole). Formally, an axis corresponds to a numerical function over the candidates, in other words to an objective function. It can be proved [4] that every problem of the considered class possesses at least one coordinate system, meaning it has a decomposition into a set of axes. In short, every such problem has some set of objective functions associated with it, one for each axis in a coordinate system for the problem.

Besides defining coordinate systems formally, [4] gives an algorithm that finds a coordinate system for a problem domain in polynomial time. The algorithm, though fast, is not guaranteed to produce the smallest possible coordinate system for the problem. Finite domains must have a minimal coordinate system, but in general even finite domains can have distinct coordinate systems of different sizes. The algorithm is not coevolutionary per se, as it examines the outcomes of tests on candidates. It is therefore applicable to the entire class of test-based problems.

1.2 Chapter Overview

To summarize, recent theoretical work on coevolutionary algorithms has elucidated a theory of underlying objectives for test-based problems which establishes a conceptual link between multiobjective optimization and coevolution. We can now view coevolution as a form of multiobjective optimization in which not all objectives are explicitly given *a priori*, but are nevertheless present theoretically and can be extracted. Although this theory originated in coevolutionary algorithms research, it is focused on problem structure and indifferent to which search algorithm is used.

---

[6] What we are here calling *tests*.

[7] Specifically, problems with a finite number of candidates and a finite number of binary-outcome tests.

Coordinate systems can be extracted from any test-based problem, which includes classifier induction, function regression, and game strategy learning problems.

However, there are two questions left unaddressed by this story. First, that *minimal* coordinate systems exist as theoretical objects does not guarantee they can be extracted algorithmically; previous work has given an algorithm which could find *some* coordinate system, but not necessarily a minimal one. Secondly, even if they could be found, there is no guarantee that the extracted coordinate systems are meaningful, i.e., that their structure relates to characteristic features of the problem. With these questions left open, it is possible coordinate systems are simply mathematical curiosities that have limited relevance in practice. The aim of understanding underlying objectives and establishing a bridge between test-based problems and multiobjective optimization would then not be met.

This chapter will focus on the two questions by developing and applying an exact coordinate system extraction algorithm to small instances of the game of Nim. The exact algorithm is guaranteed to identify a minimum dimensional coordinate system, but can only be applied to small problems due to its computational complexity. After giving the necessary background, we will describe the exact extraction algorithm and prove that it produces a minimal coordinate system, giving a positive answer to the question of whether minimal coordinate systems can be discovered algorithmically. We argue that the extracted coordinate system can be interpreted as *compressing* objective information of the problem, in the sense that knowing where a candidate solution lies in a coordinate system is equivalent to knowing how that candidate performs against all possible tests. Since the number of axes in a coordinate system can be no more than the number of tests, but may be significantly smaller, a minimal coordinate system is a maximally compressed view of candidate solutions' performance.[8]

In the game of Nim, where candidates are players and tests are potential opponents, we observe that a substantial compression does take place. Minimal coordinate systems for all tested instances of Nim have significantly fewer axes than tests. We also observe that the axes can be interpreted: each axis directly tests a candidate's strength for a particular game configuration. It is worth noting that, although each axis tests on a corresponding game configuration, not all game configurations correspond to axes. There are significantly fewer axes than game configurations, too. This correlation between axes and game configurations is noteworthy given that the coordinate system extraction algorithm is insensitive to details of the application domain. There is no reason to expect that an axis, while theoretically meaningful, would correspond in an intuitively meaningful way with a candidate player's ability at the game. Nevertheless, across the tested instances of Nim, extracted coordinate systems consistently represent ability for specific game configurations. These results validate that the notion of coordinate system is not just a theoretical curiosity; rather, coordinate systems give a view into performance that is both compressed in size and intuitively meaningful.

In short, this set of ideas establishes fruitful connections among multiobjective optimization, coevolution, machine learning, and game strategy learning techniques. The theoretical and algorithmic notion of coordinate systems provides a way to view a test-based problem as a sort of multiobjective problem, allowing conceptual

---

[8] Maximally compressed with respect to our set of assumptions, of course — naturally there are many ways to compress this information.

cross-fertilization among these disciplines. The results on Nim suggest that this conceptual link can go both ways: treating Nim strategy learning as a multiobjective optimization problem can yield insights into the nature of the game itself as well as into how to learn or evolve strategies to play it.

2 Preliminaries

2.1 Definition of Problem Structure

Our notion of problem structure is based on the observation that while the number of possible tests in a problem can be very large, the testing may be limited to a small number of *underlying objectives* [6]. The question we aim to address is how these underlying objectives may be identified. If this is possible, it permits accurate evaluation using only a limited number of tests. In the following, we consider how a minimal set of objectives can be identified for which the information provided is equivalent to the information provided by the set of all possible tests.

Let $\mathbb{S}$ be the set of candidate solutions in a problem; these can, for example, be classifiers, or game-playing agents. Let $\mathbb{T}$ denote the set of tests. In the example of classification these would be test points; in game playing they would be opponents. A test can be viewed both as an object $T \in \mathbb{T}$ or as a function $T : \mathbb{S} \longrightarrow \{0, 1\}$ that returns the outcome of the test for a given candidate solution. We will write the outcome of a test $T \in \mathbb{T}$ for a candidate solution $S \in \mathbb{S}$ as $T(S)$.

Definition 1 (Objective) *An* objective *is a function that assigns a value to a candidate solution which measures an aspect of its performance:* $o : \mathbb{S} \longrightarrow \mathbb{R}$.

Without loss of generality, we will assume that higher values are to be preferred. An ordered set of objectives OS can be viewed as a vector function that accepts a candidate solution and returns an outcome vector, where each element i is the outcome of objective OS_i: $OS(S) = OS_1(S), OS_2(S), \ldots, OS_n(S)$.

The set of all tests $\mathbb{T}$ can be viewed as a set of objectives; each test can be viewed as a binary objective whose value for a given candidate solution S is given by the outcome of the test $T(S)$. Since $\mathbb{T}$ is given as part of the problem formulation, the corresponding objectives will be called the *initial objectives*:

Definition 2 (Initial Objectives) $O_{\text{init}} = \mathbb{T}$.

The question of identifying a minimal set of objectives is thus reduced to the problem of compressing the initial objectives to a minimal set of objectives that provides equivalent information, as we define next.

2.2 Objective Compression

A given set of objectives $O1$ can be compressed to yield an equivalent but smaller set of objectives $O2$. For this purpose, the notion of equivalence is defined as follows.

Definition 3 (Equivalence) *We define two objective sets $O1$, $O2$ to be equivalent, written $equiv(O1, O2)$, if the following criteria hold:*

- *Information preservation. This criterion holds if a mapping f exists such that $\forall S \in \mathbb{S} : f(O2(S)) = O1(S)$. If this is the case, the objective values assigned by $O1$ can be reconstructed from the objective values assigned by $O2$.*
- *Order preservation. For the transformation to be meaningful, the second set of objectives should result in the same preference information. Though dependent on the preference function, this will generally be achieved if $\exists i : O1_i(x) > O1_i(y) \iff \exists j : O2_j(x) > O2_j(y)$, where $x, y \in \mathbb{S}$, $1 \leq i \leq |O1|$, and $1 \leq j \leq |O2|$.*

2.3 Order Preservation: Pareto-Dominance

As an example, we demonstrate that the second condition is sufficient to guarantee order preservation for the preference function of Pareto-dominance. The Pareto-dominance preference function states that a candidate solution x is preferred over another candidate solution y, or *dominates* it, with respect to the objectives in $O1$, written $x \overset{O1}{\succ} y$, if:

$$\forall i : O1_i(x) \geq O1_i(y) \wedge \exists i : O1_i(x) > O1_i(y) \text{ with } 1 \leq i \leq |O1|.$$

If two objective sets $O1$ and $O2$ are equivalent, then $x \overset{O1}{\succ} y \iff x \overset{O2}{\succ} y$. This can be seen as follows:

Assume $x \overset{O1}{\succ} y$. Then $\forall i : O1_i(x) \geq O1_i(y)$ and $\exists i : O1_i(x) > O1_i(y)$. Therefore, $\nexists i : O1_i(y) > O1_i(x)$. Thus, due to the order preservation condition, $\nexists j : O2_i(y) > O2_i(x)$, and hence $\forall j : O2_j(x) \geq O2_j(y)$.

Furthermore, since $\exists i : O1_i(x) > O1_i(y)$, the condition guarantees that $\exists j : O2_j(x) > O2_j(y)$. Since we have $\forall j : O2_j(x) \geq O2_j(y)$ and $\exists j : O2_j(x) > O2_j(y)$, it follows that $x \overset{O2}{\succ} y$, and thus the forward implication has been shown. The backward implication is analogous.

2.4 Problem Structure

Assume we are given a problem P with initial objectives O_{init}. Then P's problem structure is defined by the smallest set of objectives O_{min} that is equivalent to O_{init}:

Definition 4 (Problem Structure) $O_{\text{min}} = \arg\min_{O_i \in \mathbb{O}} |O_i|$ *such that $equiv(O_i, O_{\text{init}})$, where $O_i \in \mathbb{O}$, and $\mathbb{O}$ is the set of all possible objective sets given some representation of objectives.*

We define the *evaluation dimension* of a problem as the lowest number of dimensions d for which a correct coordinate system exists, or equivalently as the cardinality of the problem structure:

Definition 5 (Evaluation Dimension) $d_{eval}() = |O_{\text{min}}|$.

2.5 Coordinate Systems for Test-Based Problems

In the following, we present a representation for objectives in test-based problems. Based on this representation, a search algorithm will be used to identify the smallest objective set that is equivalent to the initial objectives, and which therefore represents the structure of the problem.

The objective sets we will consider take the form of *coordinate systems*. Each axis in a coordinate system represents an objective. The position of a test on an axis is determined by which candidate solutions are defeated by the test. We will write $SF(T,S)$ to indicate that test T defeats candidate solution S, meaning that it assigns a zero outcome to the candidate solution. Failing a solution S is written $SF(S)$.

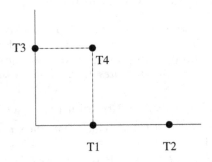

Fig. 1. Example of a coordinate system. Monotonicity: $T2$ defeats strictly more candidate solutions than $T1$. Compositionality: $T4$ defeats all and only those candidate solutions that are defeated by either $T1$ or $T3$

Definition 6 (Solution Failure) $SF(T,S) \Longleftrightarrow T(S) = 0$. *Solution failures can be grouped together in sets. We write $SF(T)$ to denote the set of all solution failures made by T: $SF(T) = \{SF(S)|S \in \$ \wedge SF(T,S)\}$. The set of all possible solution failures is named SFS: $SFS = \{SF(S)|S \in \$\}$.*

We now proceed to define the elements of a coordinate system.

Definition 7 (Axis) *An axis A represents an ordered set of increasing solution failure sets: $A = A_1, A_2, \ldots A_n$ with $A_i \subseteq SFS$ such that $\forall i < j : A_i \subset A_j$. The elements A_i of an axis will be called* coordinates.

Definition 8 (Coordinate System) *A coordinate system CS is an ordered set of axes: $CS = A^1, A^2, \ldots A^n$.*

Definition 9 (Position) *A position P in an n-dimensional coordinate system is an ordered set of coordinates, one for each axis: $P = A_i^1, A_j^2, \ldots A_k^n$.*

The coordinate systems that will be defined feature the following two properties, which motivate viewing them as coordinate systems (see Figure 1):

Definition 10 (Monotonicity) *If test T1 has a higher position on an axis than T2, this implies that T1 defeats all candidate solutions defeated by T2, in addition to one or more other candidate solutions.*

Definition 11 (Compositionality) *If test T3's position is spanned by two tests T1 and T2 which reside on different axes, i.e., T3's position is (T1,T2), then the set of candidate solutions defeated by T3 is the union of the sets of candidate solutions defeated by T1 and T2: $SF(T3) = SF(T1) \cup SF(T2)$.*

Both tests and candidate solutions can be embedded into a coordinate system.

Definition 12 (Test Embedding) *The embedding $CS(T)$ of a test T in a coordinate system CS is the position obtained by choosing the highest coordinate on each axis for which T still makes all corresponding solution failures: $CS(T) = \{A_j^i \in CS, 1 \leq i \leq n | SF(T) \supseteq A_j^i \ \wedge \ \nexists k > j : SF(T) \supseteq A_k^i\}$.*

Definition 13 (Interpretation of Test Positions) *The set $SF(P)$ of solutions defeated by a test at position P is obtained by taking the union of the solution failure sets represented by the position's coordinates: $SF(P) = \underset{1 \leq i \leq n}{\cup} P_i$.*

Definition 14 (Solution Embedding) *The embedding $CS(S)$ of a candidate solution S in a coordinate system CS is the position obtained by choosing the highest coordinate on each axis for which S is not included in the corresponding solution failures: $CS(S) = \{A_j^i \in CS, 1 \leq i \leq n | S \notin A_j^i \ \wedge \ \nexists k > j : S \notin A_j^i\}$.*

Since a coordinate system represents an objective set, $CS(S)$ may be interpreted as the objective vector for candidate solution S; each axis represents one objective, and the coordinate on the axis represents the value of the objective. In order to map coordinates into numerical values, any monotonic assignment of coordinates to values may be used; for example, the index of the coordinate on the axis can be employed such that a candidate solution with position (A_8^1, A_5^2) would have objective values $(8, 5)$.

Definition 15 (Interpretation of Solution Positions) *The set $TS(P)$ of tests solved by a candidate solution at position $P = CS(S)$ is obtained by assembling all tests whose coordinates do not exceed that of the solution: $TS(P) = \{T \in \mathbb{T} | \nexists A^i \in CS(T) | A^i \supset P_i\}$.*

Definition 16 (Correctness) *A coordinate system CS is correct for a given set of tests $TS \subseteq \mathbb{T}$, written $correct(CS, TS)$, if for each test the set of solution failures equals the set of solution failures represented by the embedding of the test: $\forall T \in TS : SF(CS(T)) = SF(T)$.*

Our central theorem states that the objectives represented by a correct coordinate system are equivalent to the initial objectives.

Theorem 1 (Correctness of coordinate systems) $correct(CS, O_{\text{init}})$ $\Longrightarrow$ $equiv(CS, O_{\text{init}})$.

Proof. The coordinate systems that have been defined above are a specific way to represent objectives for a test-based problem. We will now demonstrate that the objectives represented by a correct coordinate system are always equivalent to the initial objectives. Therefore, by restricting the search to correct coordinate systems, it is guaranteed that any coordinate systems found will be equivalent to the initial objectives. We can thus perform objective compression by searching for the smallest correct coordinate system.

Given a correct coordinate system CS, proving equivalence to the initial objectives O_{init} requires establishing the properties of information preservation and order preservation. Regarding information preservation, it is to be shown that a mapping f exists such that $\forall S \in \mathbb{S} : f(CS(S)) = O_{\text{init}}(S)$. In other words, the outcomes $\{T(S)|T \in \mathbb{T}\}$ need to be reconstructed from the coordinates of S. Thus, f can be based on the composition of (1) the embedding function $CS(S)$, which determines the position of a candidate solution in the coordinate system, and (2) the interpretation function $TS(P)$, which determines the tests solved by a candidate at position P: $TS(CS(S))$. This function returns the tests solved by S. Given all tests solved by S, the outcome of any test can be determined by seeing whether the test is part of this set. This yields the desired reconstruction function:

$$f_i(S) = \begin{cases} 1 & \text{if} \quad T_i \in TS(CS(S)) \\ 0 & \text{otherwise.} \end{cases}$$

Next, the property of order preservation is to be shown. Assume $\exists i : O_{\text{init},i}(x) > O_{\text{init},i}(y)$. Thus, $\exists T \in \mathbb{T} : T(x) > T(y)$; since we are assuming binary tests, this means $T(x) \wedge \neg T(y)$. Given that CS is a correct coordinate system, we know that $TS(CS(y))$ yields the tests solved by y, and must therefore contain a test that is not present in $TS(CS(x))$. Since the axes are monotonic, this implies y must have a higher coordinate for some axis.

Conversely, assume y has a higher coordinate than x for some axis: $TS(CS(y)) > TS(CS(x))$. Then there must exist a test solved by y that is not solved by x. Thus, y has a higher coordinate for the initial objective corresponding to this test. $\square$

3 The Exact Algorithm and Nim

3.1 Exact Algorithm

We present an algorithm that performs exact identification of problem structure, meaning that the dimensionality of the extracted coordinate system is guaranteed to be minimal. The algorithm operates by considering all possible coordinate systems in order of increasing dimensionality, and returning when a correct coordinate system has been found. This guarantees that the smallest possible coordinate system, measured in terms of its dimensionality, will be found.

As described, a coordinate system consists of axes whose coordinates represent solution failure sets. Given a set of candidate solutions, the number of all solution failure sets is the power set· of this set, and thus exponential in the size of the solution set. Considering all assignments of all solution failure sets to axes is therefore prohibitive. However, the only requirement for a coordinate system is that it be able to represent the given set of tests. Thus, the coordinate system must contain positions corresponding to the solution failure sets represented by the tests, but need not contain positions corresponding to other solution failure sets. This greatly reduces the search problem; due to this observation, we can restrict the search by considering only subsets of the solution failure sets represented by the tests.

Since the number of all possible coordinate systems grows very quickly, the exact algorithm that will be presented will in general still not be feasible for problems of realistic size. Its purpose however is to permit studying the structure of small example problems, and to provide a starting point for efficient structure approximation algorithms.

The algorithm starts from an empty coordinate system. Each axis is initially defined by two virtual tests: ORG (origin), which passes all solutions, and INF (infinity), which defeats all solutions. To search for a correct coordinate system, all tests are visited in turn. For each test, it is determined whether the test can be placed in the current coordinate system; if not, the coordinate system is adapted to permit representing the test. If this fails given the current dimensionality because one or more additional axes are required, the dimensionality is increased.

The placement of tests is stored in a state vector, representing the current placement of each test. Two search operators are employed: the operator `find_first_full_state(state)` finds the first correct coordinate system from the current state by visiting the tests in turn and placing each test correctly, adapting the coordinate system where necessary; `inc_full_state` increases the current state, and this operator is applied when the current state does not permit the construction of a correct coordinate system. By continuing the search for a correct coordinate system while the state so far permits this and increasing the state once it is found that it does not permit this, a correct coordinate system of the given dimensionality will be found if it exists. Since the dimensionality is incremented only if no correct system is found, the first coordinate system found by the algorithm is guaranteed to be minimal. The algorithm therefore returns once a correct coordinate system is found. The pseudocode of the main loop of the algorithm is given in Algorithm 1.

`calculate_options(T)` accepts a test and for each axis i determines the highest coordinate for which the test still makes all corresponding solution failures. These coordinates are written $T^i_{\min}$, and represent the embedding of the test in the partially constructed space. The successor of a coordinate $A^i_{\min}(T)$ is written $A^i_{\max}(T)$. The coordinate of a test on axis i must lie between $A^i_{\min}(T)$ and $A^i_{\max}(T)$. Potentially, a new coordinate must be created. The options for such a new coordinate are constrained by $A^i_{\min}(T)$ and $A^i_{\max}(T)$; the solution failure set it represents must be a superset of the former and a subset of the latter. Therefore, the algorithm must search the powerset of $A^i_{\max}(T) \setminus A^i_{\min}(T)$.

`init_test_state(T)` places a test at the highest possible existing coordinates, i.e., $A^1_{\min}(T), A^2_{\min}(T), \ldots, A^n_{\min}(T)$.

Algorithm 4 FIND_STRUCTURE()

1: **for** *num_dims* = 1 **to** *num_tests* **do**
2:     *init_axes*()
3:     *state* = [0, 0, 0, 0, 0]
4:     **for** *i* = 1 **to** *num_tests* **do**
5:         *calculate_options*(*T*)
6:         *init_test_state*(*i*)
7:     **end for**
8:     **while** !*done* **do**
9:         **if** *find_first_full_state*(*state*) **then**
10:             *done* = **true**!   ▷ success; break outer for loop and return
11:         **else**
12:             *ok* = *inc_full_state*
13:             **if** !*ok* **then**
14:                 *done* = **true**   ▷ failure; increment *num_dims*
15:             **end if**
16:         **end if**
17:     **end while**
18: **end for**

---

find_first_full_state(state) looks for the first feasible state from the current state by ensuring a correct placement for each test in turn. This is done by calling inc_test_state until a correct placement for a test is found.

inc_test_state(i) increments the state element corresponding to test T, and fails if this exceeds the range for this state element. If the increment succeeds, the element represents a new placement for the test. The order of the options considered is as follows: First, the test is placed at the position that its embedding would indicate, i.e., $A^1_{min}(T), A^2_{min}(T), \ldots, A^n_{min}(T)$, using init_test_state(T). The next n options consider placing the test on one of the n axes. If this succeeds, the test defines a new coordinate for the axis concerned. Finally, all combinations of solutions failures in $A^i_{max}(T) \setminus A^i_{min}(T)$ are considered.

inc_full_state moves to the next state by increasing the state vector.

3.2 The Game of Nim

We will explore the notion of problem structure and its extraction by applying the search algorithm to the Game of Nim.

Nim originates from the Chinese game Tsyanshidzi (picking stones game). The first European reference to a possible Nim-type game dates from 1503 [16]. The name is thought to stem from the German imperative 'nimm' (take), and is proposed in [1]. The game also features in Alain Resnais' 1961 movie *Last Year at Marienbad*.

The game starts by placing a number of rows of small objects such as matches on a table, where each row is called a *counter*. The players take turns, and on each turn must take one or more matches from a single row. The player to take the last match wins the game. In the *misère* version of the game, the outcome is reversed.

Nim is an *impartial game*, meaning each player has the same available moves in every position. An interesting result of combinatorial game theory is the Sprague-Grundy theorem, independently discovered by Roland P. Sprague [17] and Patrick M. Grundy [10]. The theorem states that every impartial game is equivalent to a nim position, augmented with the possibility of adding matches.

In 1901, the Harvard mathematician Charles L. Bouton presented an optimal strategy for the game of Nim [1]. To compute the strategy, the counters of a Nim position are written below one another in binary notation. Next, the columns of the numbers are summed. If all sums are even, the game position is a *safe combination*. To play optimally, the player must merely select a move that results in a safe combination.

3.3 Results

We apply the structure extraction algorithm to the Game of Nim versions with the following initial configurations: [1 3], [4], and [2 2].[9] We employ the misère version of the game, which has been noted as being more difficult to analyze. For all of these small game configurations, the structure extraction algorithm took less than a second. However, due to the high computational complexity of the exact algorithm, larger versions of the game quickly lead to high computational requirements.

To apply the structure extraction algorithm, we first generate all strategies for a game. No distinction is made between the first and second players; for both players, all combinations of the legal moves in all game configurations are considered. The complete set of strategies is then played against itself in a full squared comparison. Since some game configurations never occur for some strategies, some players will behave identically while having different strategy representations. Therefore, any first players with outcome vectors identical to those of other first players are removed, and likewise for duplicate second players. This yields an outcome matrix representing all unique first and second players, which serves as input to the structure extraction algorithm. The tests in the matrix are first sorted according to the number of solution failures they represent; since a complete search is performed, this does not affect the dimensionality of the outcome, but it may serve to consider the more likely coordinate systems first.

Results for Nim [1 3]

The first version we apply the algorithm to is the [1 3] configuration, which has one row containing a single match and one row containing three matches. There are seven non-empty game configurations. Different configurations have different numbers of legal moves; the full number of strategies is 144. However, after removing the first and second players with duplicate outcome vectors, six first player strategies (candidate solutions) and nine second player strategies (tests) remain. The resulting outcome matrix for unique strategies is shown in Table 1.

The result of applying the structure extraction algorithm to the outcome matrix for this game is a two-dimensional coordinate system; see Table 2. The first

---

[9] The notation $[x_1 x_2]$ means that this game instance has two piles of sticks, and the starting configuration has x_1 sticks in the first pile and x_2 sticks in the second pile.

Table 1. Outcome matrix of unique strategies for Nim[1 3]

	T0	T4	T8	T24	T28	T32	T48	T52	T56
S0	1	0	1	1	0	1	1	0	1
S1	0	0	0	0	0	0	0	0	0
S2	1	1	1	1	1	1	1	1	1
S3	1	1	1	0	0	0	1	1	1
S72	1	0	0	1	0	0	1	0	0
S75	0	0	0	0	0	0	1	1	1

axis consists of tests T48, T0, and T24, representing the solution failure sets $\{S1\}$, $\{S1, S75\}$, and $\{S1, S75, S3\}$, in addition to ORG and INF. Interestingly, all coordinates on the axis thus correspond exactly to actual tests, rather than being arbitrary subsets of solution failures. The same holds for the second axis, containing tests $T56$ and $T52$.

Table 2. The two-dimensional coordinate system for Nim[1 3]

dim 1	ORG	T48	T0	T24	INF
S1	1	0	0	0	0
S75	1	1	0	0	0
S3	1	1	1	0	0
S0	1	1	1	1	0
S2	1	1	1	1	0
S72	1	1	1	1	0

dim 2	ORG	T56	T52	INF
S1	1	0	0	0
S72	1	0	0	0
S0	1	1	0	0
S2	1	1	1	0
S3	1	1	1	0
S75	1	1	1	0

Since a correct coordinate systems embeds all tests onto positions that correspond to their outcomes, a question arises as to where the four tests that do not lie on the axes are located. This is shown in Table 3 (left) and visualized in Figure 2. For any of these tests, the set of solutions defeated is the union of the solution failure sets of its coordinates.

The next question is where the candidate solutions are embedded in the space. This is shown in Table 3 (right) and Figure 3. The embedding of the solutions shows how two objectives are sufficient to evaluate the solutions and express the relations

Table 3. Left: Coordinates of the tests not lying on the axes. Each test is precisely a composition of two other tests; see outcome matrix and text. Right: Coordinates of the candidate solutions

test	coordinates
T8	(T0 , T56)
T4	(T0 , T52)
T32	(T24 , T56)
T28	(T24 , T52)

test	coordinates
S0	(T24 , T56)
S1	(ORG, ORG)
S2	(T24 , T52)
S3	(T0 , T52)
S72	(T24 , ORG)
S75	(T48 , T52)

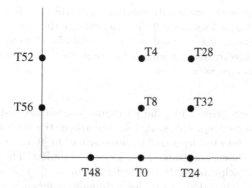

Fig. 2. Embedding of the tests into the coordination system found for Nim[1 3]

between them. For example, it is immediately seen that solution S2 is the optimal solution, as it has the highest coordinate on both dimensions. Comparing this with the initial situation in which all nine tests are objectives demonstrates the utility of objective compression: by identifying the compositionality implicitly present in the problem, the number of objectives required for evaluation is reduced from nine to two.

Apart from the potential computational advantage offered by objective compression, an interesting theoretical question is whether the automatically extracted coordinate system can tell us something about the structure of a problem. To this end, we now interpret the axes that have been identified.

Table 4 shows the actions selected by the axis tests in all of the seven different game configurations that can occur in Nim[1 3], the resulting game configuration after the action, and the outcome for the test. It is immediately seen that the tests in dimension 1 differ only in a single configuration, viz. configuration 3, which represents the game state [0 3], i.e., a single row with three matches. The first test on the axis, T48, removes all three matches and thus leaves no matches, thereby losing the game. T0 leaves two matches, so that its outcome depends on the opponent strategy. Finally, the last strategy on the axis, T24, leaves a single match, and

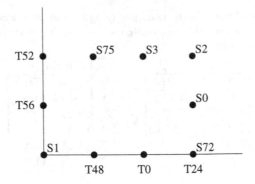

Fig. 3. Embedding of the candidate solutions into the coordination system found for Nim[1 3]. Two dimensions are sufficient to evaluate the solutions and express the relations between them, as compared to the original nine dimensions required when using all tests as objectives. This substantial reduction demonstrates the potential value of objective compression

thereby secures a sure win. Thus, the particular coordinate system that has been extracted has a clear interpretation in this dimension: the objective represents the quality of the move selected by a test in game situation [0 3].

For the second dimension (Table 5), the result is similar. The tests only differ in game situation 6, which represents the configuration [1 2]. The tests select increasingly good moves, leaving [0 2] (result depending on opponent), and [1 0] (sure win) respectively.

Table 4. Actions selected by the tests on the first axis in the seven game configurations. The axis apparently tests on the ability to select a good action in situation 3, which represents configuration [0 3]. Moving down the table, the resulting configurations show the increasing quality of the moves

	C1	C2	C3	C4	C5	C6	C7	After	Result
T48	1	1	3	1	1	1	1	[0 0]	sure loss
T0	1	1	1	1	1	1	1	[0 2]	depends
T24	1	1	2	1	1	1	1	[0 1]	sure win

Results for Nim [2 2]

Nim[2 2] has eight configurations, resulting in 288 strategies. The resulting outcome matrix contains 36 unique tests and six unique candidate solutions. The extracted minimal coordinate system is four dimensional; thus, a high degree of compression

Table 5. Actions selected by the tests on the second axis. Again, the axis concerns the behaviour of tests in a single configuration: [1 2], and the tests represent increasingly good moves for this configuration

	C1	C2	C3	C4	C5	C6	C7	After	Result
T56	1	1	3	1	1	3	1	[0 2]	depends
T52	1	1	3	1	1	2	1	[1 0]	sure win

is observed. As in Nim[1 3], all dimensions can be interpreted, and the different options for a dimension always concern a single game configuration.

Results for Nim [4]

Nim[4] has four configurations, and 24 strategies. There are six unique tests and five unique candidate solutions. The problem is two dimensional, and as with the previous two problems the dimensions have a clear interpretation. The coordinate system and the embedding of the candidate solutions is visualized in Figure 4.

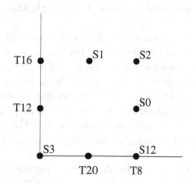

Fig. 4. Embedding of the candidate solutions into the coordination system found for Nim[4]

4 Discussion and Conclusions

We have presented an algorithm that is able to extract the *underlying objectives* of an arbitrary test-based problem. Any test-based problem has underlying objectives. Viewing each test as a separate objective yields an initial objective set. By compressing this initial set into a smaller number of objectives, meaningful underlying objectives can be obtained. The algorithm that has been presented delivers a set of objectives that are guaranteed to be minimal in number, yet which provide evaluation information that is equivalent to the initial objective set.

The optimal dimension extraction algorithm has been applied to small versions of the Game of Nim. A main finding is that by extracting the underlying objectives, the number of required objectives can be drastically reduced. In other words, the extraction algorithm substantially decreases the dimensionality of the objective space, without loss of information. For example, for a version of the game with 36 distinct tests, the initial set of 36 objectives was reduced to an equivalent objective set containing only four objectives. This demonstrates that the notion of objective set compression is not merely a theoretical possibility; in the example domain that has been explored, the number of objectives can be greatly reduced. While there is no guarantee that the rate of compression in other games will be comparable, it has been shown that a substantial reduction of the number of objectives is possible at least for some problems.

Objective compression is a new analytical tool that has several applications. First and foremost, it may serve to increase our insight into existing problems. By identifying the underlying objectives of a problem, intrinsic information about the structure of the evaluation function implicit in the problem is revealed. Our results with the Game of Nim showed that the underlying objectives can be highly interpretable; in all cases, objectives represented orthogonal dimensions of performance in the game where each subsequent position on an axis represented a better strategic choice, and where each axis corresponded to a single state that may arise during play.

A second consequence of theoretical interest is that any test-based problem is characterized by an *evaluation dimension*: the minimal number of objectives for which an objective set exists that is equivalent to the initial objective set. The evaluation dimension forms an intrinsic property of the problem. Thus, an open question for any test-based problem is what its evaluation dimension is; we expect the evaluation dimension to have implications for the complexity of evaluation and search. An open question is how the evaluation dimension or the problem structure relate to existing notions of dimensionality or complexity, such as the teaching dimension [9].

In addition to these theoretical implications, problem structure extraction may find applications in learning and search. If the relevant dimensions of performance for a problem can be identified or estimated, this may help us select which tests are to be used in evaluation. An illustration of this principle has been given in [5], which details a coevolutionary algorithm that guides selection using a coordinate system extracted from the population.

The algorithm presented here guarantees that the coordinate systems that are produced will be of minimum size, and thus have a number of objectives equal to the evaluation dimension. While this algorithm is of theoretical interest in that it can be used to identify minimal dimensional objective sets for small test problems, application to problems of practical interest will typically be computationally infeasible. Therefore, a final open question is how objective set compression may be performed efficiently when approximate solutions are acceptable. An example of an approximate dimension extraction algorithm has been given in [4], but more accurate and more efficient algorithms would form valuable extensions of the work presented here.

References

[1] Bouton, C. L. (1901–1902). Nim, a game with a complete mathematical theory. *The Annals of Mathematics, 2nd Ser.*, 3(1-4):35–39.

[2] Bucci, A. and Pollack, J. B. (2003). Focusing versus intransitivity: Geometrical aspects of coevolution. In Erick Cantú-Paz et al., editor, *Genetic and Evolutionary Computation Conference - GECCO 2003*, volume 2723 of *Lecture Notes in Computer Science*, pages 250–261. Springer.

[3] Bucci, A. and Pollack, J. B. (2005). On identifying global optima in cooperative coevolution. In Hans-Georg Beyer et al., editor, *GECCO 2005: Proceedings of the 2005 conference on Genetic and evolutionary computation*, volume 1, pages 539–544, Washington DC, USA. ACM Press.

[4] Bucci, A., Pollack, J. B., and De Jong, E. D. (2004). Automated extraction of problem structure. In Kalyanmoy Deb et al., editor, *Genetic and Evolutionary Computation Conference – GECCO 2004*, volume 3102 of *Lecture Notes in Computer Science*, pages 501–512. Springer.

[5] De Jong, E. and Bucci, A. (2006). DECA: Dimension extracting coevolutionary algorithm. In *Proceedings of the Genetic and Evolutionary Computation Conference, GECCO-06*, pages 313–320.

[6] De Jong, E. D. and Pollack, J. B. (2004). Ideal evaluation from coevolution. *Evolutionary Computation*, 12(2):159–192.

[7] Deb, K. and Saxena, D. K. (2006). Searching for Pareto-optimal solutions through dimensionality reduction for certain large-dimensional multi-objective optimization problems. In *Proceedings of the 2006 IEEE Congress on Evolutionary Computation*, pages 3353–3360. IEEE Press.

[8] Ficici, S. G. and Pollack, J. B. (2001). Pareto optimality in coevolutionary learning. In Kelemen, J. and Sosík, P., editors, *Sixth European Conference on Artificial Life (ECAL 2001)*, pages 316–325. Springer.

[9] Goldman, S. A. and Kearns, M. J. (1995). On the complexity of teaching. *Journal of Computer and System Sciences*, 50(1):20–31.

[10] Grundy, P. M. (1939). Mathematics and games. *Eureka*, 2:6–8. Reprinted in Eureka 27 (1964) 9-11.

[11] Hillis, D. W. (1990). Co-evolving Parasites Improve Simulated Evolution in an Optimization Procedure. *Physica D*, 42:228–234.

[12] Knowles, J. D., Watson, R. A., and Corne, D. W. (2001). Reducing Local Optima in Single-Objective Problems by Multi-objectivization. In Zitzler, E., Deb, K., Thiele, L., Coello, C. C., and Corne, D., editors, *First International Conference on Evolutionary Multi-Criterion Optimization*, volume 1993 of *LNCS*, pages 268–282. Springer, Berlin.

[13] Noble, J. and Watson, R. A. (2001). Pareto Coevolution: Using Performance Against Coevolved Opponents in a Game as Dimensions for Pareto Selection. In L. Spector et al., editor, *Proceedings of the Genetic and Evolutionary Computation Conference, GECCO-2001*, pages 493–500, San Francisco, CA. Morgan Kaufmann Publishers.

[14] Potter, M. A. and Jong, K. A. D. (2000). Cooperative coevolution: An architecture for evolving coadapted subcomponents. *Evolutionary Computation*, 8(1):1–29.

[15] Saxena, D. K. and Deb, K. (2007). Non-linear dimensionality reduction procedures for certain large-dimensional multi-objective optimization problems: Em-

ploying correntropy and a novel maximum variance unfolding. In Obayashi, S., Deb, K., Poloni, C., Hiroyasu, T., and Murata, T., editors, *Proceedings of the 4th International Conference on Evolutionary Multi-Criterion Optimization*, volume 4403 of *Lecture Notes in Computer Science*, pages 772–787.

[16] Singmaster, D. (1996).   Chronology   of   recreational   mathematics. http://anduin.eldar.org/ problemi/singmast/recchron.html.

[17] Sprague, R. P. (1935–1936). Über mathematische kampfspiele. *Tôhoku Mathematical Journal*, 41:438–444.

[18] Watson, R. and Pollack, J. B. (2001). Coevolutionary dynamics in a minimal substrate. In L. Spector et al., editor, *Proceedings of the Genetic and Evolutionary Computation Conference, GECCO-2001*, San Francisco, CA. Morgan Kaufmann Publishers.

On Handling a Large Number of Objectives A Posteriori and During Optimization

Dimo Brockhoff[1], Dhish Kumar Saxena[2], Kalyanmoy Deb[2], and Eckart Zitzler[1]

[1] Computer Engineering and Networks Laboratory (TIK),
ETH Zurich, 8092 Zurich, Switzerland {brockhoff,zitzler}@tik.ee.ethz.ch
[2] Kanpur Genetic Algorithms Laboratory (KanGAL)
Indian Institute of Technology Kanpur
Kanpur, PIN 208016, India {dksaxena,deb}@iitk.ac.in

Summary. Dimensionality reduction methods are used routinely in statistics, pattern recognition, data mining, and machine learning to cope with high-dimensional spaces. Also in the case of high-dimensional multiobjective optimization problems, a reduction of the objective space can be beneficial both for search and decision making. New questions arise in this context, e.g., how to select a subset of objectives while preserving most of the problem structure. In this chapter, two different approaches to the task of objective reduction are developed, one based on assessing explicit conflicts, the other based on principal component analysis (PCA). Although both methods use different principles and preserve different properties of the underlying optimization problems, they can be effectively utilized either in an *a posteriori* scenario or during search. Here, we demonstrate the usability of the conflict-based approach in a decision-making scenario after the search and show how the principal-component-based approach can be integrated into an evolutionary multicriterion optimization (EMO) procedure.

1 Introduction

The field of multiobjective evolutionary algorithms has been rapidly growing over the last decade, and there have been numerous publications dealing with two- and three-dimensional problems [10]. On the other hand, studies addressing high-dimensional problems are rare [28, 9]. The reason is that many-objective problems lead to further difficulties with respect to decision making, visualization, and computation aspects. Nevertheless, from a practical point of view, it is desirable with most applications to include as many objectives as possible, without the need to further specify preferences among the different criteria. This causes the considered problems to be practically challenging or even intractable; e.g., it has been shown experimentally that state-of-the-art algorithms, working well on low-dimensional problems, such as NSGA-II and SPEA2, have difficulties in finding well-representative sets of Pareto-optimal solutions on various test functions with many objectives [33].

Some of the additional problems occurring with many objectives are obvious; e.g., the decision-making process becomes harder when more objectives are involved. More objectives imply that more objective values per solution have to be considered, and the visualization of the objective values also becomes difficult for more than three objectives. Furthermore, the computation time needed to evaluate a single solution may increase considerably with more objectives. Even with the availability of sufficient computing resources, some search methods are practically not usable for a high number of objectives. For example, most algorithms based on the hypervolume indicator [18] have running times exponential in the number of objectives [34, 35, 3, 19]. Moreover, there is a fundamental issue involved with nondomination based optimization algorithms which is worth discussing. With more objectives the probability that any two arbitrary solutions are nondominated to each other increases, because there are more objectives in which a trade-off (one is better in one objective but worse in any other objective) can occur. While dealing with a finite-sized population-based approach, the proportion of nondominated solutions in the population increases. As most evolutionary algorithms provide more emphasis to the nondominated solutions, a large proportion of the old population gets emphasized, thereby not leaving much room for new solutions to be included in the population. This, in effect, reduces the selection pressure for the better solutions in the population and the search process slows down.

Assuming that the number of given objectives for a problem is large, the question arises whether all objectives are necessary in terms of both finding a good approximation of the Pareto front and the decision-making process. Here, we assume that the decision-making process is postponed until after the search is performed by multiobjective evolutionary algorithms. We present two different approaches to handling a large number of objectives by eliminating redundant objectives systematically. To this end, we give a brief overview of existing methods for dimensionality reduction in Section 2 together with a comparison between the two methods presented later. In Section 3, we present a conflict-based objective reduction method [4], and in Section 4 an objective reduction method based on principal component analysis [15]. The first method aims at preserving (most of) the dominance structure, whereas the second method utilizes a correlation-based notion of conflict to interpret the principal components. Although both methods use different principles and preserve different properties of the underlying optimization problems, they can be effectively utilized both in an *a posteriori* scenario and during search by reducing the number of objectives. Here, we demonstrate the usability of the conflict-based approach in a decision-making scenario after the search and show how the principal-component-based approach can be integrated into an evolutionary multicriterion optimization (EMO) procedure. Many other methodologies are possible and more effort must be spent on addressing the important issue of many-objective problems.

2 Objective Reduction Methods: Overview

There is a broad interest in dimensionality reduction methods in statistics, pattern recognition, data mining, and machine learning [27]. Since pre-processing of data is a common task in real-world problems, various dimensionality reduction techniques have been proposed and successfully used, e.g., in biology [11] and text processing [1]. The general idea of dimensionality reduction methods is to reduce large feature spaces to smaller feature spaces, where the variables under consideration are called *features*. Two distinct approaches to reduce the dimensionality of the feature space can be distinguished; they are often referred to as *feature extraction* and *feature selection*.

Given a high-dimensional data set with many "features", the task in feature extraction is to find a new feature space the data can be embedded into, and the size of which is as small as possible. In other words, feature extraction tries to extract a set of (arbitrary) features to explain the data. The emerging features are often new and defined as combinations of the original ones. Methods for this task of feature extraction are, e.g., principal component analysis (PCA) [25] and independent component analysis [24]. In contrast to the feature extraction approach, the task in feature selection is to find the smallest subset of the *given* features, representing the given data best. The task of finding a smallest subset of features is, in general, $\mathcal{NP}$-hard when formalized as an optimization problem [8]. Therefore, an exhaustive search is necessary to solve some instances of feature selection problems optimally. In practice, various methods based on greedy heuristics as well as evolutionary algorithms have been proposed and applied to feature selection problems [26, 12, 32].

The further sections will present two dimensionality reduction approaches, especially developed for the case of multiobjective optimization problems. The features are here the objectives, and we try to find a selection of objectives describing the original problem. Since new objectives—potentially defined as combinations of the given ones—are not easy to handle in the decision making, we focus on finding subsets of the given objectives, best (re-)formulating the original problem, i.e., feature selection approaches. In contrast, a feature extraction method for test-based multiobjective problems is presented in this volume, pp. 357–376. The approach, presented in the next section (Section 3), is based on a definition of conflicting objectives (also presented in Section 3) and is a pure feature selection method, whereas the approach presented in Section 4 employs a feature extraction method, namely PCA, to arrive at a subset of the given objectives.

The two approaches presented pursue different goals. On the one hand, the conflict-based approach aims at finding minimum objective sets yielding the same (or a slightly changed) dominance structure as the original problem induces. This procedure is applicable as an *a posteriori* approach to assist in the decision making after a multiobjective optimization procedure. The integration of this approach into the optimization itself will be addressed in future work. On the other hand, the PCA-based approach tries to detect the objectives which span the Pareto-optimal front, assuming that the considered problems possess a low-dimensional Pareto-optimal front although many objectives are contained in the original problem. Thus, this approach can be used during an optimization process; and, starting with all given objectives, the procedure systematically eliminates redundant objectives and finally determines the trade-off frontier corresponding to the required number of objectives.

Before we present the two objective reduction approaches in detail, we briefly address the general problem of scaling objective values. Especially in real world problems, one often has to consider objectives which are not measured on comparable length scales. Those disproportionately scaled objectives have different influences on both the objective reduction methods presented in this chapter. On the one hand, the δ error in the conflict-based approach directly depends on the objective scaling, i.e., objectives with large values, in principle, cause larger errors than objectives with small values. On the other hand, in the PCA-based approach, disproportionately

scaled objectives may lead to large differences in their variances, eventually bias-
ing the eigenvalues and eigenvectors (in favour of objectives with larger variances).
This would render the entire analysis erroneous. Since both objective reduction approaches highly depend on the objective values, we recommend scaling the objectives
to comparable values whenever it is possible. For instance, in the second approach,
we recommend using the correlation matrix instead of the covariance matrix, as
described in Section 4.3. Further research will be necessary to study the influences
of objective scaling on the proposed approaches in detail and to develop objective
reduction methods which are as insensitive to a scaling of the objectives as possible.

3 A Conflict-Based Objective Reduction Method

To present a recently proposed conflict-based objective reduction method [4], we
start with the introduction of the underlying objective conflict definition on the
basis of simple examples (Section 3.1). Section 3.2 continues with the definition
of minimum objective sets and the presentation of the two problems δ-MOSS and
k-EMOSS. Afterwards, we present both exact and greedy algorithms for the two problems and show the usability of the approach for high-dimensional test problems in
Section 3.3.

3.1 Introductory Examples and Objective Conflicts

Example 1. Assume we have a minimization problem with three solutions in the
decision space X that are pairwisely incomparable. Solution $\mathbf{x}_1 \in X$ is better than
the other two with respect to objective f_1; with regard to objective f_2, solution
$\mathbf{x}_2 \in X$ has the best value; and solution $\mathbf{x}_3 \in X$ has the best f_3-value. Figure 1
shows the value path plot[3] of the three solutions $\mathbf{x}_1$, $\mathbf{x}_2$, and $\mathbf{x}_3$. Assuming that
$\mathbf{x}_1, \mathbf{x}_2$, and $\mathbf{x}_3$ represent either the entire search space or the Pareto-optimal set,
the original objective set $\mathcal{F} := \{f_1, f_2, f_3\}$ is conflicting according to [13] as there
is no single optimal solution but three Pareto-optimal ones. For the same reason of
incomparable solution pairs, the objective set is also conflicting according to [31].
Last, every possible objective pair f_i, f_j with $i, j \in \{1, 2, 3\}, i \neq j$, "exhibits evidence
of conflict" as defined in [29].

Although the three conflict definitions in [13, 31, 29] mislead to the assumption
that all objectives are necessary to represent the dominance relation, the objective
set $\{f_1, f_2, f_3\}$ contains redundant information as defined by Gal and Leberling [21]:
the objective f_2 can be omitted, and all solutions remain incomparable to each other
with regard to the objective set $\{f_1, f_3\}$, i.e., the dominance relation on the search
space stays unaffected.

A conflict definition, based on the dominance relation between solutions and
indicating that objectives are redundant, i.e., their omission does not affect the
dominance relation, was recently proposed in [4] and uses a generalization of the
weak Pareto dominance relation, induced by a given (sub)set $\mathcal{F}$ of objectives:

$$\preceq_{\mathcal{F}} := \{(\mathbf{x}, \mathbf{y}) \mid \mathbf{x}, \mathbf{y} \in X \wedge \forall f_i \in \mathcal{F} : f_i(\mathbf{x}) \leq f_i(\mathbf{y})\}$$

---

[3] As used in [29].

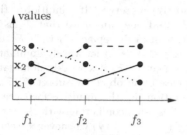

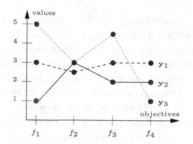

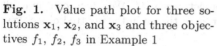

Fig. 1. Value path plot for three solutions x_1, x_2, and x_3 and three objectives f_1, f_2, f_3 in Example 1

Fig. 2. Value path plot for three solutions y_1, y_2, and y_3 and four objectives f_1, f_2, f_3, f_4 in Example 2

Definition 1 *Let $\mathcal{F}_1, \mathcal{F}_2 \subseteq \mathcal{F}$ be two sets of objectives. We call $\mathcal{F}_1$ conflicting with $\mathcal{F}_2$ iff $\preceq_{\mathcal{F}_1} \neq \preceq_{\mathcal{F}_2}$.*

If a subset $\mathcal{F}' \subseteq \mathcal{F}$ of the given objective set $\mathcal{F}$ is not conflicting with $\mathcal{F}$, we can omit all objectives in $\mathcal{F} \setminus \mathcal{F}'$ without changing the underlying dominance structure of the problem; the objectives in $\mathcal{F} \setminus \mathcal{F}'$ are redundant. In practice, one is often interested in a further objective reduction at the cost of slight changes in the dominance structure. This poses the question of how such a structural change can be quantitatively measured.

Example 2. Consider a set of solutions y_1, y_2, and y_3 with four objectives, $f_i \in \mathcal{F}$, $1 \leq i \leq 4$ as in Fig. 2. Let $\mathcal{F}' := \{f_3, f_4\}$. We observe that by reducing the set of objectives to $\mathcal{F}'$, the dominances change: on the one hand $x_1 \preceq_{\mathcal{F}'} x_2$; on the other hand $x_1 \not\preceq_{\mathcal{F}} x_2$. In this sense, we make an error: the objective values of x_1 had to be smaller by an additive term of $\delta = 0.5$ such that $x_1 \preceq_{\mathcal{F}} x_2$ would actually hold. This δ value can be used as a measure to quantify the difference in the dominance structure induced by $\mathcal{F}'$ and $\mathcal{F}$. By computing the δ values for all solution pairs x, y, we can determine the maximum error. The meaning of the maximum δ value is that whenever we wrongly assume that $x \preceq_{\mathcal{F}'} y$, we also know that x is not worse than y in all objectives by an additive term of δ. For $\mathcal{F}' := \{f_3, f_4\}$, the maximum error is $\delta = 0.5$; for $\mathcal{F}' := \{f_2, f_4\}$, the maximum δ is 4.

In the following, we formalize the definition of error in [4], according to the above example. The background for that is provided by the (additive) ε-dominance relation[4] [37] and a generalization of the notion of conflict in Definition 1.

Definition 2 *Let $\mathcal{F}_1$ and $\mathcal{F}_2$ be two objective sets. We call*

- *$\mathcal{F}_1$ δ-nonconflicting with $\mathcal{F}_2$ iff $\left(\preceq_{\mathcal{F}_1} \subseteq \preceq_{\mathcal{F}_2}^{\delta} \right) \wedge \left(\preceq_{\mathcal{F}_2} \subseteq \preceq_{\mathcal{F}_1}^{\delta} \right)$;*
- *$\mathcal{F}_1$ δ-conflicting with $\mathcal{F}_2$ iff $\neg \left(\mathcal{F}_1 \ \delta\text{-nonconflicting with } \mathcal{F}_2 \right)$.*

The above definition of δ-conflict contains Definition 1 for the case $\delta = 0$. The absence of δ-conflict between objective sets can be useful, in practice, if a problem formulation needs to be changed by considering a different objective set but

---

[4] $\preceq_{\mathcal{F}'}^{\varepsilon} := \{ (x, y) \mid x, y \in X \wedge \forall i \in \mathcal{F}' \subseteq \mathcal{F} : f_i(x) - \varepsilon \leq f_i(y) \}$

the underlying problem structure has to be (mostly) preserved. If a multiobjective optimization problem uses the objective set $\mathcal{F}_1$ and one can prove that $\mathcal{F}_1$ is δ-nonconflicting with another objective set $\mathcal{F}_2$, one can easily replace $\mathcal{F}_1$ with $\mathcal{F}_2$ and can be sure that in the new formulation, for any $\mathbf{x}, \mathbf{y} \in X$, $\mathbf{x}$ either weakly dominates $\mathbf{y}$ with respect to $\mathcal{F}_2$ or $\mathbf{x}$ ε-dominates $\mathbf{y}$ with respect to $\mathcal{F}_2$ if $\mathbf{x}$ weakly dominates $\mathbf{y}$ with respect to $\mathcal{F}_1$ and $\varepsilon = \delta$. In the special case of an objective subset $\mathcal{F}' \subseteq \mathcal{F}$, δ-nonconflicting with all objectives $\mathcal{F}$, the definition fits the intuitive measure of error in Example 2. If an objective subset $\mathcal{F}' \subset \mathcal{F}$ is δ-nonconflicting with the set $\mathcal{F}$ of all objectives, $\mathbf{x}$ δ-dominates $\mathbf{y}$, i.e., $\forall i \in \mathcal{F} : f_i(\mathbf{x}) - \delta \leq f_i(\mathbf{y})$ whenever $\mathbf{x}$ weakly dominates $\mathbf{y}$ with respect to the reduced objective set $\mathcal{F}'$. We, then, can omit all objectives in $\mathcal{F} \setminus \mathcal{F}'$ without making a larger error than δ in the omitted objectives.

3.2 Minimum Objective Sets

Based on the above conflict definitions, we will now formalize the notion of δ-minimal and δ-minimum objective sets as in [4] and present a condition under which an objective reduction is possible. Furthermore, we present the two problems δ-MOSS and k-EMOSS, already presented in [4].

Definition 3 *Let $\mathcal{F}$ be a set of objectives and $\delta \in \mathbb{R}$. An objective set $\mathcal{F}' \subseteq \mathcal{F}$ is denoted as*

- *δ-minimal with respect to $\mathcal{F}$ iff (i) $\mathcal{F}'$ is δ-nonconflicting with $\mathcal{F}$, (ii) $\mathcal{F}'$ is δ'-conflicting with $\mathcal{F}$ for all $\delta' < \delta$, and (iii) there exists no $\mathcal{F}'' \subset \mathcal{F}'$ that is δ-nonconflicting with $\mathcal{F}$;*
- *δ-minimum with respect to $\mathcal{F}$ iff (i) $\mathcal{F}'$ is δ-minimal with respect to $\mathcal{F}$, and (ii) there exists no $\mathcal{F}'' \subset \mathcal{F}$ with $|\mathcal{F}''| < |\mathcal{F}'|$ that is δ-minimal with respect to $\mathcal{F}$.*

A δ-minimal objective set is a subset of the original objectives that cannot be further reduced without changing the associated dominance structure with an error of at most δ. A δ-minimum objective set is the smallest possible set of original objectives that preserves the original dominance structure except for an error of δ. By definition, every δ-minimum objective set is δ-minimal, but not all δ-minimal sets are at the same time δ-minimum.

Definition 4 *A set $\mathcal{F}$ of objectives is called δ-redundant if and only if there exists $\mathcal{F}' \subset \mathcal{F}$ that is δ-minimal with respect to $\mathcal{F}$.*

This definition of δ-redundancy represents a necessary and sufficient condition for the omission of objectives while the obtained dominance relation preserves most of the initial dominance relation according to the definition of error in Example 2.

Both the δ-MOSS and the k-EMOSS problem from [4] formalize the question of finding objective subsets $\mathcal{F}'$ for given solution sets $A \subseteq X$ to detect whether the objective values of the given solutions include redundancy. The δ-MOSS problem asks for a δ-minimum objective set with respect to the given objective set for a given error δ. The k-EMOSS problem asks for an objective set of size k with the smallest error δ according to the entire objective set. In the following, we denote the number of objectives as $M := |\mathcal{F}|$.

Definition 5 *Given a multiobjective optimization problem, the problem δ-MINIMUM OBJECTIVE SUBSET (δ-MOSS) is defined as follows.*

Instance: The objective vectors $f(\mathbf{x}_1), \ldots, f(\mathbf{x}_m) \in \mathbb{R}^M$ of the solutions $\mathbf{x}_1, \ldots, \mathbf{x}_m \in A \subseteq X$ and a $\delta \in \mathbb{R}$.

Task: Compute a δ-minimum objective subset $\mathcal{F}' \subseteq \mathcal{F}$ with respect to $\mathcal{F}$.

Definition 6 *Given a multiobjective optimization problem, the problem MINIMUM OBJECTIVE SUBSET OF SIZE k WITH MINIMUM ERROR (k-EMOSS) is defined as follows.*

Instance: The objective vectors $f(\mathbf{x}_1), \ldots, f(\mathbf{x}_m) \in \mathbb{R}^M$ of the solutions $\mathbf{x}_1, \ldots, \mathbf{x}_m \in A \subseteq X$ and a $k \in \mathbb{N}$.

Task: Compute an objective subset $\mathcal{F}' \subseteq \mathcal{F}$ which has size $|\mathcal{F}'| \leq k$ and is δ-nonconflicting with $\mathcal{F}$ with the minimal possible δ.

In the unlikely case that we know the objective vectors of all solutions within our search space, we can compute the exact number of non-δ-redundant objectives within a given problem with algorithms for δ-MOSS. Otherwise, if we know the objective vectors only for a small sample of the solution space, algorithms for δ-MOSS and k-EMOSS can be used to simplify the decision-making process after the search by computing objective subsets with certain properties. We can, for example, use only the generated nondominated solutions as inputs for the algorithms to compute a subset of objectives describing the dominance structure between the search points in our Pareto front approximation. If we take all solutions, generated during the optimization into account, the outcome of the algorithms tells the decision maker which objectives are necessary to describe the dominance structure between the solutions in the sampled section of the search space.

3.3 Algorithms and Results

After the definition of an objective subset's error regarding the dominance structure and the presentation of the two problems δ-MOSS and k-EMOSS, we present both exact and greedy algorithms for them, while dealing with two questions: (i) what is the exact intrinsic number of objectives in a problem and (ii) how can it be approximated if only samples of the search space are considered?

Regarding question (i), we present an exact algorithm to examine the minimum number of non-redundant objectives for a given problem. Regarding question (ii), since the entire search space is usually unknown and the problem of computing a minimum objective set itself is $\mathcal{NP}$-hard, we present greedy algorithms which can cope with samples of the search space of appropriate size and provide an example of how to use their outputs to (ii) gain information about the intrinsic number of objectives in a problem even if the entire search space is unknown.

An Exact Algorithm

To solve both the δ-MOSS and the k-EMOSS problem exactly, we present here Algorithm 5, already proposed in [4]. Since the above problems are known to be $\mathcal{NP}$-hard

---

Algorithm 5 An exact algorithm for δ-MOSS and k-EMOSS

1: Init:
2:　　$C := \emptyset, \quad S_C := \emptyset$
3: **for all** pairs $\mathbf{x}, \mathbf{y} \in A$, $\mathbf{x} \neq \mathbf{y}$ of solutions **do**
4:　　　$S_{\{(\mathbf{x},\mathbf{y})\}} := \emptyset$
5:　　　**for all** objective pairs $i, j \in \mathcal{F}$, where not necessary $i \neq j$ **do**
6:　　　　compute $\delta_{ij} := \delta_{\min}(\{i\} \cup \{j\}, \mathcal{F})$ with respect to $\mathbf{x}, \mathbf{y}$
7:　　　　$S_{\{(\mathbf{x},\mathbf{y})\}} := S_{\{(\mathbf{x},\mathbf{y})\}} \sqcup (\{i\} \cup \{j\}, \delta_{ij})$
8:　　　**end for**
9:　　　$S_{C \cup \{(\mathbf{x},\mathbf{y})\}} := S_C \sqcup S_{\{(\mathbf{x},\mathbf{y})\}}$
10:　　　$C := C \cup \{(\mathbf{x}, \mathbf{y})\}$
11: **end for**
12: Output for δ-MOSS: $(s_{\min}, \delta_{\min})$ in S_C with minimal size $|s_{\min}|$ and $\delta_{min} \leq \delta$
13: Output for k-EMOSS:(s, δ) in S_C with size $|s| \leq$ k and minimal δ

---

[5], the exact algorithm has a running time exponentially in the number of objectives. Thus, it can solve only small problem instances in reasonable time and is therefore more of theoretical interest although it has some advantages; cf. the next section. The basic idea is to consider all solution pairs $(\mathbf{x}, \mathbf{y})$ successively and store in the set S_C all minimal objective subsets $\mathcal{F}'$ together with the minimal δ' value such that $\mathcal{F}'$ is δ'-nonconflicting with the set $\mathcal{F}$ of all objectives when taking into account only the solution pairs in C considered so far.

The subfunction $\delta_{\min}(\mathcal{F}_1, \mathcal{F}_2)$ computes the minimal δ error for two solutions $\mathbf{x}, \mathbf{y} \in X$ such that $\mathcal{F}_1$ is δ-nonconflicting with $\mathcal{F}_2$ with respect to $\mathbf{x}, \mathbf{y}$. Due to space limitations, we cannot show here how this minimal δ can be computed in time $O(M \cdot m^2)$ and refer the reader to [5]. Furthermore, Algorithm 5 computes the union $\sqcup$ of two sets of objective subsets with simultaneous deletion of non-δ'-minimal pairs $(\mathcal{F}', \delta')$:

$$S_1 \sqcup S_2 := \{(\mathcal{F}_1 \cup \mathcal{F}_2, \max\{\delta_1, \delta_2\}) \mid (\mathcal{F}_1, \delta_1) \in S_1 \wedge (\mathcal{F}_2, \delta_2) \in S_2$$
$$\wedge \; \nexists(\mathcal{F}_1', \delta_1') \in S_1, (\mathcal{F}_2', \delta_2') \in S_2 \colon (\mathcal{F}_1' \cup \mathcal{F}_2' \subset \mathcal{F}_1 \cup \mathcal{F}_2 \wedge \max\{\delta_1', \delta_2'\} \leq \max\{\delta_1, \delta_2\})$$
$$\wedge \; \nexists(\mathcal{F}_1', \delta_1') \in S_1, (\mathcal{F}_2', \delta_2') \in S_2 \colon (\mathcal{F}_1' \cup \mathcal{F}_2' \subseteq \mathcal{F}_1 \cup \mathcal{F}_2 \wedge \max\{\delta_1', \delta_2'\} < \max\{\delta_1, \delta_2\})\}$$

The correctness proof of Algorithm 5—as well as the proof of its running time of $O(m^2 \cdot M \cdot 2^M)$—can also be found in [5]. Note, that the exact algorithm can be easily parallelized, as the computation of the sets $S_{\{(\mathbf{x},\mathbf{y})\}}$ are independent for different pairs $(\mathbf{x}, \mathbf{y})$. It also can be accelerated if line 9 of Algorithm 5 is tailored to either the δ-MOSS or the k-EMOSS problem by including a pair $(\mathcal{F}', \delta')$ into $S_{C \cup \{(\mathbf{x},\mathbf{y})\}}$ only if $\delta' \leq \delta$ and $|\mathcal{F}'| \leq$ k, respectively.

Greedy Strategies

To solve the δ-MOSS and k-EMOSS problems for the entire search space exactly is unrealistic due to both the size of the search space and the running time of the

---

Algorithm 6 A greedy algorithm for δ-MOSS

1: Init:
2:     compute the relations $\preceq_i$ for all $1 \le i \le M$ and $\preceq_{\mathcal{F}}$
3:     $\mathcal{F}' := \emptyset$
4:     $R := A \times A \backslash \preceq_{\mathcal{F}}$
5: **while** $R \ne \emptyset$ **do**
6:     $i^* = \underset{i \in \mathcal{F} \backslash \mathcal{F}'}{\operatorname{argmin}} \{ |(R \cap \preceq_i) \backslash \left(\preceq^0_{\mathcal{F}' \cup \{i\}} \cap \preceq^{\delta}_{\mathcal{F} \backslash (\mathcal{F}' \cup \{i\})} \right)| \}$
7:     $R := (R \cap \preceq_{i^*}) \backslash \left(\preceq^0_{\mathcal{F}' \cup \{i^*\}} \cap \preceq^{\delta}_{\mathcal{F} \backslash (\mathcal{F}' \cup \{i^*\})} \right)$
8:     $\mathcal{F}' := \mathcal{F}' \cup \{i^*\}$
9: **end while**

Algorithm 7 A greedy algorithm for k-EMOSS

1: Init:
2:     $\mathcal{F}' := \emptyset$
3: **while** $|\mathcal{F}'| < $ k **do**
4:     $\mathcal{F}' := \mathcal{F}' \cup \underset{i \in \mathcal{F} \backslash \mathcal{F}'}{\operatorname{argmin}} \{\delta_{\min} (\mathcal{F}' \cup \{i\}, \mathcal{F}) \text{ with respect to } A\}$
5: **end while**

---

presented exact algorithm. Two ways to bypass this are (i) to reduce the input size and (ii) to develop approximation algorithms with lower runtimes. We will first present simple approximation algorithms for the two problems and afterwards discuss what information can be gained with the presented algorithms to assist in the decision-making step.

Given a set $A \subseteq X$ of solutions and a $\delta > 0$, Algorithm 6, as an approximation algorithm for δ-MOSS, computes an objective subset $\mathcal{F}'$, δ-nonconflicting with the set $\mathcal{F}$ of all objectives in a greedy way. Starting with an empty set $\mathcal{F}'$, Algorithm 6 chooses in each step the objective f_i which yields the smallest set $\preceq_{\mathcal{F}'} \cap \preceq_i$ without considering the relationships in $\preceq^0_{\mathcal{F}' \cup \{i\}} \cap \preceq^{\delta}_{\mathcal{F}}$ until $\mathcal{F}'$ is δ-nonconflicting with $\mathcal{F}$. For the correctness proof of Algorithm 6 and the proof of its running time of $O(\min\{M^3 \cdot m^2, M^2 \cdot m^4\})$ we once again refer the reader to [5]. Note that Algorithm 6 does not necessarily yield a δ-minimum or even a δ-minimal objective set with respect to $\mathcal{F}$.

Algorithm 7 is an approximation algorithm for k-EMOSS. It supplies always an objective subset of size k but does not guarantee our finding find the set with minimal δ. The greedy algorithm needs time $O(m^2 \cdot M^3)$ since at most k $\le M$ loops with M calls of the $\delta_{\min}$ subfunction are needed. One call of the $\delta_{\min}$ function needs time $\Theta(m^2 \cdot M)$ (cf. [5]), and all other operations need time $O(1)$ each. Note that Algorithm 7 can be accelerated in a concrete implementation as the while loop can be aborted if either $|\mathcal{F}'| = $ k or $\delta_{\min}(\mathcal{F}', \mathcal{F}) = 0$.

Results

With the presented greedy algorithms, we can now approximate the optimal solutions for δ-MOSS and k-EMOSS for larger instances than can be solved with the exact

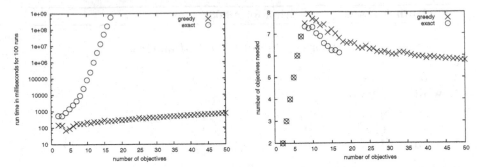

Fig. 3. Comparison between the greedy and the exact algorithm on 0-MOSS for a problem with 32 solutions and random objectives on a linux computer (SunFireV60X with 3060 MHz) over 100 runs: summed running times (left); averaged sizes of the computed minimum / minimal sets (right). See [6] for details

algorithm; cf. Fig. 3. Nevertheless, computing a minimum objective set for the entire search space is still unrealistic. The question remains about how the algorithms presented can be used to assist in the decision making. That the algorithms are useful in reducing the vast amount of information considered in the decision-making step is shown in Fig. 4. The first plot in Fig. 4(a) shows a value path plot of a Pareto front approximation for the 0-1-knapsack problem with 20 objectives and 50 solutions, computed by the indicator-based evolutionary algorithm IBEA [36]. A decision maker who has to choose one out of the 50 solutions would have difficulties when dealing with all 1,000 objective values. A run of the greedy Algorithm 6 with $\delta = 0$ indicates that nine objectives are sufficient to describe the dominance relation between the given 50 solutions (Fig. 4(b)). When using the exact Algorithm 5 we can find an even smaller objective set, yielding the same dominance relation (Fig. 4(c)). This indicates for the decision maker that the 350 objective values, i.e., only seven out of the 20 objectives, contain the same information on the dominance relation between the 50 solutions as the 1,000 values of Fig. 4(a). If the decision maker wants to further reduce the number of objectives, Algorithm 5 can be applied to the k-EMOSS problem with $k = 6, 5, \ldots, 1$. This demonstrates that for the given example a reduction to six objectives causes an error of $\delta = 129$. Further reduction cause errors of 289 ($k = 5$), 363 ($k = 4$), 447 ($k = 3$), 485 ($k = 2$), and 552 ($k = 1$).

If not only the decision about which of the various nondominated solutions to take is important but also the increase of knowledge on the problem itself, a different approach might be useful which has been discussed recently in detail in [7]. We exemplify this approach by applying it to a radar waveform optimization problem. In a recently proposed study of designing waveforms for a pulsed Doppler radar [23], the author describes nine objectives, characterizing different waveforms in terms of median/minimum range and velocity extent of detectable targets (objectives 1–8) and the required time to transmit the total waveform (objective 9). According to the author, the objectives are not totally independent, e.g., objectives 1 and 5 should not conflict and objective pairs 1/3, 5/7, 2/4, and 6/7 "tend to have a degree of correlation, i.e., they may not conflict strongly" ([23], page 709). The definition of conflict, proposed in the previous section, allows us to test the mentioned prediction of conflicts between the objectives analytically. To this end, after normalizing the

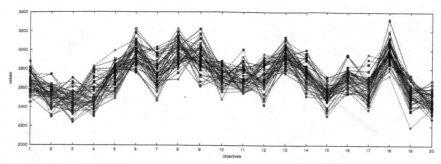

(a) Original problem formulation with 20 objectives

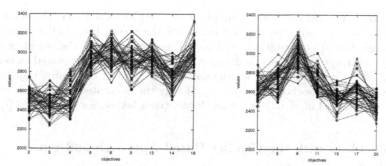

(b) The same solutions reduced to the objective subset computed by the greedy algorithm

(c) The solutions from (a) reduced to the objective subset computed by the exact algorithm

Fig. 4. Visualization of the capability of the presented algorithms for 0-MOSS on the 0-1-knapsack problem with 20 objectives. The plots show the objective values for the 50 solutions computed by an IBEA run. Figure (a) shows the values for the complete set of 20 objectives. The other figures show the objective subsets, 0-nonconflicting with the entire objective set, computed by the greedy algorithm (b), and the exact algorithm (c)

objective vectors, the set of 22,844 known nondominated solutions is reduced to a significantly smaller set of 107 solutions by computing the ε-nondominated set with $\varepsilon = 0.062^5$. Afterwards, the δ errors between all pairs of single objectives are computed. According to Definition 1, all objective pairs conflict with each other.

Surprisingly, the smallest error occurs between objectives 4 and 9, and the second smallest between objective pair 1/7, in contrast to the prediction of [23]; the left part of Fig. 5 shows the computed δ errors graphically. Using the computed δ errors between the objectives, a tree-like visualization of the objectives can also be derived to gain information on the problem (Fig. 5, right part); cf. [7] for details.

---

[5] A reduction was necessary due to the large running times of the algorithms on $\gg 20,000$ points. The reduction to 107 points, however, was arbitrary. Note, that larger sets of up to 5,000 points yield similar results.

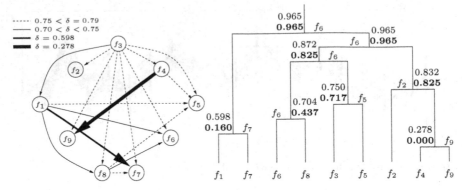

Fig. 5. Analysis of the radar waveform problem: (left) Visualization of the minimum delta error between objective pairs; the thicker the arrow, the smaller the error. Errors larger than 0.8 are omitted for clarity. The arrows point to the objective with the larger δ error; (right) tree visualization of the δ errors. The annotated values on the inner nodes correspond to the minimal δ error between the paired objectives, and the overall δ error (bold) respectively, if only the annotated objective is taken into account instead of all objectives on the subtree's leaves. For details, see [7].

3.4 Difficulties with the Conflict-Based Approach and Future Directions

Due to the fact that finding a minimum objective subset is an $\mathcal{NP}$-hard problem, the running time of the exact algorithm is not practical. Although the running times of the greedy heuristics are much smaller, their running times are also a problem when many objectives ($\gg 50$) and reasonable solution sets (≥ 200) are taken into account. While the relation-based approach was introduced as an *a posteriori* method here, a first preliminary attempt to use the approach within an EMO procedure was recently done in [7], but the integration of the objective reduction method into an evolutionary algorithm still remains future work.

4 A PCA-Based Objective Reduction Method

4.1 Difficulties of EMO Methodologies in Handling Many Objectives

Before we discuss the objective reduction procedure used during an EMO simulation run, we want to draw the reader's attention towards Figure 6, which highlights quantitatively the number of EMO publications by the number of objectives [10], until 2001. The overwhelming majority have used only two objective functions, most probably for the ease of their solution principles. Some have used three to nine objectives, and only a few have tried beyond ten objectives. The studies using more than ten objectives have mostly employed a single-objective optimization method, by converting the multiobjective optimization problem into a single objective. On the whole, such a sparse and fragmentary (more details are cited in [15]) research

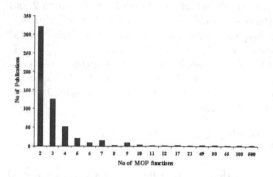

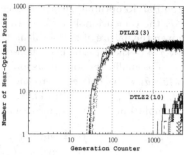

Fig. 6. EMO publications by number of objectives (until 2001), taken from [10]

Fig. 7. Illustration of difficulty in 10-objective MOP with NSGA-II.

effort also reflects the difficulties involved with the EMO approach towards many-objective problems.

While the difficulties associated with a large number of objectives were highlighted in Section 1 (more details may be found in [15]), it would be interesting for readers to witness a concrete example — the vulnerability of EMO methodology in general (and of the elitist nondominated sorting GA or NSGA-II [14] in particular) to large dimensional Multiobjective Optimization Problems (MOPs). We defer exposing this vulnerability to first explain the test problems which we have designed and utilized for all demonstrations in the remainder of the chapter. These test problems, to be to referred as DTLZ5(I, M), are built upon the well-known M-objective DTLZ5 problems [16], the changes introduced being in terms of the parameter θ_i, and the additional constraints [15] to help maintain the claimed scalability of the latter. DTLZ5(I, M), where I denotes the dimensionality of the Pareto-optimal surface (in terms of the number of objectives) and M denotes the total number of objectives in the problem, is as follows:

$$
\left.
\begin{aligned}
&\text{Min} \quad f_1(\mathbf{x}) = (1 + g(\mathbf{x}_M)) \cos(\theta_1) \cos(\theta_2) \cdots \cos(\theta_{M-2}) \cos(\theta_{M-1}), \\
&\text{Min} \quad f_2(\mathbf{x}) = (1 + g(\mathbf{x}_M)) \cos(\theta_1) \cos(\theta_2) \cdots \cos(\theta_{M-2}) \sin(\theta_{M-1}), \\
&\text{Min} \quad f_3(\mathbf{x}) = (1 + g(\mathbf{x}_M)) \cos(\theta_1) \cos(\theta_2) \cdots \sin(\theta_{M-2}), \\
&\quad\vdots \qquad \vdots \\
&\text{Min} \quad f_{M-1}(\mathbf{x}) = (1 + g(\mathbf{x}_M)) \cos(\theta_1) \sin(\theta_2), \\
&\text{Min} \quad f_M(\mathbf{x}) = (1 + g(\mathbf{x}_M)) \sin(\theta_1), \\
&\text{where } g(\mathbf{x}_M) = \sum_{x_i \in \mathbf{x}_M} (x_i - 0.5)^2, \\
&\qquad \theta_i = \begin{cases} \frac{\pi}{2} x_i, & \text{for } i = 1, \dots, (I-1), \\ \frac{\pi}{4(1+g(\mathbf{x}_M))} (1 + 2g(\mathbf{x}_M) x_i), & \text{for } i = I, \dots, (M-1), \end{cases} \\
&\qquad 0 \le x_i \le 1, \quad \text{for } i = 1, 2, \dots, n.
\end{aligned}
\right\} \quad (1)
$$

The total number of variables are $n = M + k - 1$, where $k = |\mathbf{x}_M| = 10$ is prescribed. The problem is so designed that the Pareto-optimal surface will correspond to (i) a zero value of the g function, in turn implying $x_i = 0.5$ for $i = M, \dots, (M+k-1)$, (ii) a fixed value of $\pi/4$ for the variables x_1 to x_{M-1}, and (iii) independent values for the variables x_1 to x_{I-1}, by virtue of which they are responsible for the dimensionality of the Pareto-optimal surface. Hence, by simply setting I to an integer between 2 and M, the dimensionality (I) of the Pareto-optimal surface can be changed. Further, given the nature of the problem, the Pareto-optimal surface (solutions being denoted by f^*) to these problems is non-convex and follows the relationship, $\sum_{i=1}^{M}(f_i^*)^2 = 1$.

For an illustration, let us look at the $I = 2$ case. For the Pareto-optimal solutions, while θ_1 varies between zero and $\pi/2$, $\theta_i = \pi/4$ for all $i = 2, \ldots, M-1$. Also, $x_i = 0.5$ for $i = M, \ldots, (M + 9)$. Thus, the Pareto-optimal front is one dimensional (w.r.t. variables) and two dimensional (w.r.t. objectives):

$$\left. \begin{array}{l} f_M = \sin(\theta_1), f_j = \cos(\theta_1)/(\sqrt{2})^{M-2}; \text{ for } j = 1 \\ f_j = \cos(\theta_1)/(\sqrt{2})^{M-j}; \forall j = 2, \ldots, M - 1 \end{array} \right\} \tag{2}$$

However, it is observed that for $M > 3$ there exist other solutions having $g > 0$ which are also Pareto-optimal solutions and which violate the dimensionality claim made above. To remedy this problem (details may be found in [15]), we introduce $M - 1$ constraints to the original DTLZ5(2,M) problem, in the form

$$\left. \begin{array}{l} f_M^2 + 2^{M-2}f_j^2 \geq 1, \text{for } j = 1 \\ f_M^2 + 2^{M-j}f_j^2 \geq 1, \text{for all } j = 2, \ldots, M - 1. \end{array} \right\} \tag{3}$$

Let us now turn to Figure 7, where a comparison is drawn between DTLZ5(10,10) (or DTLZ2(10)) and DTLZ5(3,3) (or DTLZ2(3)) in terms of percentage of solutions converging to the Pareto-optimal surface. Here, results corresponding to five runs (5,000 generations in each run) for each problem, with different initial populations of size 200, are plotted. The SBX crossover [13] with a probability of 0.9 and index of 5 and polynomial mutation [13] with a probability of $1/n$ and index of 50 are used. It can be seen that while for the 3-objective version of the problem a significantly high percentage of solutions converge (having $g(\mathbf{x}_M) \leq 0.01$) to the front, for the 10-objective version, approximately only 4% solutions come to the Pareto-optimal front. Although these results are specific to the test problems considered here, they reflect the general underlying difficulty with a large number of objectives — in which case the proportion of nondominated solutions in the population increases (as discussed in Section 1, [15]) to eventually stagnate the search process. With this understanding, the results here can be treated as representing the general characteristic of large-dimensional problems where only a small percentage of solutions converge to the Pareto-optimal surface. This feature can well be termed the "curse of dimensionality" in evolutionary multiobjective optimization.

4.2 Redundancy Among Objectives: Existence, Form, and Past Research

Simulation results in the previous section indicated the vulnerability of EMO methodologies to those large-dimensional M-objective MOPs that have an M-dimensional Pareto-optimal front.

However, there may exist many-objective problems in practice that have *redundant* objectives; that is, although the problem has M objectives, the Pareto-optimal front involves a much lower-dimensional interaction. Such a reduction in the dimensionality may happen, either gradually from the region of random solutions towards the Pareto-optimal region, or, the entire search space may have such a structure. The former type of problem is quite likely to exist in practice. In this case, the nonconflicting nature of objectives does not exist for all solutions in the search space. For randomly picked solutions, all the objectives may be conflicting and there may be an M-dimensional interaction; however, for solutions close to the Pareto-optimal front (special solutions), the optimality conditions of some objectives may be similar and behave in a nonconflicting manner. On the other hand, the latter case implies that there exist some objectives in the problem formulation which are nonconflicting to each other. Often, such information about the nature of variation of objective values is not intuitive to a designer/decision maker. The DTLZ5(I,M) problem constructed earlier offers a way to simulate this second scenario with the flexibility of controlling the reduction in dimension.

Given the fact that not many studies have been performed in the total realm of high-dimensional multiobjective problems, it is only natural, not to expect many studies on the determination of redundancy in many-objective problems. Those catalogued are largely suggestive in nature in terms of proposed definitions of redundancy and the possible ways to address it. Some of these studies are by Gal and Leberling [21], Gal and Hanne [20], and Agrell [2]. Another recent study [22] used a correlation matrix to identify the redundant objectives, if any, in a four-objective optimization problem.

Here, to target determination of redundant objectives, we propose a principal component analysis (PCA) procedure coupled with the NSGA-II method. The proposed combined procedure works iteratively from the interior of the search space and iteratively moves towards the Pareto-optimal region and adaptively attempts to find the correct lower-dimensional interactions.

4.3 Proposed PCA-NSGA-II Procedure

Given two random variables X_1, $X_2 \in R^{\mathbf{N}}$ ($X_1 = [x_{11}, x_{12} \ldots x_{1\mathbf{N}}]$, $X_2 = [x_{21}, x_{22} \ldots x_{2\mathbf{N}}]$) with expected values μ_{X_1} and μ_{X_2} and standard deviations σ_{X_1} and σ_{X_2}, the correlation between them can be defined as $\rho_{X_1,X_2} = \frac{Covariance(X_1,X_2)}{\sigma_{X_1} \sigma_{X_2}}$ $= \frac{E((X_1-\mu_{X_1}) (X_2-\mu_{X_2}))}{\sigma_{X_1} \sigma_{X_2}}$. Given $\mathbf{M}$ such random variables, the correlation between all pairs of variables can be computed as above, and correlation matrix $\mathbf{R}_{\mathbf{M} \times \mathbf{M}}$ can be composed, where $\mathbf{R}_{i,j}$ would imply ρ_{X_i,X_j}. Here, two features are worth highlighting. One is the centring of data (where the mean of each variable is subtracted before multiplication) for computation of the covariance and the second is division of the covariance by the standard deviation (referred to as standardization), for computation of the correlation. An equivalent but simpler representation of the correlation matrix can be obtained by transforming the given random variables to centred and standardized form. Let $\bar{X}_i = \frac{(X_i - \mu_{X_i})}{\sigma_{X_i}}$ represent the ith random variable in both centred and standardized form. Given M such random variables, we can define a new $M \times N$ matrix $\mathbf{X} = [\bar{X}_1 \ \bar{X}_2 \ldots \bar{X}_M]^T$. In this case, the correlation matrix $\mathbf{R}_{\mathbf{M} \times \mathbf{M}}$ is given by $\frac{1}{N} \mathbf{X} \mathbf{X}^T$. In principal component analysis (PCA), it is either the covariance matrix or the correlation matrix that is eigen-decomposed and analysed. PCA on covariance matrices causes problems when variables show large differences in variances, in which case variables with large variances get larger weights (and ones with small variance get insignificant weights) and can dominate the whole covariance matrix and, hence, the eigenvalues and eigenvectors. Large differences in the variables' variances can naturally arise due to their different scales (units). The remedy to this problem of different scales and large differences in variances of the variables can be found in the use of the correlation matrix, which involves standardizing the variables to have variance equal to 1 (over and above, setting the mean to 0, which is also done in the covariance matrix). Hence, we recommend and employ the correlation matrix as opposed to the covariance matrix, for our study.

In the context of evolutionary multiobjective optimization, the dimension M, which basically represents the 'measurement types' can be taken to represent the objective functions, while N, which basically represents the time samples, can be taken to represent the population members. To illustrate the customization of the PCA procedure to the domain of evolutionary multiobjective optimization, we consider the DTLZ5(2,3) problem. With $M = 3$ (3-objective problem), $N = 200$ (population size) is taken and NSGA-II is run over 5,000 generations. The correlation matrix R obtained corresponding to the final population is shown in Table 1. It can be observed from this matrix that the first and third objectives are negatively correlated, that is, they are in conflict with each other. The same is true for the second and third objectives. Hence, while the first and second objectives are nonconflicting[6], each of them is in conflict with the third objective. Thus, in this simple case it can be concluded from this matrix that either the first or the second objective is redundant in this problem.

Table 1. PCA-NSGA-II: Formulation and Solution Module illustrated on DTLZ5(2,3)

Correlation Matrix : R	1.000000	0.999998	-0.916090
	0.999998	1.000000	-0.916109
	-0.916090	-0.916109	1.000000
Eigenvalues of	E1	E2	E3
1. R i. actual	2.889	0.000	0.111
ii. by ratio	0.962898	0.000	0.037101
2. R^2 i. actual	8.345	0.000	0.012
ii. by ratio	0.998518	0.0000	0.001482
Eigenvectors	V1–PCA1	V2–PCA3	V3–PCA2
	0.583104	0.707070	0.400052
	0.583108	-0.707143	0.399917
	-0.565664	-0.000080	0.824636

However, for a large number of objectives and in more complex problems, such a clear analysis may not be possible from an $M \times M$ matrix of real numbers. To address this, we prescribe a three-step procedure, which would eventually help us identify a reduced set (if possible) of objectives — those which are most important for forming the Pareto-optimal front.

---

[6] Note that in the remainder of this chapter we will use a correlation-based idea of conflicting objectives instead of the conflict definition given in Section 3.

Eigenvalue Analysis for Dimensionality Reduction

The eigenvalues of the correlation matrix $\mathbf{R}$ for DTLZ5(2,3) and their corresponding eigenvectors are shown in Table 1. Here, the first principal component (eigenvector $(0.583104, 0.583108, -0.565664)^T$, corresponding to the largest eigenvalue and accounting for 96.2898% of the variance) is designated as 'PCA1'. The first component of this vector denotes the contribution of the first objective function towards this vector, the second component denotes the contribution of the second objective, and so on. For a three-objective problem, like this one, the three contributions could easily be seen as the direction cosines defining a directed-ray in the three-dimensional objective space (this analogy would naturally also extend to higher dimensional problems). Now, the task here is to extract the information about conflict or no-conflict between objectives, based on the relative signs of their contributions. It is understood that the eigenvectors in PCA are only defined up to their sign (that is, if $\mathbf{V}$ is an eigenvector then so is $\mathbf{-V}$); hence, objectives with contributions with the same sign could be interpreted as increasing or decreasing together, when moving in the direction of the eigenvector. Similarly, objectives with opposite sign contributions could be interpreted as being in conflict with respect to the eigenvector. Thus, if we consider the objectives corresponding to the most positive and most negative contributions towards an eigenvector, they would mark the two most conflicting objectives with respect to this direction. To this effect, in the above example f_2 and f_3 could be observed to be the two most critically conflicting objectives with respect to PCA1.

Acknowledging that each principal component accounts for a certain fraction of the total variance of a data set, we can define a Threshold Cut (TC) as the proportion of the total variance to be accounted for and we can preset it for a given study. We can then start with analysing the first principal component, and then the subsequent ones, until their cumulative contribution exceeds TC. Hence, setting an appropriate TC is very important. If too high a TC (close to 1) is set, even some redundant objectives may be picked as important ones, thereby defeating the purpose of the PCA analysis. On the other hand, if too low a value of TC is set, even some important objectives may be ignored, thereby rendering the whole study erroneous. However, we understand that in order to make a reliable study, a TC value of 0.95 may be a balanced option. The choice of the TC can also be based on the relative magnitudes of the eigenvalues. If the reduction in two consecutive eigenvalues is more than a predefined proportion, no further principal component may be considered. This decision can also be based on the choice of the decision maker. If the decision maker wishes to include certain preferred objectives in the EMO procedure, the principal component analysis can be continued till all such preferred objectives are chosen. Many such possibilities exist and are worth experimenting with, but we prescribe a TC of 0.95. With this value, for DTLZ5(2,3) being considered, only the first principal component needs to be analysed, since PCA1 itself contributes 96.28% of the total variance (0.9628, by fraction). Hence, we need not consider any subsequent PCAs, and on the basis of larger conflict can declare the second and third objectives as important objectives to this problem. This way, the first objective is determined to be redundant and an EMO procedure can be applied to solve the two-objective problem (f_2 and f_3), instead of our using all three objectives.

Effect of Multiple Principal Components

For the DTLZ5(2,3) being considered, a TC of 0.95 necessitated consideration of PCA1 alone, along with which identifying the two most conflicting objectives naturally led to identification of the redundant objective. However, for large-objective problems (say, $M = 50$), not only many PCAs may be required to be considered but also the notion of identifying only the two most conflicting objectives along a particular PCA may not be adequate. Thus, the need is to build upon the understanding developed above, to be able to lay a dimensionality-reduction procedure which is compact (requiring us to account for as small a number of PCAs as possible) and effective (ensuring identification of the right set of objectives as important).

For compactness, we prescribe usage of matrix $\mathbf{R}^2$ instead of $\mathbf{R}$ for eigen-decomposition; in the case of $\mathbf{R}^2$, the eigenvalues get squared and more emphasized (evident in Table 1), while the eigenvectors remain unchanged. As the variance contribution of a principal component relates to the ratio of the corresponding eigenvalue of the total, fewer principal components would now have to be considered to meet the preset TC.

For effectiveness, we suggest the following procedure for interpretation of eigenvalues and eigenvectors of matrix $\mathbf{R}^2$:

1. As the first principal component accounts for a significant proportion of the total variance in the data set, we would want to capture any signal of a conflicting objective. Hence, for the first principal component, along with the objective corresponding to the most-positive element, we consider as important any and all objectives which correspond to a negative component, however small. If in some case all the elements along PCA1 are positive, we pick up the objectives corresponding to the first two most positive elements.

2. We refer to subsequent principal components only if the fraction of variance accounted for by the first principal component is less than TC. In such a case, we consider the second principal component, then the third, and so on, until the cumulative variance accounted exceeds TC. Given any subsequent principal component, we first check if the proportion of the total variance that it accounts for is greater than 10% (0.1, by fraction) or not.

   a) If not, we only choose the objective corresponding to the highest absolute element in the eigenvector.

   b) If yes, we consider various cases:

      i. If all the elements of the eigenvector are identical in sign, we choose only the objective corresponding to the highest magnitude.

      ii. If the value of the highest positive element (p) is less than the absolute value of the most negative element (n), we check if $p \geq 0.9|n|$. If yes, then we choose the two objectives corresponding to p and n; otherwise we choose only the objective corresponding to n.

      iii. Similarly, we also consider the possibility of the absolute value of the most-negative element (n) being less than the highest positive element (p), in which case we further check if $|n| \geq 0.8p$. If yes, we choose the two objectives corresponding to p and n; otherwise we only choose the objective corresponding to p.

At the outset, the criteria for selection of objectives in the above paragraph may appear as ad hoc fixations. Let us now mention the logic in the same. The first principal component, which accounts for the most significant proportion of the total variance (of the data set), is given the highest weight and based on the prescriptions above, it will provide us at least two important objectives. Further, a principal component which accounts for less than 10% of the total variance is interpreted in accordance with its low importance and is prescribed to provide only one objective as important. We utilize, the principal components with intermediate importance to provide either one or more important objectives depending on how comparable the most positive and the most negative contributions of corresponding objectives are along these directions. The fixations like $p \geq 0.9|n|$ and $|n| \geq 0.8p$ (as in the above paragraph) are just tests of the comparableness. For instance, testing for $p \geq 0.5|n|$ instead of $p \geq 0.9|n|$, or testing for $|n| \geq 0.5p$ instead of $|n| \geq 0.8p$, is more likely to also infer the objectives corresponding to p (apart from n) and n (apart from p) as important, in their respective cases. Hence, here, the user can be flexible in such fixations depending on how robust the procedure is expected to be. Extracting more and more objectives as important at this stage (based on importance along important directions) would make the procedure more robust on one hand; on the other hand, it may render the dimensionality reduction procedure slower, as more iterations may be required to reach the minimal set of important objectives. Our experience with many test problems and some real-world problems suggests that our prescriptions above are fairly grounded. The readers should note that the inaccuracy in results presented in this chapter basically emanate from the limitations of the PCA technique itself, which we discuss in a later section. Interestingly, the same procedure, when employed while utilizing a nonlinear dimensionality reduction technique, namely, maximum variance unfolding [30], provides fully accurate results. Hence, the prescriptions made above can be fairly relied upon while remaining open to suitable amendments.

Final Reduction Using the Correlation Matrix

It is expected that interpretation of principal components (eigenvalue analysis) based on the preceding procedure would lead to identification of many of the redundant objectives. To consider whether further reduction in the number of objectives is possible, we now return to a *reduced* correlation matrix (only columns and rows corresponding to non-redundant objectives adjudged so by eigenvalue analysis) and investigate if there still exists a set of objectives having identical positive or negative correlation coefficients with other objectives and having positive correlations among themselves. This will suggest that any one member from such a group is enough to establish the conflicting relationships with the remaining objectives.

Consider an identically correlated set of S objectives, each being represented by f_i, $i = 1 \ldots S$. Further, assume that V principal components were utilized for eigenvalue analysis, each represented by v_j, $j = 1 \ldots V$, and each accounting for a proportion e_j, $j = 1 \ldots V$, towards the total variance of the data set. Let the contribution of f_i along v_j be represented by f_{ij}. Then, compute for each of the identically correlated objectives f_i the value $c_i = \sum_{j=1}^{V} \|(f_{ij} \cdot e_j)\|$. Pick from the set $\{c_i \,|\, i = 1 \ldots S\}$ of scalars the highest value, say c_k. Then the objective f_k can be considered as a representative of the set of S objectives. This selection criterion physically implies that an objective which contributes most along the important principal components collectively is deemed fit to be the representative. It should also be mentioned that once NSGA-II is run for a sufficiently large number of generations, the correlation matrix stabilizes and correlation patterns turn invariant over a number of generations. Hence, the inferences drawn from the reduced correlation matrix can be trusted.

Overall PCA-NSGA-II Procedure

We are now ready to present the overall PCA-NSGA-II procedure.

Step 1: Set an iteration counter $t = 0$ and an initial set of objectives $\mathbb{I}_0 = \{1, 2, \ldots, M\}$.

Step 2: Initialize a random population for all objectives in the set $\mathbb{I}_t$, run an EMO, and obtain a population P_t.

Step 3: Perform a PCA analysis on P_t using $\mathbb{I}_t$ to choose a reduced set of objectives $\mathbb{I}_{t+1}$ using the predefined TC. Steps of the PCA analysis are as follows:
  1. Compute the correlation matrix.
  2. Compute eigenvalues and eigenvectors and choose non-redundant objectives using the procedure discussed in Section 4.3.
  3. Reduce the number of objectives further, if possible, by using the correlation coefficients of the non-redundant objectives found in item 2 above, applying the procedure discussed in Section 4.3.

Step 4: If $\mathbb{I}_{t+1} = \mathbb{I}_t$, stop and declare the obtained front. Else set $t = t + 1$ and go to Step 2.

Thus, starting with all M objectives, the above procedure iteratively finds a reduced set of objectives by analysing the obtained nondominated solutions by an EMO procedure. When no further objective reduction is possible, the procedure stops and declares the final set of objectives and corresponding nondominated solutions.

We realize that the above procedure of dimensionality reduction has not much of a meaning for those problems which have an exactly M-dimensional Pareto-optimal front. In such a scenario, it is expected that the proposed algorithm will deem all objectives to be important in the very first iteration, thereby not performing any dimensionality reduction. However, even in this case the proposed procedure will establish a relative order of importance of the objectives by the PCA analysis, which may provide additional information to a decision maker.

4.4 Simulation Results with the PCA-NSGA-II Procedure

We now show the simulation results obtained with the iterative PCA-NSGA-II procedure described above on a number of test problems having a varying number of objectives. Having realized the curse of dimensionality in Section 4.1, to solve DTLZ5(I,M) problems with $I > 3$, we use a population size of 800 and run NSGA-II for 10,000 generations. It is expected that such a choice of these parameters will offer a reasonable computational effort towards convergence of the population. PCA is then applied to the nondominated population so obtained. The rest of the parameters remain the same as in Section 4.1.

Problem DTLZ5(2,10)

Let us now discuss the implementation of the complete PCA-NSGA-II procedure, as highlighted in Section 4.3 on DTLZ5(2,10) problem. The inferences drawn by

Table 2. DTLZ5(2,10): Iter1(i). Eigenvalue Analysis

Iter. 1 : PCA1 (58.83 % variance)	f_7	f_{10}
PCA2 (28.26 % variance)	f_1	
PCA3 (06.53 % variance)		f_8
PCA4 (03.27 % variance)		f_8

Table 3. DTLZ5(2,10): Iter1(ii). Reduced Correlation Analysis

	f_1	f_7	f_8	f_{10}
f_1	+	-	+	-
f_7	-	+	+	-
f_8	+	+	+	-
f_{10}	-	-	-	+

Table 4. DTLZ5(2,10): Iter2(i). Eigenvalue Analysis

Iter. 2 : PCA1 (94.58 % variance)	f_7	f_{10}
PCA2 (4.28 % variance)		f_8

Table 5. DTLZ5(2,10): Iter2(ii). Reduced Correlation Analysis

	f_7	f_8	f_{10}
f_7	+	+	-
f_8	+	+	-
f_{10}	-	-	+

Table 6. DTLZ5(2,10): Iter2(iii). Selection criterion for Reduced Correlation Matrix

	e1:0.9458	e2:0.0428	
f_7	+0.543	-0.275	c7=0.5253
f_8	+0.457	+0.672	c8=0.4610
	PCA1	PCA2	

eigenvalue analysis in Iteration 1, as tabulated above, suggests that a total of four objectives are important. Further, the reduced correlation matrix, as shown above, suggests that no further reduction is possible as each objective is differently correlated with the other. Consequently, Iteration 2 is performed using this reduced set of four objectives, namely f_1, f_7, f_8 and f_{10}. The eigenvalue analysis, highlights three objectives, f_7, f_8, and f_{10} to be important. Further, the reduced correlation matrix shows f_7 and f_8 to be identically correlated. Table 6 is based on the guidelines in Section 4.3 and evaluates f_7 to be a better representative than f_8, amongst the two.

While the overall results are tabulated in Table 7, the corresponding populations are shown in Figures 8, 9, and 10.

Table 7. DTLZ5(2,10)

Iter. 1	f_1	f_7	f_8	f_{10}
Iter. 2		f_7		f_{10}
Iter. 3		f_7		f_{10}

Table 8. DTLZ5(2,20)

Iter. 1	f_2	f_5	f_{10}	f_{12}	f_{16}	f_{18}	f_{20}
Iter. 2		f_5				f_{18}	f_{20}
Iter. 3						f_{18}	f_{20}
Iter. 4						f_{18}	f_{20}

Table 9. DTLZ5(2,30)

Iter. 1	f_1	f_8	f_{30}
Iter. 2		f_8	f_{30}
Iter. 3		f_8	f_{30}

Higher-Objective DTLZ5(2,M) Problems

To further investigate the performance of the proposed procedure on a larger number of objectives, we apply it to 20, 30, and 50 objective problems, whose results are summarized in Table 8, Table 9, and Table 10, respectively. In each case, for the set of objectives declared critical by PCA-NSGA-II procedure, the Pareto-optimal front obtained by running NSGA-II matches with that expected theoretically, from Equation 2. However, for brevity we show such conformance only for DTLZ5(2,20), in Figures 11 and 12.

Consider a situation where, given an M objective optimization problem, neither the cardinality of the Pareto-optimal front nor the indices of the objectives contributing to the front is known. An exhaustive attempt to find the front would require us to consider all possible combinations $\left(\binom{M}{i}\right)$ of objectives for a particular cardinality (i), which in turn would have to be varied from 1 to M. Consequently, as many as $\sum_{i=0}^{M} \binom{M}{i}$ or, equivalently, 2^M combinations would need to be checked. On the other hand, our proposed procedure uses a fraction of these computations to make a good estimate of the Pareto-optimal front. Though it is applicable only to quasi-convex objectives, readers may find the work in [17] interesting, in pursuit

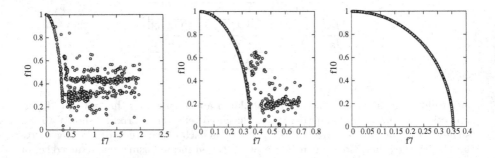

Fig. 8. Population obtained after the first iteration of PCA-NSGA-II on DTLZ5(2,10)

Fig. 9. Population obtained after the second iteration of PCA-NSGA-II on DTLZ5(2,10)

Fig. 10. Population obtained after the third iteration of PCA-NSGA-II on DTLZ5(2,10)

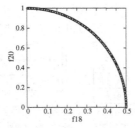

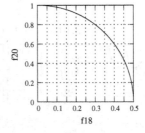

Fig. 11. Population obtained after the third iteration of PCA-NSGA-II on 20-objective DTLZ5(2,20)

Fig. 12. Theoretical Pareto-optimal front in f_{20}-f_{18} plane for 20-objective DTLZ5(2,20)

of deciding Pareto optimality for a point in the decision space of a MOP using only subsets of the set of objectives.

Table 10. DTLZ5(2,50)

Iter. 1	f_{13} f_{50}
Iter. 2	f_{13} f_{50}

Table 11. DTLZ2(3)

Iter. 1	f_1 f_2 f_3

Table 12. DTLZ2(5)

Iter. 1	f_1 f_2 f_3 f_4 f_5

DTLZ2(M) or DTLZ5(M,M) Problems

Unlike a reduced dimensionality for the Pareto-optimal frontier in DTLZ5(I,M) with $I < M$, DTLZ5(M,M), also referred to as DTLZ2(M), problems preserve the dimensionality of the Pareto-optimal front. This implies that, in DTLZ2(M) problems, the entire Pareto-optimal front involves all M objectives. While Table 11 shows all the three objectives to be critical for DTLZ2(3), all the five objectives are found critical for DTLZ2(5), as shown in Table 12. Hence, the results obtained reflect the problem characteristics. Further, the fact that the obtained results neatly highlight the case of 'no redundancy', induces in us further confidence in the proposed procedure, that it is not 'greedily tuned' to figure out redundancy when there isn't any. For DTLZ2(3), the Pareto-optimal front obtained by NSGA-II with three objectives is mapped over the theoretical front based on Equation 1, and is shown in Figure 13.

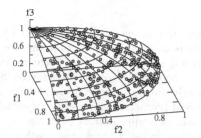

Fig. 13. DTLZ2(3): Pareto-optimal front with all 3-objectives

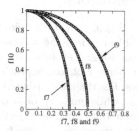

Fig. 14. Illustration of limitation of PCA

4.5 Difficulties with the PCA Approach and Future Directions

While PCA yields a smaller dimensional linear subspace that best represents the full data according to a minimum square error criterion, it may be ineffective in revealing the underlying dimensionality when the data points live on a nonlinear manifold (manifolds are spaces that are locally linear but unlike Euclidean subspaces, can be globally nonlinear) or when the data structure is non-Gaussian. The strength of our proposed PCA-NSGA-II algorithm emerges from the fact that we can relate the most important directions in the data set (in terms of variance) to the importance of objectives given a multiobjective optimization problem. Now if the determination of important directions in a data set is erroneous, the inferences drawn about the importance of objectives, and hence the determination of redundant objectives, will be meaningless. Hence, it would be worthwhile to assess situations in which PCA is likely to extract erroneous directions. Such situations can be best examined under the question "Does the data live in a low-dimensional subspace?" or "Does the data live on a low-dimensional sub-manifold?".

To highlight difficulties with standard PCA, let us begin with a concrete example of DTLZ5(2,10) for which PCA-NSGA-II declared f_7 and f_{10} as critical objectives and the rest as redundant. Let us investigate these results in light of two facts. Fact 1 relates to the property of DTLZ5(2,M) problems where the Pareto-optimal front corresponds to the last objective and any other one. In this context, declaration of f_7 and f_{10} as critical is right. Fact 2 relates to the criteria of judging an objective set as critical. PCA-NSGA-II is expected to declare those objectives as critical which apart from being in conflict with each other also account for variances larger than those declared redundant. From Figure 14, f_9 and f_{10} can be seen to account for the largest variance amongst the set of ten objectives. This, in fact, is generalizable to all DTLZ5(2,M) problems, where objectives with indices M and $(M - 1)$ will collectively account for the largest variance in the data set of all given objectives. Hence, the last two objectives must come up as the critical objectives given any DTLZ5(2,M) problem, which is not the case with the PCA utilized for this chapter. The remedy lies in the nonlinear version of PCA, some preliminary results of which can be found in [30].

5 Conclusions

In this chapter, we have addressed the important issue of tackling multiobjective problems when the number of objectives is large. We argued that the use of domination-based evolutionary approaches is not computationally efficient in handling such many-objective problems. Hence, we proposed two objective reduction approaches. For the conflict-based approach, which aims at preserving the dominance structure while omitting objectives, we discussed how exact and greedy algorithms can eliminate redundant objectives in an *a posteriori* approach, i.e., after an EMO procedure generates a Pareto front approximation, to assist the decision maker. For the second approach, we showed how it can be used during an EMO procedure, namely NSGA-II, to detect the crucial objectives in large-dimensional MOPs which degenerate to possess a lower-dimensional Pareto-optimal frontier. Both methods presented indicate that EMO methodologies could be a worthy option for generating smaller-dimensional Pareto-optimal solutions when solving problems with many objectives. Hopefully, this study will motivate further research for devising more reliable and efficient methods of dimensionality reduction and eventually facilitate solutions to large-dimensional multiobjective optimization problems which are common in practice.

References

[1] *The Journal of Machine Learning Research: Special Issue on Variable and Feature Selection*. MIT Press, 2003.

[2] P. J. Agrell. On redundancy in multi criteria decision making. *European Journal of Operational Research*, 98(3):571–586, 1997.

[3] N. Beume and G. Rudolph. Faster S-Metric Calculation by Considering Dominated Hypervolume as Klee's Measure Problem. Technical Report CI-216/06, Sonderforschungsbereich 531 Computational Intelligence, Universität Dortmund, July 2006.

[4] D. Brockhoff and E. Zitzler. Are All Objectives Necessary? On Dimensionality Reduction in Evolutionary Multiobjective Optimization. In *Parallel Problem Solving from Nature (PPSN-IX)*, volume 4193 of *LNCS*, pages 533–542. Springer, 2006. ISBN 3-540-38990-3.

[5] D. Brockhoff and E. Zitzler. Dimensionality Reduction in Multiobjective Optimization with (Partial) Dominance Structure Preservation: Generalized Minimum Objective Subset Problems. TIK Report 247, Institut für Technische Informatik und Kommunikationsnetze, ETH Zürich, April 2006.

[6] D. Brockhoff and E. Zitzler. Dimensionality Reduction in Multiobjective Optimization: The Minimum Objective Subset Problem. In *Proceedings of the Operations Research 2006 conference*, 2007. to appear.

[7] D. Brockhoff and E. Zitzler. Offline and Online Objective Reduction in Evolutionary Multiobjective Optimization Based on Objective Conflicts. TIK Report 269, Institut für Technische Informatik und Kommunikationsnetze, ETH Zürich, Apr. 2007.

[8] M. Charikar, V. Guruswami, R. Kumar, S. Rajagopalan, and A. Sahai. Combinatorial feature selection problems. In *IEEE Symposium on Foundations of Computer Science*, pages 631–640, 2000. URL citeseer.ist.psu.edu/376451.html.

[9] C. Coello Coello and A. Hernández Aguirre. Design of Combinational Logic Circuits through an Evolutionary Multiobjective Optimization Approach. *Artificial Intelligence for Engineering, Design, Analysis and Manufacture*, 16(1):39–53, 2002.

[10] C. A. Coello Coello, D. A. Van Veldhuizen, and G. B. Lamont. *Evolutionary Algorithms for Solving Multi-Objective Problems*. Kluwer Academic Publishers, New York, 2002.

[11] J. J. Dai, L. Lieu, and D. Rocke. Dimension Reduction for Classification with Gene Expression Microarray Data. *Statistical Applications in Genetics and Molecular Biology*, 5(1), 2006.

[12] M. Dash and H. Liu. Feature selection for classification. *Intelligent Data Analysis*, 1(3):131–156, 1997.

[13] K. Deb. *Multi-objective optimization using evolutionary algorithms*. Wiley, Chichester, UK, 2001.

[14] K. Deb, S. Agrawal, A. Pratap, and T. Meyarivan. A fast elitist non-dominated sorting genetic algorithm for multi-objective optimization: NSGA-II. In *Parallel Problem Solving from Nature (PPSN-VI)*, pages 849–858, 2000.

[15] K. Deb and D. Saxena. Searching For Pareto-Optimal Solutions Through Dimensionality Reduction for Certain Large-Dimensional Multi-Objective Optimization Problems. In *IEEE Congress on Evolutionary Computation (CEC 2006)*, pages 3352–3360, 2006.

[16] K. Deb, L. Thiele, M. Laumanns, and E. Zitzler. Scalable Test Problems for Evolutionary Multi-Objective Optimization. In A. Abraham, R. Jain, and

R. Goldberg, editors, *Evolutionary Multiobjective Optimization: Theoretical Advances and Applications*, pages 105–145. Springer, 2005. ISBN 1-85233-787-7.

[17] M. Ehrgott and S. Nickel. On the number of criteria needed to decide Pareto-optimality. *Mathematical Methods of Operations Research*, 55(3):329–345, 2002.

[18] M. Emmerich, N. Beume, and B. Naujoks. An EMO Algorithm Using the Hypervolume Measure as Selection Criterion. In *Evolutionary Multi-Criterion Optimization (EMO 2005)*, volume 3410 of *LNCS*, pages 62–76. Springer, 2005.

[19] C. M. Fonseca, L. Paquete, and M. López-Ibáñez. An Improved Dimension-Sweep Algorithm for the Hypervolume Indicator. In *IEEE Congress on Evolutionary Computation (CEC 2006)*, pages 1157–1163, 2006.

[20] T. Gal and T. Hanne. Consequences of dropping nonessential objectives for the application of MCDM methods. *European Journal of Operational Research*, 119:373–378, 1999.

[21] T. Gal and H. Leberling. Redundant objective functions in linear vector maximum problems and their determination. *European Journal of Operational Research*, 1(3):176–184, 1977.

[22] T. Goel, R. Vaidyanathan, R. T. Haftka, N. Queipo, W. Shyy, and K. Tucker. Response surface approximation of Pareto-optimal front in multi-objective optimization. Technical report, Department of Mechanical and Aerospace Engineering, University of Florida, USA, 2004.

[23] E. J. Hughes. Radar Waveform Optimization as a Many-Objective Application Benchmark. In *Evolutionary Multi-Criterion Optimization (EMO 2007)*, volume 4403 of *LNCS*, pages 700–714, 2007.

[24] A. Hyvärinen, J. Karhunen, and E. Oja. *Independent Component Analysis*. John Wiley & Sons, 2001.

[25] I. T. Jolliffe. *Principal component analysis*. Springer, 2002. ISBN 0-387-95442-2.

[26] P. Langley. Selection of relevant features in machine learning. In *Proceedings of the AAAI Fall Symposium on Relevance*, pages 140–144, 1994.

[27] H. Liu and H. Motoda, editors. *Feature Extraction, Construction and Selection: A Data Mining Perspective*. Kluwer Academic Publishers, Norwell, MA, USA, 1998.

[28] B. Paechter, R. C. Rankin, A. Cumming, and T. C. Fogarty. Timetabling the Classes of an Entire University with an Evolutionary Algorithm. In *Parallel Problem Solving From Nature (PPSN-V)*, pages 865–874. Springer, 1998.

[29] R. C. Purshouse and P. J. Fleming. Conflict, Harmony, and Independence: Relationships in Evolutionary Multi-criterion Optimisation. In *Evolutionary Multi-Criterion Optimization (EMO 2003)*, volume 2632 of *LNCS*, pages 16–30. Springer, 2003.

[30] D. K. Saxena and K. Deb. Non-linear Dimensionality Reduction Procedures for Certain Large-Dimensional Multi-Objective Optimization Problems: Employing Correntropy and a Novel Maximum Variance Unfolding. In *Evolutionary Multi-criterion Optimization (EMO 2007)*, 2007.

[31] K. C. Tan, E. F. Khor, and T. H. Lee. *Multiobjective Evolutionary Algorithms and Applications*. Springer, London, UK, 2005.

[32] H. Vafaie and K. De Jong. Robust feature selection algorithms. In *Tools with Artificial Intelligence (TAI '93)*, pages 356–363, 1993.

[33] T. Wagner, N. Beume, and B. Naujoks. Pareto-, Aggregation-, and Indicator-based Methods in Many-objective Optimization. In *Evolutionary Multi-*

Criterion Optimization (EMO 2007), volume 4403 of *LNCS*, pages 742–756. Springer, 2007.

[34] L. While. A New Analysis of the LebMeasure Algorithm for Calculating Hypervolume. In *Evolutionary Multi-Criterion Optimization (EMO 2005)*, volume 3410 of *LNCS*, pages 326–340. Springer, 2005.

[35] L. While, L. Bradstreet, L. Barone, and P. Hingston. Heuristics for Optimising the Calculation of Hypervolume for Multi-objective Optimisation Problems. In *IEEE Congress on Evolutionary Computation (CEC 2005)*, pages 2225–2232, 2005.

[36] E. Zitzler and S. Künzli. Indicator-Based Selection in Multiobjective Search. In *Parallel Problem Solving from Nature (PPSN-VIII)*, September 2004.

[37] E. Zitzler, L. Thiele, M. Laumanns, C. M. Foneseca, and V. Grunert da Fonseca. Performance Assessment of Multiobjective Optimizers: An Analysis and Review. *IEEE Transactions on Evolutionary Computation*, 7(2):117–132, 2003.

Index

Natural Computing Series

Printed in the United States
By Bookmasters

Printed in the United States
By Bookmasters